Lecture Notes in Mathematics

Volume 2310

This series reports on new developments in all areas of mathematics and their applications - quickly, informally and at a high level. Mathematical texts analysing new developments in modelling and numerical simulation are welcome. The type of material considered for publication includes:

1. Research monographs.
2. Lectures on a new field or presentations of a new angle in a classical field.
3. Summer schools and intensive courses on topics of current research.

Texts which are out of print but still in demand may also be considered if they fall within these categories. The timeliness of a manuscript is sometimes more important than its form, which may be preliminary or tentative.

Titles from this series are indexed by Scopus, Web of Science, Mathematical Reviews, and zbMATH.

Herbert Lange • Rubí E. Rodríguez

Decomposition of Jacobians by Prym Varieties

Springer

Herbert Lange
Department Mathematik
University of Erlangen-Nuremberg
Erlangen, Germany

Rubí E. Rodríguez
Department of Mathematics
Universidad de La Frontera
Temuco, Chile

ISSN 0075-8434   ISSN 1617-9692   (electronic)
Lecture Notes in Mathematics
ISBN 978-3-031-10144-1   ISBN 978-3-031-10145-8   (eBook)
https://doi.org/10.1007/978-3-031-10145-8

This Springer imprint is published by the registered company Springer Nature Switzerland AG
The registered company address is: Gewerbestrasse 11, 6330 Cham, Switzerland

Preface

Abelian varieties and in particular Jacobians as well as theta functions have been studied since the nineteenth century. Whereas Jacobian varieties are used as a device to investigate algebraic curves, theta functions occur in several branches of mathematics and also in other natural sciences such as physics. Their relation became clear only much later: theta functions can be considered as sections of line bundles on abelian varieties.

Here, decomposable Jacobians correspond to reducible theta functions, that is, products of theta functions in less variables, which were intensively studied at the end of the nineteenth century. The decomposition of Jacobians started with the introduction of Prym varieties by Mumford in the 1960s. They contribute to the understanding of curves and special abelian varieties.

The present volume contains, apart from some basics, several new results and also applications, such as new proofs of the bigonal and trigonal constructions. It should be of interest for most algebraic geometers and several arithmetical geometers or number theorists.

Erlangen, Germany
Temuco, Chile

Herbert Lange
Rubí E. Rodríguez

Acknowledgements

We feel very grateful to our friend Sevin Recillas for his great ideas used in the book, which unfortunately he could not see in print. We thank Angel Carocca and Benjamin Moraga for proofreading parts of the manuscript and Hans-Joachim Schmid for helping us with the pictures.

Contents

Notations

Throughout the book we use the following notations. Given a finite group G, we denote by $\widetilde{C}$ a (smooth projective) curve with an action of G. Given this, we denote

- by C the quotient curve $C := \widetilde{C}/G$ with projection map $f : \widetilde{C} \to C$;
- by $\widetilde{J}$ the Jacobian of $\widetilde{C}$, $\widetilde{J} := J\widetilde{C}$;
- by J the Jacobian of C, $J := JC$;
- for any subgroup $H \subset G$ by C_H the quotient curve $C_H := \widetilde{C}/H$ (instead of writing $\widetilde{C}_H$ which at some points seemed too complicated).

We also use the following notations for any abelian variety A:

- if B is a subgroup of A, we denote by B^0 the connected component of B containing 0;
- if a finite group G acts on A, we denote by A^G the subgroup of A fixed by G, $A^G := \{x \in A \mid gx = x \text{ for all } g \in G\}$. If G is generated by one element say g, we write A^g instead of $A^{\langle g \rangle}$;
- for any abelian variety A and any integer n we denote by n_A multiplication by n on A: $n_A = n \cdot id_A$. Sometimes we write n instead of n_A;
- if A and B are isogenous abelian varieties, we write $A \sim B$;
- by $\equiv$ we denote algebraic equivalence.

For any finite group G we denote:

- by $|G|$ the order of G;
- for any subgroup H of the group G, by ρ_H the representation of G induced by the trivial representation of H;
- for any two complex representations U and V of G, by $\langle U, V \rangle$ the inner product of the characters of U and V. For two rational representations W_1 and W_2 we define $\langle W_1, W_2 \rangle := \langle W_1 \otimes \mathbb{C}, W_2 \otimes \mathbb{C} \rangle$.

Finally denote by

- $\mathbb{C}_1$ the circle group, that is, $\mathbb{C}_1 = \{z \in \mathbb{C} \mid |z| = 1\}$.

Chapter 1
Introduction

Theta functions in several variables were introduced by Göpel and Rosenhain and first intensively studied by Riemann and Weierstrass. It was Riemann who stated the theorem that every abelian function can be rationally represented by theta functions. A rigorous proof of this result was given by Picard and Poincaré.

Decomposable theta functions, meaning theta functions which are products of theta functions in less variables, were extensively investigated at the end of the nineteenth century. They are related to reducible abelian integrals, that is, such integrals which come from abelian integrals of smaller genus (see [18, Chapter 11] and [19, Chapter 15]).

Much later it was understood that theta functions can be considered as sections of line bundles on abelian varieties. Hence, in a sense it is equivalent to decompose theta functions and to decompose the corresponding abelian varieties, and this is our aim here. To be more precise, we consider smooth projective curves $\widetilde{C}$ with an action of a finite group G and quotient map $f : \widetilde{C} \to C = \widetilde{C}/G$. The action induces an action of G on the Jacobian $\widetilde{J} := J\widetilde{C}$, which in turn gives some isogenies from $\widetilde{J}$ to products of abelian subvarieties of $\widetilde{J}$.

We distinguish two such decompositions, the first called the *isotypical decomposition*, which is unique. Its factors correspond bijectively to the irreducible rational representations of the group G, although some of the factors might be zero. The factor of $\widetilde{J}$ corresponding the irreducible rational representation W is called *the isotypical component of $\widetilde{J}$ associated* to W. In general we denote it by A_W.

Some of these factors decompose further. In fact, let V be an irreducible complex representation, Galois associated to W, such that

$$n_W := \frac{\dim V}{s_V} \geq 2,$$

where s_V denotes the Schur index of V; then there is an abelian subvariety B_W of A_W and an isogeny

H. Lange, R. E. Rodríguez, *Decomposition of Jacobians by Prym Varieties*, Lecture Notes in Mathematics 2310, https://doi.org/10.1007/978-3-031-10145-8_1

$$B_W^{n_W} \sim A_W.$$

The abelian subvariety B_W is highly non-unique. In fact, there are many such subvarieties of A_W, all isogenous to each other. Any resulting decomposition of $\widetilde{J}$ is called *a group algebra decomposition* of $\widetilde{J}$ and any B_W a *group algebra component associated to* W.

Our main method to compute A_W and a component B_W is given by the following theorem. For this we denote, for any subgroup H of G, by ρ_H the representation of G induced by the trivial representation of H. Moreover, we use a more general notion of a Prym variety than the original one, introduced by Mumford in [27]. For any finite cover of curves $g : X \to Y$, the *Prym variety* $P(g) = P(X/Y)$ is by definition the complement of the abelian subvariety $g^* JY$ in JX with respect to the canonical polarization of JX. Then see Corollary 3.5.10 for the following theorem. There we do not call it a theorem, because it is a special case of a more general result.

Theorem *Suppose there are subgroups $H \subset N \subseteq G$ such that*

$$\rho_H \simeq \rho_N \oplus W$$

for some irreducible rational representation W of G. Then

(i) *There is an abelian subvariety B_W which is given explicitly by a primitive idempotent associated to W such that*

$$P(C_H/C_N) \sim B_W$$

where $P(C_H/C_N)$ denotes the Prym variety of the cover $C_H := \widetilde{C}/H \to C_N := \widetilde{C}/N$.

(ii) *If V is an irreducible complex representation of G Galois associated to W with Schur index s_V and*

$$n = \frac{\dim V}{s_V},$$

then there is an isogeny

$$P(C_H/C_N)^n \sim A_W$$

of the n-th power of $P(C_H/C_N)$ onto the isotypical component A_W associated to W.

In many cases the theorem can be applied to work out the isotypical decomposition as well as a group algebra decomposition of $\widetilde{J}$ in terms of products of Prym varieties. In some cases we need a slightly more general concept, namely, the Prym

variety of a pair of covers. We also give an example where even the more general
Prym variety does not apply.

As an implication we get new proofs of the classical bigonal and trigonal
constructions. The importance of these proofs lies in the fact that they generalize
to give isogenies between Prym varieties of different subcovers of $\tilde{C}$. In several
cases we compute the degree of the corresponding isogeny, which equals one in the
classical cases.

For example, the classical trigonal construction, due to Recillas [30], is an
isomorphism between the Prym variety of an étale doble cover of a trigonal curve
and the Jacobian of an associated tetragonal curve. Now the Galois closure of a non-
Galois tetragonal cover is a Galois cover with group either the dihedral group D_4 of
order 8, or the alternating group $\mathcal{A}_4$ of degree 4, or the symmetric group $\mathcal{S}_4$ of degree
4. In each case we obtain an isogeny between Prym varieties which generalizes
the classical isomorphism. Moreover, the generalization works for any non-Galois
degree 4 cover. A similar statement is true for the classical bigonal construction due
to Pantazis [28]. Moreover, here we give explicit conditions for its validity in terms
of the covers.

In Chap. 2 we recall some preliminary material on abelian varieties, mainly in
order to set the notations. We use some results of [3] as well as on the action of a
Hecke algebra associated to a subgroup of G as given in [8] and [10].

Chapter 3 deals with Prym varieties. In Sect. 3.1 we recall some basics of finite
covers of curves including the existence of Galois covers of curves for any finite
group G in terms of the signature of the cover (see [33]). Sect. 3.2 gives the
definition and basics on Prym varieties. In Sect. 3.4 we define Prym varieties of
a pair of covers, following [20]. Finally, in Sect. 3.5 we deduce the results on Galois
covers of curves needed in subsequent chapters. In particular here we prove the
theorem mentioned above.

Chapter 4 deals with covers of degree 2 and 3. Since every cover of degree 2 is
cyclic, there is only the decomposition of $\tilde{J}$ into the product of the pullback of J and
the Prym variety $P(f)$. For degree 3 covers, there is also the non-Galois cover. Here
the Galois closure is a degree 6 cover with an action of $\mathcal{S}_3 = D_3$. We decompose its
Jacobian following [31].

The next chapter was actually the starting point for writing this book. In 2004 the
second author and our common friend Sevin Recillas wrote the paper [32] which
never was published due to the tragic death of Sevin right after the paper was
finished. We thought that the contents of that paper are worth being published, in
particular since now the general theory has developed quite a bit and consequently
some of the proofs have simplified. We hope the reader will agree with this.

There are five cases of covers of degree 4, two Galois covers and three non-Galois
cases. Sections 5.1 and 5.2 deal with the cyclic covers of degree 4 and the case of the
Galois cover with group the Klein group of order 4, respectively. If the cover is non-
Galois, there are three possibilities for the group of its Galois closure: it is either the
dihedral group D_4 of order 8, the alternating group $\mathcal{A}_4$, or the symmetric group $\mathcal{S}_4$ of
degree 4. Sections 5.3, 5.5, and 5.7 deal with these covers, respectively. In Sect. 5.4
we study the bigonal construction. In particular we give a new proof of Pantazis'

theorem which has the advantage of precise hypotheses. In Sects. 5.6 and 5.8 we give new proofs of the trigonal constructions for $\mathcal{A}_4$ and $\mathcal{S}_4$ covers, respectively. The generalizations of theorems mentioned above are also given in Sect. 5.6 in the $\mathcal{A}_4$-case and in Sect. 5.8 in the $\mathcal{S}_4$ case.

In the sixth chapter, we consider some series of covers, the covers with cyclic group $\mathbb{Z}_n$ in Sect. 6.1, with dihedral group D_n of order $2n$ in Sect. 6.2 and the semidirect product of $\mathbb{Z}_3$ by an arbitrary power of Klein groups. In Sect. 6.2 we follow [6] and in Sect. 6.3 [5]. Section 6.3 gives an example for a Jacobian isogenous to an arbitrary number of Jacobians.

Finally, in the last chapter, we worked out the decomposition of Jacobians for some special groups which we found worthwhile. Section 7.1 deals with the simple group $\mathcal{A}_5$ of order 60. Section 7.2 deals with actions of two groups which admit an irreducible representation of Schur index 2. First is the group Q_8 of Hamiltonian quaternions and then a semidirect product of Q_8 by the group of order 3. The second group gives the only case in this book where an isotypical component of the action (which equals a group algebra component) is not isogenous to a Prym variety of a pair of subquotients nor to a pair of two such subquotients.

In the last section, we consider the problem of giving example of Jacobians isogenous to a product of elliptic curves. We give a proof of the criterion of Ekedahl-Serre and finish with several examples.

Finally, following a suggestion of the referees, let us say a few words why the results and constructions are important and and for whom they should be interesting.

First of all, we mentioned already above the relation with the classical theory of decompositions of theta functions as a product of theta functions and need not say more about it.

To every smooth projective variety, one can associate in a canonical way two abelian varieties, its Albanese and dually its Picard variety (see [3]). They coincide in the case of a curve and are then called the Jacobian of the curve. Abelian varieties and in particular Jacobians originally were introduced to study the curves, roughly speaking as a sort of abelianization of the curves.

In general it is not easy to determine these abelian varieties explicitly, even in the curve case. The best possibility for doing this is when the curve admits a non-trivial group of automorphisms. It induces an action on the Jacobian of the curve in a natural way. If the group is sufficiently big, one can use the action even to compute a period matrix for the Jacobian. In general the action decomposes the Jacobian up to isogeny.

A slightly more general type of abelian varieties than Jacobians are Prym varieties of finite covers. So it seems naturally important to decompose the Jacobians up to isogeny into a product of these, since they are more accessible than general abelian varieties. So our results give a lot of interesting examples which should be of interest for many algebraic geometers and in particular those ones working on algebraic curves or Riemann surfaces. In fact the group of geometers working explicitly on the decomposition of abelian varieties is already fairly large. Our bibliography is by no means complete.

Another subject where Abelian varieties play a role are parts of arithmetic geometry. Let us mention only Shimura varieties, which are a generalization of some special abelian varieties. Also here our examples might find some interest.

Chapter 2
Preliminaries and Basic Results

In the first chapter, we recall some basics on abelian varieties. We try to give full definitions. For the proof of the results mentioned, we take the following point of view. For results we need and which are proved in [3], we only give a precise quotation. For the other results, which mostly appeared in some journal, we give a proof.

In Sects. 2.1 to 2.5, we recall the basics on abelian varieties, polarizations, endomorphisms, the Weil form associated to an ample line bundle, norm endomorphisms of an abelian subvariety, associated symmetric idempotents, and pairs of complementary abelian subvarieties.

In Sect. 2.6 we give some relations for the induced polarizations on abelian subvariety and its complement. In the case of a principal polarization, where this is well known, so we quote the corresponding results of [3]. In the case of a non-principally polarized abelian variety, we give a weak result (Theorem 2.6.10).

In Sect. 2.7 we recall Poincaré's complete reducibility theorem with action of a finite group, and in Sect. 2.8 we compile what we need on complex and rational representations of finite groups. We give proofs for the less well-known results.

In Sect. 2.9 we recall the isotypical decomposition and the group algebra decompositions of an abelian variety with the action of a finite group. Moreover we study the induced action on the tangent space.

For any subgroup H of a finite group acting on an abelian variety A, one can associate an abelian subvariety A_H of A in a natural way on which there is no corresponding subgroup action. However the Hecke algebra associated to H acts on it. In the last section, we study this action.

H. Lange, R. E. Rodríguez, *Decomposition of Jacobians by Prym Varieties*, Lecture Notes in Mathematics 2310, https://doi.org/10.1007/978-3-031-10145-8_2

2.1   Line Bundles on Abelian Varieties

An *abelian variety of dimension g* is by definition an algebraic complex torus, $A = V/\Lambda$ where V is a complex vector space of dimension g and Λ a lattice of maximal rank in V. Here "algebraic" means that A admits a projective embedding. So A is a smooth projective variety admitting the structure of an abelian group. A *homomorphism of abelian varieties* $f : A_1 \to A_2$ is by definition a holomorphic map f which is compatible with the group structures, which satisfies

$$f(x + y) = f(x) + f(y) \qquad \text{for all } x, y \in A_1.$$

The set of homomorphisms of A_1 to A_2 is an abelian group denoted by $\text{Hom}(A_1, A_2)$. If $A_i = V_i/\Lambda_i$ for $i = 1$ and 2, every homomorphism $f : A_1 \to A_2$ induces a vector space homomorphism $F : V_1 \to V_2$, and we get an injective homomorphism of abelian groups

$$\rho_a : \text{Hom}(A_1, A_2) \to \text{Hom}_{\mathbb{C}}(V_1, V_2), \qquad f \mapsto F,$$

called the *analytic representation* of $\text{Hom}(A_1, A_2)$. The restriction F_{Λ_1} of F to the lattice Λ_1 is $\mathbb{Z}$-linear and determines F and thus f completely. Hence we get an injective homomorphism

$$\rho_r : \text{Hom}(A_1, A_2) \to \text{Hom}_{\mathbb{Z}}(\Lambda_1, \Lambda_2), \qquad f \mapsto F_{\Lambda_1},$$

called the *rational representation of* $\text{Hom}(A_1, A_2)$. Clearly in the case of $A_1 = A_2$, ρ_a and ρ_r are ring homomorphisms.

A special type of homomorphisms is particularly important: a homomorphism $f : A_1 \to A_2$ is called an *isogeny* if it is surjective with finite kernel. Note that $f : A_1 \to A_2$ is an isogeny if and only if $\rho_a(f)$ is an isomorphism. The *degree* of the isogeny f is defined to be the order of its kernel, $\deg f = |\text{Ker } f|$, and one easily checks

$$\deg f = |\Lambda_2/\text{Im}(F_{\Lambda_1})|.$$

In this section we recall some properties of the group $\text{Pic}(A)$ of line bundles on $A = V/\Lambda$. Let $\pi : V \to A$ denote the canonical projection. Since every line bundle on V is trivial and Λ is the fundamental group of A, any line bundle $L \in \text{Pic}(A)$ can be described by a factor of automorphy, that is, a 1 cocycle of Λ with values in $H^0(\mathcal{O}_V^*)$. Hence pullback via π gives an isomorphism

$$\text{Pic}(A) = H^1(\mathcal{O}_A^*) \simeq H^1(\Lambda, H^0(\mathcal{O}_V^*)).$$

This implies that the global sections of L can be considered as some sort of theta functions, which depend of course on the choice of a basis of V.

At the moment we need only the Appell-Humbert theorem, which gives a description of Pic(A). For this recall that the first Chern class $c_1(L)$ of L is the image of $L \in \text{Pic}(A) = H^1(\mathcal{O}_A^*)$ under the boundary homomorphism $c_1 : H^1(\mathcal{O}_A^*) \to H^2(A, \mathbb{Z})$ of the exponential sequence of A and hence can be considered as a hermitian form on V whose imaginary part $\text{Im}\, c_1(L)$ is an integer valued alternating form on the lattice Λ (see [3, Section 2.1]).

Let $\mathbb{C}_1$ denote the circle group and $H = c_1(L)$ a hermitian form on V. A *semicharacter for H* is a map $\chi : \Lambda \to \mathbb{C}_1$ satisfying

$$\chi(\lambda + \mu) = \chi(\lambda)\chi(\mu)\exp(\pi i\, \text{Im}\, H(\lambda, \mu)) \quad \text{for all} \quad \lambda, \mu \in \Lambda.$$

Let $\mathcal{P}(\Lambda)$ denote the set of all pairs (H, χ) with H the first Chern class of a line bundle on A and χ a semicharacter for H. Recall furthermore that $\text{Pic}^0(A)$ denotes the subgroup of Pic(A) of line bundles which are algebraically equivalent to 0, and

$$\text{NS}(A) = \text{Pic}(A)/\text{Pic}^0(A)$$

is the Néron-Severi group of A. Then the **Appell-Humbert theorem** [3, Section 2.2] is the following statement:

For any abelian variety $A = V/\Lambda$ there is a canonical isomorphism of exact sequences

$$
\begin{array}{ccccccccc}
1 & \longrightarrow & \text{Hom}(\Lambda, \mathbb{C}_1) & \longrightarrow & \mathcal{P}(\Lambda) & \longrightarrow & \text{NS}(A) & \longrightarrow & 0 \\
& & \simeq \downarrow & & \simeq \downarrow & & \| & & \\
1 & \longrightarrow & \text{Pic}^0(A) & \longrightarrow & \text{Pic}(A) & \longrightarrow & \text{NS}(A) & \longrightarrow & 0
\end{array}
\tag{2.1}
$$

2.2 Polarized Abelian Varieties

A *polarization* on an abelian variety A is by definition the first Chern class $c_1(L)$ of an *ample* line bundle L on A. The line bundle L is ample if and only if the hermitian form $H = c_1(L)$ is positive definite. Hence a complex torus is an abelian variety if and only if it admits a polarization. The pair (A, H) or slightly less precisely the pair (A, L) is called a *polarized abelian variety*.

Let (A, L) be a polarized abelian variety of dimension g with associated hermitian form H. The alternating form $\text{Im}(H)$ is integer valued on the lattice Λ. Let $d_1, \ldots, d_g$ denote the elementary divisors of $\text{Im}(H)$, ordered such that $d_1 \leq d_2 \leq \cdots \leq d_g$. If L is ample, all d_i are positive integers, and the theorem of **Riemann-Roch** says

$$h^0(L) = \prod_{i=1}^{g} d_i \quad \text{and} \quad h^i(L) = 0 \text{ for } i > 0. \tag{2.2}$$

The tuple $(d_1, \ldots, d_g)$ is called the *type* of L or of the corresponding polarization. A polarization of type $(1, \ldots, 1)$ is called a *principal polarization*.

Given an abelian variety, its *dual abelian variety* $\widehat{A}$ is defined as follows: consider the $\mathbb{C}$-vector space $\overline{\Omega} := \operatorname{Hom}_{\overline{\mathbb{C}}}(V, \mathbb{C})$ of $\mathbb{C}$-antilinear forms $\ell : V \to \mathbb{C}$. As a real vector space, $\overline{\Omega}$ is canonically isomorphic to $\operatorname{Hom}_{\mathbb{R}}(V, \mathbb{R})$, with isomorphism given by $\ell \mapsto \operatorname{Im} \ell$. The canonical $\mathbb{R}$-bilinear form

$$\langle \, , \, \rangle : \overline{\Omega} \times V \to \mathbb{R}, \quad \langle \ell, v \rangle := \operatorname{Im} \ell(v)$$

is non-degenerate. This implies that

$$\widehat{\Lambda} := \{\ell \in \overline{\Omega} \mid \langle \ell, \Lambda \rangle \subseteq \mathbb{Z}\}$$

is a lattice in $\overline{\Omega}$. The quotient

$$\widehat{A} := \overline{\Omega}/\widehat{\Lambda}$$

is an abelian variety of dimension g, the *dual abelian variety* of A. Identifying V with the space of $\mathbb{C}$-antilinear forms on $\overline{\Omega}$, the non-degeneracy of $\langle \, , \, \rangle$ implies that we can identify

$$\widehat{\widehat{A}} = A.$$

The following description of $\widehat{A}$ will often be used (see [3, Proposition 2.4.1]): there is a canonical isomorphism $\widehat{A} \simeq \operatorname{Pic}^0(A)$ induced by the isomorphism

$$\overline{\Omega} \to \operatorname{Hom}(\Lambda, \mathbb{C}_1), \quad \ell \mapsto \exp(2\pi i \langle \ell, \cdot \rangle).$$

So we can identify:

$$\widehat{A} = \operatorname{Pic}^0(A). \tag{2.3}$$

It is clear how to define the dual homomorphism $\widehat{f} : \widehat{B} \to \widehat{A}$ of a homomorphism $f : A \to B$. Using (2.3) it is given by the pullback of line bundles.

Proposition 2.2.1

(a) *If* $0 \to A \to B \to C \to 0$ *is an exact sequence of abelian varieties, the dual sequence* $0 \to \widehat{C} \to \widehat{B} \to \widehat{A} \to 0$ *is also exact.*

(b) *If* $f : A \to B$ *is an isogeny of degree* d, *the dual map* $\widehat{f} : \widehat{B} \to \widehat{A}$ *is also an isogeny of degree* d. *To be more precise, if* K *is the kernel of the isogeny* f, *the dual group* $\widehat{K} = \operatorname{Hom}(K, \mathbb{C}_1)$ *is the kernel of the dual isogeny* $\widehat{f}$.

For a proof see [3, Propositions 2.4.2 and 2.4.3].

For any point $a \in A$, let $t_a : A \to A$ denote the *translation by a*. If L is any line bundle on A, the line bundle $t_a^* L \otimes L^{-1}$ has first Chern class zero. Hence using the identification (2.3), we get a map

$$\phi_L : A \to \widehat{A}, \quad a \mapsto t_a^* L \otimes L^{-1} \tag{2.4}$$

which according to [3, Theorem of the Square 2.3.3] is a homomorphism. Since the isogeny ϕ_L determines and is determined by $c_1(L)$, we often denote by ϕ_L also the polarization defined by L.

For any $L \in \text{Pic}(A)$, we denote

$$K(L) := \text{Ker}(\phi_L : A \to \widehat{A})$$

called *the kernel of the polarization* ϕ_L. Clearly $K(L)$ depends only on the first Chern class of L. Let (A, L) be a polarized abelian variety of type $(d_1, \ldots, d_g)$. Then we have for $K(L)$

$$K(L) \simeq \left(\bigoplus_{i=1}^{g} \mathbb{Z}/d_i\mathbb{Z} \right)^2 \tag{2.5}$$

Hence ϕ_L is an isogeny of degree d^2 with $d = d_1 \cdots d_g$. In particular, for a principal polarization, ϕ_L is an isomorphism, by which we can identify $A = \widehat{A} = \text{Pic}^0(A)$.

Proposition 2.2.2 *Let L be a polarization of type $(d_1, \ldots, d_g)$ on an abelian variety A. Then there is a unique polarization $\widehat{L}$ on the dual abelian variety $\widehat{A}$ satifying the property*

$$\phi_{\widehat{L}} \circ \phi_L = (d_1 d_g)_A := d_1 d_g \cdot id_A.$$

We call $\widehat{L}$ (respectively $\phi_{\widehat{L}}$) the *dual polarization of L* (respectively ϕ_L). For the proof of the proposition, see [3, Proposition 14.4.1].

2.3 Endomorphisms of Abelian Varieties

With respect to addition, the algebra $\text{End}(A)$ of endomorphisms of an abelian variety A of dimension g is a free abelian group of rank $\leq 4g^2$. This implies that

$$\text{End}_{\mathbb{Q}}(A) := \text{End}(A) \otimes \mathbb{Q}$$

is a finite dimensional $\mathbb{Q}$-vector space. Here we want to study some properties of these rings which are induced by a polarization on A.

Let $A = V/\Lambda$ be an abelian variety and L be a polarization of type $(d_1, \ldots, d_g)$ on A. The number d_g is called the *exponent* of the polarization. It is the exponent of the finite abelian group $K(L)$.

Then there is a unique isogeny $\psi_L : \widehat{A} \to A$ such that $\psi_L \phi_L = e(L)_A$ and $\phi_L \psi_L = e(L)_{\widehat{A}}$, where $e(L)_A$ and $e(L)_{\widehat{A}}$ denote the multiplications by the exponent $e(L)$ on A and $\widehat{A}$, respectively. So ϕ_L has an inverse in $\mathrm{Hom}_{\mathbb{Q}}(\widehat{A}, A) = \mathrm{Hom}(\widehat{A}, A) \otimes \mathbb{Q}$, namely,

$$\phi^{-1} = \frac{1}{e(L)} \psi_L.$$

Every $\alpha \in \mathrm{End}_{\mathbb{Q}}(A)$ can be written as rf with an endomorphism f of A and $r \in \mathbb{Q}$. The dual of α is defined as

$$\widehat{\alpha} := r \widehat{f}.$$

This definition does not depend on the choice of r and f.

The map

$${}' : \mathrm{End}_{\mathbb{Q}}(A) \to \mathrm{End}_{\mathbb{Q}}(A), \quad \alpha' := \phi_L^{-1} \widehat{\alpha} \phi_L$$

is an anti-involution on $\mathrm{End}_{\mathbb{Q}}(A)$ (i.e., it satisfies $(\alpha\beta)' = \beta'\alpha'$ for all $\alpha, \beta \in \mathrm{End}_{\mathbb{Q}}(A)$), called the *Rosati-involution* with respect to the polarization L (although it is an anti-involution).

For any abelian variety A, the analytic and rational representations

$$\rho_a : \mathrm{End}(A) \to \mathrm{End}_{\mathbb{C}}(V) \quad \text{and} \quad \rho_r : \mathrm{End}(A) \to \mathrm{End}_{\mathbb{Z}}(\Lambda)$$

extend to representations of $\mathrm{End}_{\mathbb{Q}}(A)$ denoted by the same symbol. The extended map $\rho_r \otimes 1 : \mathrm{End}_{\mathbb{Q}}(A) \otimes \mathbb{C} \to \mathrm{End}_{\mathbb{C}}(\Lambda \otimes \mathbb{C}) \simeq \mathrm{End}_{\mathbb{C}}(V)$ and the analytic representation are related as follows:

$$\rho_r \otimes 1 \simeq \rho_a \oplus \overline{\rho_a}. \tag{2.6}$$

The Rosati involution $f \mapsto f'$ with respect to a polarization L is the adjoint operator with respect to the hermitian form $H = c_1(L)$ (see [3, Proposition 5.1.1]); that is,

$$H(\rho_a(f)(v), w) = H(v, \rho_a(f')(w)) \quad \text{for all} \quad v, w \in V.$$

For any $\alpha \in \mathrm{End}_{\mathbb{Q}}(A)$, the characteristic polynomials P_α^r of $\rho_r(\alpha)$ and P_α^a of $\rho_a(\alpha)$ satisfy

$$P_\alpha^r(n) = P_\alpha^a(n) \cdot \overline{P_\alpha^a}(n) = \deg(n_A - \alpha) \quad \text{for all} \quad n \in \mathbb{Z} \tag{2.7}$$

where $\deg(\alpha) = r \deg(f)$ if $\alpha = rf$ as above and $\deg(f) = 0$ if f does not have a finite kernel [3, Proposition 5.1.2]. Suppose

$$P_\alpha^r(t) = \sum_{i=0}^{2g} (-1)^i r_i t^{2g-i} \quad \text{and} \quad P_\alpha^a(t) = \sum_{i=0}^{g} (-1)^i a_i t^{g-i}$$

with coefficients $r_i \in \mathbb{Q}$, $r_0 = 1$ and $a_i \in \mathbb{C}$, $a_0 = 1$. The *rational trace* and *norm* and the *analytic trace* and *norm* of α are defined by

$$\text{tr}_r(\alpha) = r_1, \quad \text{tr}_a(\alpha) = a_1, \quad \text{Nm}_r(\alpha) = r_{2g}, \quad \text{Nm}_a(\alpha) = a_g.$$

We use the following result [3, Theorem 5.1.8]

Theorem 2.3.1 *The map* $(\alpha, \beta) \mapsto \text{tr}_r(\alpha'\beta)$ *is a positive definite symmetric bilinear form on the* $\mathbb{Q}$-*vector space* $\text{End}_\mathbb{Q}(A)$.

This has the following important consequences:

The group of automorphisms of a polarized abelian variety (A, L) is finite.

(2.8)

(see [3, Corollary 5.1.9]). Here an automorphism $f \in \text{End}(A)$ is an automorphism of the polarized abelian variety (A, L) if it respects the polarization given by L, that is, if f^*L is algebraically equivalent to L or, equivalently, if $f^*(c_1(L)) = c_1(L)$.

For any positive integer n, let $A[n]$ denote the subgroup of n-division points of A. For any automorphism f of the polarized abelian variety (A, L), we have [3, Corollary 5.1.10]

$$\text{if} \quad f|_{A[n]} = id_{A[n]} \text{ for some } n \geq 3, \quad \text{then} \quad f = id_A. \tag{2.9}$$

2.4 The Weil Form on $K(L)$

The Weil form is a bimultiplicative non-degenerate complex valued form on the finite group $K(L)$ for any ample line bundle L. It was introduced by Weil in [35] in the special case where L is the square of the canonical principal polarization of a Jacobian J and hence $K(L) = J[2]$.

Let (A, L) be a polarized abelian variety and suppose $A = V/\Lambda$. We consider the first Chern class of L as a hermitian form H on V whose imaginary part is integer valued on the lattice Λ. If $\pi : V \to A$ is the canonical projection, we denote

$$\Lambda(L) := \pi^{-1}(K(L)).$$

so that

$$K(L) = \Lambda(L)/\Lambda.$$

Then the *Weil form* $e^L : K(L) \times K(L) \to \mathbb{C}^*$ is defined as

$$e^L(a_1, a_2) := \exp(-2\pi i \operatorname{Im} H(\alpha_1, \alpha_2))$$

for all $a_1, a_2 \in K(L)$ and $\pi(\alpha_i) = a_i$ and thus $\alpha_i \in \Lambda(L)$ for $i = 1, 2$.

The Weil form equals the commutator map on the theta group of L (see [3, Proposition 6.3.1]), but we don't need this fact. The map e^L is a bimultiplicative form in the following sense: it satisfies for all $a_1, a_2, a \in K(L)$:

$$e^L(a_1 + a_2, a) = e^L(a_1, a)e^L(a_2, a) \quad \text{and} \quad e^L(a_2, a_1) = e^L(a_1, a_2)^{-1}.$$

The form e^L is non-degenerate for any ample L: if $e^L(a, b) = 1$ for all $b \in K(L)$, then $a = 0$. This implies that for any subgroup $K \subset K(L)$, there is an *orthogonal complement* $K^\perp$. A subgroup $K \subset K(L)$ is called *isotropic* if $e^L(a_1, a_2) = 1$ for all $a_1, a_2 \in K$. We need the following result [3, Corollary 2.4.4].

Proposition 2.4.1 *For an isogeny $f : A \to B$ of abelian varieties and an ample line bundle $L \in \mathrm{Pic}(A)$, the following statements are equivalent:*

(i) $L = f^*M$ *for some* $M \in \mathrm{Pic}(B)$.
(ii) $\mathrm{Ker}\, f$ *is an isotropic subgroup of $K(L)$ with respect to e^L.*

Clearly M in condition i) is also an ample line bundle, so that $f : (A, L) \to (B, M)$ is an isogeny of polarized abelian varieties. The proposition has an important corollary:

Corollary 2.4.2 *Let L be an ample line bundle on A.*

(a) *If K is a maximal isotropic subgroup of $K(L)$ with respect to e^L and $\pi : A \to B := A/K$ is the canonical map, then there is a principal polarization M on B such that $L = \pi^*M$.*
(b) *If L is of type $(d_1, \ldots, d_g)$, then there there is a line bundle N on A such that $L = N^{d_1}$; N is of type $(1, \frac{d_2}{d_1}, \ldots, \frac{d_g}{d_1})$.*

For the proof of (a), use that $K(L) = \Lambda(L)/\Lambda$, and use [3, Lemma 3.1.4 and Remark 3.1.5]. The statement (a) only means that $c_1(L) = \pi^*c_1(M)$. But clearly then there is even a line bundle M with $L = \pi^*M$. For the proof of (b), use that all d_1-division points of $A[d_1]$ are contained in $K(L)$.

2.5 Symmetric Idempotents

In this section we describe the abelian subvarieties of a polarized abelian variety in terms of the endomorphism algebra of the abelian variety and derive some consequences.

Let (A, L) be a polarized abelian variety and B an abelian subvariety of A with canonical embedding $\iota : B \hookrightarrow A$. Let

$$e(B) := e(\iota^* L)$$

denote the exponent of the restricted polarization $L|_B$. It is called the *exponent of B* (with respect to the polarization L). Then

$$\psi_{\iota^* L} := e(B)\phi_{\iota^* L}^{-1} : \widehat{B} \to B$$

is an isogeny. The *norm-endomorphism* of A associated to B (with respect to L) is the composition

$$N_B = \iota\,\psi_{\iota^* L}\,\widehat{\iota}\,\phi_L : A \xrightarrow{\phi_L} \widehat{A} \xrightarrow{\widehat{\iota}} \widehat{B} \xrightarrow{\psi_{\iota^* L}} B \xrightarrow{\iota} A. \tag{2.10}$$

In other words, it is given by the commutative diagram

$$
\begin{array}{ccc}
A & \xrightarrow{\phi_L} & \widehat{A} \\
{\scriptstyle \iota}\big\uparrow & & \big\downarrow{\scriptstyle \widehat{\iota}} \\
B & \xleftarrow[\psi_{\iota^* L}]{} & \widehat{B}.
\end{array}
$$

The element

$$\epsilon_B := \frac{1}{e(B)} N_B$$

is a symmetric idempotent of the algebra $\mathrm{End}_{\mathbb{Q}}(A)$; that is, satisfying $\epsilon_B' = \epsilon_B$ and $\epsilon_B^2 = \epsilon_B$.

Conversely, if ϵ is a symmetric idempotent of $\mathrm{End}_{\mathbb{Q}}(A)$, there is a positive integer n such that $n\epsilon \in \mathrm{End}(A)$. Define

$$A^\epsilon := \mathrm{Im}(n\epsilon).$$

Then A^ϵ is an abelian subvariety of A, and the definition does not depend of the choice of n. The smallest such integer is the exponent of the abelian subvariety A^ϵ. For the following theorem, see [3, Theorem 5.3.2].

Theorem 2.5.1 *The assignments $B \mapsto \epsilon_B$ and $\epsilon \mapsto A^\epsilon$ are inverse to each other and give a bijection between the sets of*

(a) *Abelian subvarieties of A,*
(b) *Symmetric idempotents of $\mathrm{End}_{\mathbb{Q}}(A)$.*

Corollary 2.5.2 *Let (A, L) be a polarized abelian variety.*

(*a*) *For any abelian subvariety B of A, there is an abelian subvariety P, uniquely determined by the polarization, such that the addition map*

$$\mu : B \times P \to A$$

is an isogeny.

(*b*) *The map given by $B \mapsto P$ is a fixed-point free involution on the set of all abelian subvarieties of A.*

Proof If ϵ is the symmetric idempotent corresponding to B, the element $1_A - \epsilon$ is also a symmetric idempotent of $\mathrm{End}_{\mathbb{Q}}(A)$, defining P. The relation

$$\epsilon + (1_A - \epsilon) = 1_A$$

implies that the addition map $A \times P \to A$ is an isogeny.

The map is an involution, since $1_A - (1_A - \epsilon) = \epsilon$. The involution is fixed-point free, since the relation $\epsilon = 1_A - \epsilon$ implies $2\epsilon = 1_A$ which is impossible. □

The abelian subvariety P is called the *complementary abelian subvariety* of B. In general it depends on the polarization. The corollary implies that conversely B is the complementary abelian subvariety of P. So it makes sense to call the pair (B, P) of abelian subvarieties *a pair of complementary abelian subvarieties* of A with respect to the polarization L.

Another immediate consequence of Theorem 2.5.1 is the following corollary. For the proof see [3, Corollary 5.3.3].

Corollary 2.5.3 *For $\varphi \in \mathrm{End}(A)$ and $B = \mathrm{Im}\,\varphi$, the following statements are equivalent:*

(i) $\varphi = N_B$.

(ii) $\varphi' = \varphi$ and $\varphi^2 = e(B)\varphi$, *where φ' and $e(B)$ denote the Rosati involution and exponent with respect to the polarization L.*

In general it is not easy to apply this corollary, since in general it is not easy to compute the exponent of an abelian subvariety. In particular, the exponents of B and its complementary abelian subvariety may be different from each other. For an example see [3, Section 12.1].

In the case of a principally polarized abelian variety (A, L), we have a better result, for which we use the following definition. An endomorphism f of A is called *primitive* if $f = ng$ for some $g \in \mathrm{End}(A)$ implies $n = \pm 1$. Equivalently, f is primitive if and only if its kernel does not contain a subgroup $A[n]$ for some $n \geq 2$. Then, according to [3, 5.3.4] we have the following criterion.

Theorem 2.5.4 *Let (A, L) be a principally polarized abelian variety. For $f \in \mathrm{End}(X)$, the following statements are equivalent:*

(i) $f = N_B$ *for some abelian subvariety B of A.*

(ii) *The following three conditions hold*

    (a) f *is either primitive or* $f = 0$;

    (b) $f' = f$;

    (c) $f^2 = ef$ *for some positive integer* e.

If (ii) holds, then the integer e in (ii)(c) is the exponent $e(B)$ of the abelian subvariety B. An almost immediate consequence is the following corollary for the proof of which we refer to [3, Corollary 12.1.2].

Corollary 2.5.5 *Let* (B, P) *be a pair of complementary abelian subvarieties of a principally polarized abelian variety* (A, L). *Then the exponents satisfy*

$$e(B) = e(P).$$

Hence it makes sense to speak about the *exponent of the pair* (B, P) in the principally polarized case. Another almost immediate consequence of Theorems 2.5.1 and 2.5.4 is (see [3, Section 5.3])

Corollary 2.5.6 *Let* (B, P) *be a pair of complementary abelian subvarieties of exponent* e *of the (not necessarily principally) polarized abelian variety* (A, L). *Then*

$$B = \operatorname{Im} N_B = (\operatorname{Ker}(e_A - N_B))^0 \quad and \quad P = \operatorname{Im}(e_A - N_B) = (\operatorname{Ker} N_B)^0.$$

Proof The assumption on the pair (B, P) means for the corresponding symmetric idempotents $\epsilon_P = 1_A - \epsilon_B$. Since both abelian subvarieties are assumed to have the same exponent e, this means for the norm-endomorphisms $N_P = e_A - N_B$. This implies $N_B N_P = N_P N_B = 0$, which immediately gives the assertion. $\qquad\square$

To any automorphism α of a polarized abelian variety, one can associate two abelian subvarieties in the following way.

First let $\alpha : A \to A$ be an arbitrary endomorphism of an abelian variety and L a polarization of A. The pullback polarization $\alpha^* L$ is given by the commutative diagram

$$
\begin{array}{ccc}
A & \xrightarrow{\;\phi_L\;} & \widehat{A} \\[4pt]
{\scriptstyle \alpha}\Big\uparrow & & \Big\downarrow{\scriptstyle \widehat{\alpha}} \\[4pt]
A & \xrightarrow[\;\phi_{\alpha^* L}\;]{} & \widehat{A}
\end{array}
$$

Hence, if α is an automorphism of A, then α is an automorphism of the polarized abelian variety (A, L) if and only if

$$\widehat{\alpha}\,\phi_L\,\alpha = \phi_L. \tag{2.11}$$

In this case it follows from Eq. (2.9) that α is of finite order. Suppose α is of order n. Then we have

$$0 = 1_A - \alpha^n = (1_A - \alpha) \sum_{i=0}^{n-1} \alpha^i = (\sum_{i=0}^{n-1} \alpha^i)(1_A - \alpha).$$

As usual we denote for any subgroup H of an abelian variety A by H^0 the connected component of H containing 0. Consider the abelian subvarieties

$$B := \mathrm{Im}(\sum_{i=0}^{n-1} \alpha^i) = \mathrm{Ker}(1_A - \alpha)^0 \quad \text{and} \quad P := \mathrm{Im}(1_A - \alpha) = (\mathrm{Ker}(\sum_{i=0}^{n-1} \alpha^i))^0.$$

$$(2.12)$$

For any polarized abelian variety (A, L), we denote by $\mathrm{Aut}(A, L)$ the group of automorphisms α of A respecting the polarization, that is, satisfying $\alpha^* L$ algebraically equivalent to L.

Proposition 2.5.7 *Let* $\alpha \in \mathrm{Aut}(A, L)$ *be of order* n *and* L *a principal polarization. Then* (B, P) *is a pair of complementary abelian varieties of exponent* n.

Proof We will apply Theorem 2.5.4 to $\varphi = \sum_{i=0}^{n-1} \alpha^i$. Equation (2.11) is equivalent to

$$\alpha' = \phi_L \widehat{\alpha} \phi_L = \alpha^{-1} = \alpha^{n-1}$$

from which we get for $(\alpha^i)' = \alpha^{n-i}$ for $i = 0, \ldots, n-1$ and thus

$$\varphi' = (\sum_{i=0}^{n-1} \alpha^i)' = \sum_{i=0}^{n-1} \alpha^i = \varphi.$$

Moreover,

$$\varphi^2 = (\sum_{i=0}^{n-1} \alpha^i)^2 = n \sum_{i=0}^{n-1} \alpha^i = n\varphi.$$

Since $\sum_{i=0}^{n-1} \alpha^i$ clearly is primitive, Theorem 2.5.4 implies

$$N_B = \sum_{i=0}^{n-1} \alpha^i$$

and moreover B is of exponent n.

Let for a moment P' be the complementary abelian subvariety of B with respect to the principal polarization of A. By Corollary 2.5.5 we have $e(P') = e(B)$, and we may apply Corollary 2.5.6 to give

$$P' = (\text{Ker } N_B)^0 = (\text{Ker } \sum_{i=0}^{n-1} \alpha^i)^0 = P$$

where the last equality follows from (2.12). This completes the proof of the proposition. □

Remark 2.5.8 In general we have by the uniqueness of the complementary abelian subvariety that

$$P = \text{Im}(1_A - \alpha) = \text{Im}(n_A - \sum_{i=0}^{n-1} \alpha^i).$$

But by Theorem 2.5.4 in general, $1_A - \alpha$ is not the norm-endomorphism of P.

However, if α is an involution of (A, L), then $N_P = 1_A - \alpha$, since then $(1-\alpha)^2 = 2(1_A - \alpha)$.

In general, in terms of the analytic representation, B is the image in A of the eigenspace of 1 of $\rho_a(\alpha)$, and P is the image of the space generated by the other eigenspaces of $\rho_a(\alpha)$.

Proposition 2.5.9 *Let (A, L) be a principally polarized abelian variety and $\alpha \in \text{Aut}(A, L)$ be of order n. Then the set of fixed points of α restricted to P is*

$$\text{Fix}(\alpha|_P) = \text{Ker}(1 - \alpha) \cap P \subset P[n].$$

Proof If $y \in \text{Fix}(\alpha|_P)$, then $0 = (1 + \alpha + \alpha^2 + \cdots + \alpha^{n-1})(y) = ny$, which gives the assertion. □

Since $B + P = A$, we get immediately the following corollary, which we will apply in Chapter 4.

Corollary 2.5.10 *Let (A, L) be a principally polarized abelian variety and $\alpha \in \text{Aut}(A, L)$ be of order n. Then*

(i) $\text{Ker}(1_A - \alpha) = B + P_0$ with $P_0 = \text{Ker}(1_A - \alpha) \cap P \subset P[n]$.
(ii) $\text{Ker}(\sum_{i=0}^{n-1} \alpha^i) = P + B[n]$.

2.6   Abelian Subvarieties of a Polarized Abelian Variety

Let (A, L) be a polarized abelian variety with associated isogeny $\phi_L : A \to \widehat{A}$ and (B, P) be a pair of abelian subvarieties, complementary with respect to the polarization L. In this section we study the restriction of the polarization L to B and P, denoted by L_B and L_P, respectively.

2.6.1   The Principally Polarized Case

Suppose that L defines a principal polarization on the abelian variety A. Then this was done already in [3, Section 12.1]. Here we recall only the results without (the easy) proofs. Let $\iota : B \to A$ be the canonical embedding and $\widehat{\iota} : \widehat{B} \to \widehat{A} = A$ its dual map. Then [3, Proposition 12.1.3] gives

Proposition 2.6.1 $P = (\operatorname{Ker} N_B)^0 = \operatorname{Ker}\widehat{\iota} \simeq \widehat{(A/B)}$.

As an almost immediate consequence, we get for the kernel $K(L_B) = \operatorname{Ker} \phi_{L_B}$ (see [3, Corollary 12.1.4]).

Corollary 2.6.2   *As subgroups of A, we have the following equalities:*

$$K(L_B) = \iota^{-1}P = B \cap P = K(L_P).$$

Proposition 2.6.3   *If $\dim P \geq \dim B$ and the induced polarization L_B is of type $(d_1, \ldots, d_r)$, then the induced polarization L_P is of type $(1, \ldots, 1, d_1, \ldots, d_r)$, with $\dim P - \dim B$ numbers 1.*

For the proof see [3, Corollary 12.1.5].

Proposition 2.6.4   *Let (A, L) be a polarized abelian variety (with a not necessarily principal polarization L) and (B, P) be a pair of complementary abelian subvarieties. Let $\mu : B \times P \to A$ the isogeny given by the addition map. Then the pullback polarization $\mu^* L$ on $B \times P$ splits; that is*

$$\phi_{\mu^* L} = \phi_{L_B} \times \phi_{L_P} : B \times P \to \widehat{B \times P} = \widehat{B} \times \widehat{P}.$$

For the proof see [3, Corollary 5.3.6].
Finally, [3, Lemma 12.1.7] gives the following lemma.

Lemma 2.6.5   *Let (B, P) be a pair of complementary abelian subvarieties of exponent n in the principally polarized abelian variety (A, L). Then*

$$\operatorname{Ker} N_B \cap A[n] = \operatorname{Ker} N_B \cap \operatorname{Ker} N_P = B[n] + P[n] \subset A[n].$$

2.6.2   The Case of an Arbitrary Polarization

In the case of a non-principally polarized abelian variety, not much is known about the type of the restricted polarization. For example, in general the exponents of an abelian subvariety and its complement do not always coincide. For an example see

[3, Section 12.1]. We have only a weak result, Theorem 2.6.10 below. Here we follow [24].

Let (A, L) be a polarized abelian variety and ϵ_B and N_B the symmetric idempotent and norm-endomorphism of A associated to B and similarly ϵ_P and N_P for the complementary abelian subvariety P. Let e be the exponent of B. We make the following:

Assumption e is also the exponent of P; that is, we assume

$$N_B = e\epsilon_B \quad \text{and} \quad N_P = e\epsilon_P.$$

Since B is the complementary abelian subvariety of P, it suffices to prove the following lemmas concerning B and P for B alone.

Let

$$\iota_B : B \hookrightarrow A \quad \text{and} \quad \pi_B : A \to A/B$$

denote the canonical inclusion and projection and similarly ι_P and π_P for the complementary abelian subvariety P. The addition $\mu : B \times P \to A$ satisfies $\mu = \iota_B + \iota_P$, and we have the following exact sequence

$$0 \to B \cap P \to B \times P \xrightarrow{\mu} A \to 0. \tag{2.13}$$

Lemma 2.6.6 *The finite subgroup $B \cap P$ is contained in $B[e]$ and in $P[e]$.*

Proof For $x \in B \cap P$ we have $N_B(x) = ex$, since $x \in B$, and $N_B(x) = 0$ since $x \in P$. This gives the assertion.   □

Recall the notation for the restricted polarizations

$$L_B = L|_B \quad \text{and} \quad L_P = L|_P$$

and that $K(L_B)$ is the kernel of the isogeny $\phi_{L_B} : B \to \widehat{B}$ and similarly for $K(L_P)$.

Lemma 2.6.7

$$|K(L_B)| \cdot |K(L_P)| = |B \cap P|^2 \cdot |K(L)|.$$

Proof According to Proposition 2.6.4, the polarization $\mu^* L$ splits; that is, $\phi_{\mu^* L} = \phi_{L_B} \times \phi_{L_P}$, which gives

$$|K(\mu^* L)| = |K(L_B)| \cdot |K(L_P)|.$$

On the other hand, the exact sequence (2.13) implies $\deg \mu = |B \cap P|$ and thus

$$|K(\mu^*L)| = |\operatorname{Ker}(\widehat{\mu}\phi_L\mu)| = \deg(\mu)^2 \cdot |K(L)| = |B \cap P|^2 \cdot |K(L)|.$$

Combining both equations gives the assertion.   □

Lemma 2.6.8 *We have the following equality of abelian subvarieties of $\widehat{A}$:*

$$\phi_L(B) = \widehat{A/P} \subset \widehat{A} \quad and \quad \phi_L(P) = \widehat{A/B} \subset \widehat{A}.$$

Proof The norm-endomorphism N_B (see (2.10)) factorizes as $\iota_B \circ \gamma$ with $\gamma : A \to B$. Dualizing gives the factorization

$$\widehat{N_B} : \widehat{A} \xrightarrow{\widehat{\iota_B}} \widehat{B} \xrightarrow{\widehat{\gamma}} \widehat{A}.$$

With Proposition 2.2.1 this gives $\operatorname{Ker} \widehat{\iota_B} = \widehat{A/B} \subset \ker \widehat{N_B}$. On the other hand, by Corollary 2.5.6 we have $P \subset \operatorname{Ker} N_B$, and since the symmetry of N_B with respect to the Rosati involution says $\phi_L N_B = \widehat{N_B}\phi_L$, this implies

$$\phi_L(P) \subset \operatorname{Ker} \widehat{N_B}.$$

This implies the second assertion, since both abelian subvarieties are the connected component containing the origin of $\operatorname{Ker} \widehat{N_B}$. The first assertion follows by interchanging P and B.   □

We give two descriptions of $K(L_P)$ ($= \operatorname{Ker}(\phi_{L_P} : P \xrightarrow{\iota_P} A \xrightarrow{\phi_L} \widehat{A} \xrightarrow{\widehat{\iota_P}} \widehat{P})$) and $K(L_B)$. For the first one, note that $\operatorname{Ker} \widehat{\iota_P} = \widehat{A/P}$ and, by Lemma 2.6.8, $\widehat{A/P} = \phi_L(B)$. Hence we get

$$K(L_P) = \bigcup_{x \in K(L)} (B + x) \cap P \quad \text{and similarly} \quad K(L_B) = \bigcup_{x \in K(L)} (P + x) \cap B.$$
$$(2.14)$$

In particular,

$$B \cap P \subset K(L_B) \cap K(L_P).$$

For the second description, consider the isogenies

$$\alpha : P \xrightarrow{\iota_P} A \xrightarrow{\pi_B} A/B \quad \text{and} \quad \beta : B \xrightarrow{\iota_B} A \xrightarrow{\pi_P} A/P$$

with

$$\operatorname{Ker} \alpha = B \cap P \subset P \quad \text{and} \quad \operatorname{Ker} \beta = B \cap P \subset B.$$

Moreover, Lemma 2.6.8 gives the isogenies

$$\varphi_P := \phi_L|_P : P \to \widehat{A/B} \quad \text{and} \quad \varphi_B := \phi_L|_B : B \to \widehat{A/P}.$$

With these notations we have

Lemma 2.6.9 *The following sequences are exact:*

$$0 \to B \cap P \hookrightarrow K(L_P) \xrightarrow{\alpha} \operatorname{Ker}\widehat{\varphi_P} \to 0 \ with \ K(L) \cap P \simeq \operatorname{Ker}\widehat{\varphi_P} \subset A/B.$$

$$(2.15)$$

$$0 \to B \cap P \hookrightarrow K(L_B) \xrightarrow{\beta} \operatorname{Ker}\widehat{\varphi_B} \to 0 \ with \ K(L) \cap B \simeq \operatorname{Ker}\widehat{\varphi_B} \subset A/B.$$

$$(2.16)$$

Proof It suffices to prove (2.15). The dual isogeny of α factorizes as

$$\widehat{\alpha} : \widehat{A/B} \xrightarrow{\widehat{\pi_B}} \widehat{A} \xrightarrow{\widehat{\iota_P}} \widehat{P}.$$

According to Lemma 2.6.8, $\phi_L(P) = \widehat{A/B}$ which implies $\phi_{L_P} = \widehat{\iota_P}\phi_L\iota_P = \widehat{\alpha}\phi_L\iota_P$. Hence ϕ_{L_p} factorizes as $\phi_{L_P} : P \xrightarrow{\varphi_P} \widehat{A/B} \xrightarrow{\widehat{\alpha}} \widehat{P}$. Dualizing and using $\widehat{\phi_{L_P}} = \phi_{L_P}$, we get the factorization

$$\phi_{L_P} : P \xrightarrow{\alpha} A/B \xrightarrow{\widehat{\varphi_P}} \widehat{P}.$$

This gives the exact sequence (2.15). The assertion on $\operatorname{Ker}\widehat{\varphi_P}$ follows from the equation $\operatorname{Ker}\widehat{\varphi_P} \simeq \operatorname{Ker}\varphi_P = K(L) \cap P$.   □

In particular, Lemma 2.6.9 gives

$$|K(L_P)| = |B \cap P| \cdot |K(L) \cap P| \quad \text{and} \quad |K(L_B)| = |B \cap P| \cdot |K(L) \cap B|. \quad (2.17)$$

Inserting this into Lemma 2.6.7 gives the following theorem.

Theorem 2.6.10 *Let B and P be a pair of complementary abelian subvarieties of the polarized abelian variety* (A, L). *Then*

$$|K(L)| = |K(L) \cap B| \cdot |K(L) \cap P|.$$

Now we consider for any pair (B, P) in (A, L) as above the finite groups of connected components of the kernels of N_P and N_B (see Corollary 2.5.6).

$$G_P := \operatorname{Ker} N_B / P \quad \text{and} \quad G_B := \operatorname{Ker} N_P / B = \operatorname{Ker}(e_A - N_B)/B.$$

Lemma 2.6.11 *There are canonical exact sequences:*

$$0 \to B \cap P \to B[e] \to G_P \to 0 \quad \text{and} \quad 0 \to B \cap P \to P[e] \to G_B \to 0.$$

Proof By definition, $P \subset \operatorname{Ker} N_B$. Hence N_B descends to an isogeny $\overline{N}_B : A/P \to B$ with kernel G_P. From $N_B|_B = e_B$ we conclude that the composite isogeny

$$B \xrightarrow{\pi_P \iota_B} A/P \xrightarrow{\overline{N}_B} B$$

equals multipliplication by e. Taking kernels gives the first exact sequence. The second one follows by interchanging B and P.   □

We denote by

$$N_B[e] : A[e] \to B[e] \quad \text{and} \quad (e_A - N_B)[e] : A[e] \to P[e]$$

the restrictions of N_B and $N_P = e_A - N_B$ to the e-torsion points. Note that $N_B[e]$ and $(e_A - N_B)[e]$ are nilpotent:

$$(N_B[e])^2 = ((e_A - N_B)[e])^2 = 0$$

and we have

$$G_P \subset (A/P)[e] \quad \text{and} \quad G_B \subset (A/B)[e].$$

To be more precise, we have

Proposition 2.6.12 *We have the following equalities of finite groups:*

$$\operatorname{Im} N_B[e] = B \cap P \quad \text{and} \quad G_P = \operatorname{Coker} N_B[e].$$

$$\operatorname{Im}(e_A - N_B)[e] = B \cap P \quad \text{and} \quad G_B = \operatorname{Coker}(e_A - N_B)[e].$$

Proof It suffices to prove the first two equalities. Note first that

$$\operatorname{Im} N_B[e] \subset B \cap P.$$

To see this, let $b \in \operatorname{Im} N_B[e]$, so $b = N_B(a)$ for some $a \in A[e]$. Then $ea = 0$ and hence $b = (N_B - e_A)(a) \in P$.

In order to show equality, we compute the indices as subgroups of $B[e]$. By Lemma 2.6.11 the index of $B \cap P$ in $B[e]$ is $|G_P|$. The index of $\operatorname{Im} N_B[e]$ in $B[e]$ is

$$\frac{|B[e]|}{|\operatorname{Im} N_B[e]|} = \frac{|B[e]|}{|A[e]|} \cdot |\operatorname{Ker} N_B[e]| = \frac{|\operatorname{Ker} N_B[e]|}{|P[e]|}.$$

Moreover, since $G_P \subset (A/P)[e]$, the exact sequence $0 \to P \to \mathrm{Ker}\, N_B \to G_P \to 0$ remains exact after restricting to the e-torsion points. This implies the first equality. The second equality follows from this and Lemma 2.6.11.   □

Finally we compute the kernels of the duals of the isogenies φ_P and φ_B of above:

$$\widehat{\varphi_P} : A/B \to \widehat{P} \quad and \quad \widehat{\varphi_B} : A/P \to \widehat{B}.$$

Proposition 2.6.13 *We have the following equalities of subgroups of A/B and A/P, respectively:*

$$\mathrm{Ker}\, \widehat{\varphi_P} = \pi_B(K(L)) \quad and \quad \mathrm{Ker}\, \widehat{\varphi_B} = \pi_P(K(L)).$$

In particular $\widehat{\varphi_P}$ and $\widehat{\varphi_P}$ are injective if L is a principal polarization.
Proof Again it suffices to show the first equality. The inclusion $\pi_B(K(L)) \subset \mathrm{Ker}\, \widehat{\varphi_P}$ follows from the commutative diagram

$$
\begin{array}{ccc}
A & \xrightarrow{\phi_L} & \widehat{A} \\
{\scriptstyle \pi_B}\downarrow & & \downarrow{\scriptstyle \widehat{\iota_P}} \\
A/B & \xrightarrow{\widehat{\varphi_P}} & \widehat{P}.
\end{array}
$$

Hence it will be enough to show that these two groups have the same order. Using (2.15) we have

$$| \mathrm{Ker}\, \varphi_P| = | \mathrm{Ker}\, \widehat{\varphi_P}| = |K(L) \cap P|.$$

On the other hand, clearly

$$|\pi_B(K(L))| = \frac{|K(L)|}{|K(L) \cap B|}.$$

The assertion follows from Theorem 2.6.10.   □

2.7 Poincaré's Reducibility Theorem

Let G be a finite group acting on an abelian variety A. A is called G-*decomposable* if there are non-trivial G-stable abelian subvarieties A_1 and A_2 such that the addition $\mu : A_1 \times A_2 \to A$ is an isogeny. Otherwise A is called G-*simple*. *Poincaré's complete reducibility theorem with G-action* is the following theorem (see [3, Theorem 13.5.2]).

Theorem 2.7.1 *There are G-simple abelian subvarieties $A_1, \ldots, A_r$ of A such that the addition map*

$$\mu : A_1 \times \cdots \times A_r \to A$$

is a G-equivariant isogeny. This decomposition is unique up to G-equivariant isogenies and permutations.

For the special case $G = \{1_A\}$, we get the usual Poincaré's complete reducibility theorem. Note that the abelian subvarieties A_i are not uniquely determined, they are determined only up to isogeny. For the proof one uses that there always is a G-invariant polarization on A and then applies Theorem 2.5.1.

2.8   Complex and Rational Representations of Finite Groups

In this section we recall from [7] some results on complex and rational representations of a finite group G. Concerning the notation, we follow [34].

Let V be an irreducible complex representation of G (of finite dimension). We denote by L its *field of definition* and by K its *character field,* that is, the field generated over the rationals by the values of the characters of V. Then $K \subseteq L$, and the index

$$s = s(V) = [L : K]$$

is called the *Schur index of V*.

For any subgroup H of G, we denote by ρ_H the representation induced by the trivial representation of H and by V^H the subspace of V fixed by H. For any two complex representations U and V of G we denote by $\langle U, V \rangle$ the usual inner product of the characters of the representations U and V. Then we get from Frobenius reciprocity theorem that

$$\langle \rho_H, V \rangle = \dim V^H. \tag{2.18}$$

Both fields K and L are Galois over $\mathbb{Q}$. For any $\gamma \in \mathrm{Gal}(L/\mathbb{Q})$, the representation V^γ is conjugate to V if and only if γ is contained in the subgroup $\mathrm{Gal}(L/K)$ of $\mathrm{Gal}(L/\mathbb{Q})$. The direct sum

$$U := \bigoplus_{\gamma \in \mathrm{Gal}(L/K)} V^\gamma \simeq sV$$

is defined over K, where s is the Schur index of V.

There is a unique irreducible rational representation W of G such that

$$W \otimes K \simeq \bigoplus_{\gamma \in \mathrm{Gal}(K/\mathbb{Q})} U^{\gamma} \quad \text{and} \tag{2.19}$$

$$W \otimes L \simeq \bigoplus_{\delta \in \mathrm{Gal}(L/\mathbb{Q})} V^{\delta} \simeq s \bigoplus_{\gamma \in \mathrm{Gal}(K/\mathbb{Q})} V^{\gamma}. \tag{2.20}$$

Conversely, any irreducible rational representation is of this form. The representations V and W are called *Galois associated* in this case.

Observe that it follows immediately that the character of the rational irreducible representation W with Galois associated complex irreducible representation V is given by

$$\chi_W(g) = s \sum_{\gamma \in \mathrm{Gal}(K/\mathbb{Q})} \mathrm{tr}_{K/\mathbb{Q}} \chi_V(g) = s \sum_{\gamma \in \mathrm{Gal}(K/\mathbb{Q})} \mathrm{tr}_{K/\mathbb{Q}} \chi_V(g^{-1}),$$

where the last equality follows since $\chi_W(g)$ is a rational number and $\chi_V(g^{-1}) = \overline{\chi_V(g)}$.

Recall that the complex and rational group algebras $\mathbb{C}[G]$ and $\mathbb{Q}[G]$ are semisimple algebras of finite dimension. As such they admit unique decompositions

$$\mathbb{C}[G] = R_1 \times \cdots \times R_s \quad \text{and} \quad \mathbb{Q}[G] = Q_1 \times \cdots \times Q_r \tag{2.21}$$

with simple $\mathbb{C}$-algebras R_i and simple $\mathbb{Q}$-algebras Q_j. The R_i correspond bijectively to the irreducible complex representations V_i of G. Similarly the Q_j corresponds bijectively to the irreducible rational representations W_j of G. If we denote by e_{V_i} the unique central idempotent of $\mathbb{C}[G]$ that generates R_{V_i} and by e_{W_j} the unique central idempotent of $\mathbb{Q}[G]$ that generates Q_{W_j}, then

$$1 = e_{V_1} + \cdots + e_{V_s} \quad \text{and} \quad 1 = e_{W_1} + \cdots + e_{W_r} \tag{2.22}$$

are the decompositions of $1 \in \mathbb{C}[G]$ and $1 \in \mathbb{Q}[G]$ given by (2.21). Note that these decompositions are orthogonal with respect to the inner product $\langle U, V \rangle$ of above, respectively the induced inner product on the set of rational representations.

Moreover, e_{V_i} and e_{W_j} are given as follows:

$$e_{V_i} = \frac{\dim V_i}{|G|} \sum_{g \in G} \chi_{V_i}(g^{-1}) g$$

where χ_{V_i} denotes the character of V_i and

$$e_{W_j} = \frac{\dim V_j}{|G|} \sum_{g \in G} \mathrm{tr}_{K_j/\mathbb{Q}} \chi_{V_j}(g^{-1}) g \tag{2.23}$$

where V_j is an irreducible complex representation Galois associated to W_j and $\mathrm{tr}_{K_j/\mathbb{Q}}$ denotes the trace of K_j over $\mathbb{Q}$. Furthermore, the idempotents e_{V_i} (respectively, e_{W_j}) are orthogonal to each other.

Observe that the decompositions (2.21) are given by

$$R_i = \mathbb{C}[G]e_{V_i} \qquad \text{and} \qquad Q_j = \mathbb{Q}[G]e_{W_j}$$

respectively, where G acts on each R_i via the representation $(\dim V_i)\,V_i$, and on Q_j via the representation $\left(\frac{\dim V_j}{s_j}\right) W_j$, as is well known in the first case and a consequence of (2.20) in the second one.

We need the following lemma.

Lemma 2.8.1 *Let $W_1, \ldots, W_r$ denote the irreducible rational representations of a finite group G. For each W_j choose a Galois associated representation V_j. Then for any subgroup $H \subseteq G$, the rational isotypical decomposition of the induced representation ρ_H of the trivial representation of H is*

$$\rho_H = \bigoplus_{j=1}^{r} \frac{1}{s_j} \dim(V_j^H) W_j$$

where s_j denotes the Schur index of the representation V_j.

Proof The representation ρ_H is of the form $\rho_H = \oplus_{j=1}^{r} a_j W_j$ with $a_j \in \mathbb{Z}$. We have to show that $a_i = \frac{1}{s_i} \dim V^H$ for each $i = 1, \ldots, r$.

But using Eqs. (2.18) and (2.20) and the fact that W_i and W_j are orthogonal if $j \neq i$, we have

$$\dim V_i^H = \langle \rho_H, V_i \rangle = \sum_{j=1}^{r} a_j \langle W_j, V_i \rangle$$

$$= a_i s_i \left\langle \bigoplus_{\gamma \in \mathrm{Gal}(K/\mathbb{Q})} V_i^\gamma, V_i \right\rangle$$

$$= a_i s_i \sum_{\gamma \in \mathrm{Gal}(K/\mathbb{Q})} \langle V_i^\gamma, V_i \rangle = a_i s_i$$

where the last equality follows from the fact that there is exactly one $\gamma (= \mathrm{id}_\mathbb{Q})$ such that $V_i^\gamma = V_i$ and for this γ we have $\langle V_i, V_i \rangle = 1$ while for all the others $\langle V_i^\gamma, V_i \rangle = 0$. □

Corollary 2.8.2 *Let W be an irreducible rational representation of G with Galois associated complex representation V. Then for any subgroup $H \subset G$, we have*

$$\dim W^H = [L : \mathbb{Q}] \dim V^H.$$

Proof If s is the Schur index of V, we have

$$\dim W^H = \langle \rho_H, W \rangle \quad \text{(by Frobenius reciprocity)}$$

$$= \frac{\dim V^H}{s} \langle W, W \rangle \quad \text{(by Lemma 2.8.1)}$$

$$= \frac{\dim V^H}{s} \langle s \bigoplus_{\gamma \in \mathrm{Gal}(K/\mathbb{Q})} V^\gamma, s \bigoplus_{\delta \in \mathrm{Gal}(K/\mathbb{Q})} V^\delta \rangle \quad \text{(by (2.20))}$$

$$= s \dim V^H [K : \mathbb{Q}] \langle V, V \rangle = [L : \mathbb{Q}] \dim V^H$$

where the last equation follows from (2.20) and $\langle V, V \rangle = 1$. □

For any subgroup $H \subset G$ denote

$$p_H := \frac{1}{|H|} \sum_{h \in H} h. \tag{2.24}$$

p_H is the central idempotent in $\mathbb{Q}[H]$ corresponding to the trivial representation. Since $\mathbb{Q}[H] \subset \mathbb{Q}[G]$, it is also an idempotent of $\mathbb{Q}[G]$.

Lemma 2.8.3 *For any subgroup $H \subset G$, the following rational representations of G are isomorphic:*

$$\mathbb{Q}[G]p_H \simeq \rho_H.$$

We say that the $\mathbb{Q}[G]$-module $\mathbb{Q}[G]p_H$ *affords* the representation ρ_H.
Proof Let χ denote the trivial representation of H. Then $p_H = e_\chi$ is the corresponding central idempotent of $\mathbb{Q}[H]$, and the simple algebra $M := \mathbb{Q}[H]p_H$ is the component of $\mathbb{Q}[H]$ in the isotypical decomposition corresponding to the trivial representation: $\mathbb{Q}[H]p_H$ affords the trivial representation of H.

Since $\rho_H := \mathrm{Ind}_H^G(\chi)$ is the representation of G induced by χ, it follows from [10, Proposition 11.21] that ρ_H is afforded by

$$\mathbb{Q}[G] \otimes_{\mathbb{Q}[H]} M = \mathbb{Q}[G]p_H.$$

□

Corollary 2.8.4 *In particular, for $H = \{1_G\}$, the regular representation ρ_{1_G} of G given by $\mathbb{Q}[G] \simeq \rho_{1_G}$ satisfies*

$$\rho_{1_G} = \sum_{j=1}^{r} \frac{\dim V_j}{s_j} W_j.$$

Proof It is well known that the regular representation of G given by $\mathbb{Q}[G] \simeq \rho_{1_G}$ satisfies

$$\rho_{1_G} = \sum_{V \in \mathrm{Irr}_{\mathbb{C}}(G)} \dim(V)\, V.$$

Now considering (2.20), the result follows.   □

Let W be an irreducible rational representation of G with associated central idempotent e_W, and let V be a complex irreducible representation of G Galois associated to W. The element

$$f_{H,W} := p_H e_W = e_W p_H$$

is an element of the simple algebra $\mathbb{Q}[G]e_W$ satisfying

(i) $f_{H,W}^2 = f_{H,W}$
(ii) $f_{H,W}h = h f_{H,W} = f_{H,W}$ for every $h \in H$

(i) follows since e_W is central and e_W and p_H are idempotents. (ii) is clear from the definition of $f_{H,W}$.

Multiplying (2.22) by p_H, it follows that

$$p_H = p_H e_{W_1} + \ldots + p_H e_{W_r}.$$

Since all summands are idempotents and orthogonal among themselves, this gives the following equality of left ideals in $\mathbb{Q}[G]$,

$$\mathbb{Q}[G]p_H = \mathbb{Q}[G]p_H e_{W_1} \oplus \ldots \oplus \mathbb{Q}[G]p_H e_{W_r}.$$

and therefore, for each left ideal $I_j := \mathbb{Q}[G]p_H e_{W_j}$, we have

$$I_j = \mathbb{Q}[G]e_{W_j}p_H = \mathbb{Q}[G]e_{W_j} \cap \mathbb{Q}[G]p_H. \qquad (2.25)$$

Comparing equality (2.25) with Lemma 2.8.1, we see that for each j either $\langle \rho_H, W_j \rangle = 0$, in which case $I_j = 0$ (or, equivalently, $p_H e_{W_j} = 0$), or $\langle \rho_H, W_j \rangle \neq 0$, in which case $I_j \neq 0$ (or, equivalently, $p_H e_{W_j} \neq 0$) and G acts on I_j via the representation $\dfrac{\dim V_j^H}{s_j} W_j$.

Proposition 2.8.5 *For any subgroup $H \subset G$ and any irreducible rational representation W, the following rational representations of G are isomorphic:*

$$\mathbb{Q}[G]f_{H,W} \simeq \frac{\dim V^H}{s} W.$$

Here V is a complex irreducible representation of G Galois associated to W with Schur index s.

Proof From Lemmas 2.8.1 and 2.8.3, we get

$$\bigoplus_{j=1}^{r} \frac{1}{s_j} \dim(V_j^H) W_j \simeq \rho_H \simeq \mathbb{Q}[G] p_H.$$

On the other hand, by (2.25) we have

$$\mathbb{Q}[G] f_{H,W} = \mathbb{Q}[G] p_H \cap \mathbb{Q}[G] e_W.$$

Together, this implies the assertion, since, as we saw above, G acts on $\mathbb{Q}[G] e_W$ via the representation $\frac{\dim V}{s} W$. $\qquad\square$

Proposition 2.8.6 *With the above hypotheses, the following statements are equivalent:*

(1) $f_{H,W} = 0$.
(2) $\langle \rho_H, W \rangle = 0$.
(3) $\dim V^H = 0$.

Proof (1) $\iff$ (3) follows from Proposition 2.8.5. (2) $\iff$ (3) follows from Lemma 2.8.1 and the orthogonality of the representations W_i. $\qquad\square$

2.9 The Isotypical and Group Algebra Decompositions

2.9.1 Generalities

In this subsection we recall the definition of the isotypical and the group algebra decompositions of an abelian variety with finite group action and recall two results of [3, Chapter 13].

Let A be an abelian variety with an action of a finite group G. The action induces an algebra homomorphism

$$\rho : \mathbb{Q}[G] \to \mathrm{End}_{\mathbb{Q}}(A) \tag{2.26}$$

of the rational group algebra $\mathbb{Q}[G]$ of G into $\mathrm{End}_{\mathbb{Q}}(A)$. In the notation we do not distinguish between elements of $\mathbb{Q}[G]$ and their images in $\mathrm{End}_{\mathbb{Q}}(A)$ (although the representation ρ is not necessarily faithful). Every element $\alpha \in \mathbb{Q}[G]$ defines an abelian subvariety

$$A^{\alpha} := \mathrm{Im}(\alpha) := \mathrm{Im}(m\alpha) \subseteq A,$$

where m is some positive integer such that $m\alpha$ is an endomorphism. Clearly this definition does not depend on the chosen m.

Recall that $\mathbb{Q}[G]$ is a semisimple $\mathbb{Q}$-algebra of finite dimension. As such it admits a unique decomposition

$$\mathbb{Q}[G] = Q_1 \times \cdots \times Q_r \qquad (2.27)$$

with simple $\mathbb{Q}$-algebras $Q_i = \mathbb{Q}[G]e_i$ with central idempotents e_i. Note that the Q_i corresponds bijectively to the irreducible rational representations of the group G. Let us say that e_i corresponds to the representation W_i. Then the group G acts on A^{e_i} by a multiple of the representation W_i. For the sake of abbreviation, we say that *G acts on A^{e_i} by the representation W_i.*

Let

$$1 = e_1 + \cdots + e_r \qquad (2.28)$$

be the decomposition of $1 \in \mathbb{Q}[G]$ given by (2.27). Then [3, Proposition 13.6.1] gives

Theorem 2.9.1

(a) *A^{e_i} is a G-stable abelian subvariety of A with*

$$\mathrm{Hom}_G(A^{e_i}, A^{e_j}) = 0 \quad for \quad i \neq j.$$

(b) *The addition map induces a G-equivariant isogeny*

$$\mu : A^{e_1} \times \cdots \times A^{e_r} \to A,$$

where G acts on A^{e_i} by the representation W_i.

This decomposition is called the *isotypical decomposition* of A. It is unique up to permutations, since the e_i are uniquely determined by $\mathbb{Q}[G]$. Each rational irreducible representation of G gives one of the components A^{e_i}. However several of them might be zero.

The non-zero abelian subvarieties A^{e_i} are called the *isotypical components of A* with respect to the action of G. Any sum $\sum_{j=1}^{t} A^{e_{i_j}}$ is called an *isotypical abelian subvariety*. To any rational representation W, one can associate an isotypical abelian subvariety A_W as follows: let W_i for $i = 1, \ldots, r$ be the irreducible rational representations of G with corresponding symmetric idempotents e_i. If $W = \sum n_i W_i,\ n_i \geq 0$, then

$$A_W := \sum_{i \text{ with } n_i > 0} A^{e_i}$$

is an abelian subvariety of A, called the *isotypical abelian subvariety associated to* W. Clearly not all such subvarieties are different from each other, since several of them might be zero.

In some cases the components A^{e_i} corresponding to the irreducible rational representations W_i can be decomposed further. In order to avoid indices, let W denote an irreducible rational representation of G with unit element e_W given by Eq. (2.23) and corresponding simple $\mathbb{Q}$-algebra Q. According to Schur's lemma,

$$D := \text{End}_G(W)$$

is a skew-field of finite dimension over $\mathbb{Q}$. The $\mathbb{Q}$-vector space W is a finite dimensional left-vector space over D, and we have

$$Q \simeq \text{End}_D(W). \tag{2.29}$$

Let V be a complex representation Galois associated to W. Then we have

$$n := \dim_D(W) = \frac{\dim V}{[L:K]} = \frac{\dim V}{s}. \tag{2.30}$$

There is a set of orthogonal primitive idempotents $\{q_1, \ldots, q_n\}$ in $Q \subseteq \mathbb{Q}[G]$ such that

$$e_W = q_1 + \cdots + q_n. \tag{2.31}$$

For $j = 1, \ldots, n$, let $B_j := \text{Im}(q_j)$. Then (2.31) implies that the addition map

$$\mu : B_1 \times \cdots \times B_n \to A_W \tag{2.32}$$

is an isogeny. Note that the q_j are conjugate to each other. This implies that the abelian subvarieties B_j are pairwise isogenous. We get the following decomposition, called the *group algebra decomposition* of the abelian variety A with an action of the group G. The components B_i are called *group algebra components* corresponding to the representation W. Note that all group algebra components with respect to the same representation W are pairwise isogenous to each other.

Theorem 2.9.2 *Let* $W_1, \ldots, W_r$ *denote the irreducible rational representations of* G *and* $n_i := \dim_{D_i}(W_i)$ *with* $D_i := \text{End}_G(W_i)$ *for* $i = 1, \ldots, r$. *Then there are abelian subvarieties* $B_1, \ldots B_r$ *of* A *and a* G-*equivariant isogeny*

$$B_1^{n_1} \times \cdots \times B_r^{n_r} \sim A,$$

where G *acts on* $B_i^{n_i}$ *via the representation* W_i *(with appropriate multiplicity).*

For the proof see [3, Theorem 13.6.3].

Whereas the abelian subvarieties A^{e_i} are uniquely determined, the B_j are only unique up to isogeny. Hence the dimension of B_j is uniquely determined. It may be computed from the corresponding rational representation of G as follows [8].

Proposition 2.9.3 *Let* $W_1, \ldots, W_r$ *denote the irreducible rational representation of* G, *and assume the rational representation* $\rho_r : \mathrm{End}(A) \to \mathrm{End}(\Lambda)$ *of the action of* G *on the abelian variety* $A = V/\Lambda$ *is given by*

$$\rho_r = \sum_{j=1}^{r} h_j W_j.$$

Then, if L_j *is the field of definition of the irreducible complex representation* V_j *of* G, *Galois associated to* W_j *and* s_j *its Schur index, we have, for any group algebra component* B_j *and the isotypical component* A_j *corresponding to* W_j,

(i) $\dim A_j = \frac{1}{2} h_j [L_j : \mathbb{Q}] \dim V_j$

(ii) $\dim B_j = \frac{1}{2} \langle \rho_r, W_j \rangle = \frac{1}{2} h_j s_j [L_j : \mathbb{Q}]$

Proof Since ρ_r acts on A_j and on $B_j^{n_j}$ by the representation $h_j W_j$, comparing dimensions we obtain the second equation of

$$2 \dim A_j = 2 n_j \dim(B_j) = h_j \, \mathrm{rk}(W_j) = h_j [L_j : \mathbb{Q}] \dim(V_j).$$

The first equation follows from (2.32) and the last equation from (2.20).

This gives the assertion (i) on $\dim A_j$ and the first assertion of (ii) on $\dim B_j$.

Now according to (2.30), we have $n_j = \frac{\dim V_j}{s_j}$. Inserting this completes the proof of (ii) and thus of the proposition.   □

2.9.2 Induced Action on the Tangent Space

If A is an abelian variety, say $A = V/\Lambda$ with a complex vector space V and a lattice Λ of maximal rank in V, then V can be considered in a natural way as the tangent space $T_0 A$ of A at 0. The natural map

$$\pi : V \to A$$

is just the exponential map from the Lie-theoretic point of view. Any homomorphism $f : A \to B$ of abelian varieties induces a homomorphism $d f_0 : T_0 A \to T_0 B$ on the tangent spaces in the usual way. The map $d f_0$ is called the *differential of the homomorphism* f.

Right from the definitions, it is clear that a homomorphism $f : A \to B$ of abelian varieties is an isogeny if and only if its differential $d f_0$ is an isomorphism.

Let G be a finite group action on the abelian variety $A = V/\Lambda$. We will apply the following Lemma to the tangent space V.

Lemma 2.9.4 *Let*

$$V = n_1 V_1 \oplus \cdots \oplus n_r V_r$$

be the isotypical decomposition of the complex representation V of a finite group G. Suppose there exist a positive integer s and subgroups $H \subset N$ of G such that

$$\rho_H = \rho_N \oplus \bigoplus_{i=1}^{s} (\dim V_i^H) V_i.$$

Then the fixed subspaces of V for H and N are related by

$$V^H = V^N \oplus n_1 V_1^H \oplus \cdots \oplus n_s V_s^H = V^N \oplus (n_1 V_1)^H \oplus \cdots \oplus (n_s V_s)^H$$

Proof Note first that for any subgroup F of G, the following equality holds

$$V^F = n_1 V_1^F \oplus \cdots \oplus n_r V_r^F. \tag{2.33}$$

This follows from the bilinearity of the scalar product $\langle \cdot, \cdot \rangle$ and the additivity of the restriction map Res_F^G, which give

$$V^F = \langle a \bigoplus_{i=1}^{r} \mathrm{Res}_F^G (n_i V_i), 1_F \rangle = \sum_{i=1}^{s} n_i \langle \mathrm{Res}_F^G V_i, 1_F \rangle = \sum_{i=1}^{s} n_i V_i^F.$$

Now we claim that

$$V_i^N = \begin{cases} 0 & \text{if } 1 \leq i \leq s, \\ V_i^H & \text{if } s+1 \leq i \leq r. \end{cases} \tag{2.34}$$

Since obviously $V_i^N \subseteq V_i^H$ for any $i = 1, \ldots, r$, this follows from the following calculation:

$$\dim V_i^H = \langle V_i, \rho_H \rangle$$

$$= \langle V_i, \rho_N \bigoplus \bigoplus_{j=1}^{r} (\dim V_j^H) V_j \rangle$$

$$= \dim V_i^N \bigoplus \sum_{j=1}^{r} (\dim V_j^H) \langle V_i, V_j \rangle.$$

Combining Eqs. (2.34) and (2.33), we get

$$V^H = \bigoplus_{i=s+1}^{r} n_i V_i^N \oplus \bigoplus_{i=1}^{s} n_i V_i^H = V^N \oplus n_1 V_1^H \oplus \cdots \oplus n_s V_s^H.$$

This implies the assertion.   □

2.10 Action of a Hecke Algebra on an Abelian Variety

If H is a subgroup of the finite group G, the Hecke algebra for the pair $H \subseteq G$ is the $\mathbb{Q}$-algebra generated by the double cosets of H in G. If G acts on an abelian variety A, then G does not act on the abelian subvariety A_H, the image of the projector associated to H. However, the Hecke algebra for $H \subseteq G$ acts on it. This gives a decomposition of A_H.

To be more precise, let G be a finite group acting on an abelian variety A with corresponding algebra homomorphism $\rho : \mathbb{Q}[G] \to \operatorname{End}_{\mathbb{Q}}(A)$, where again we denote corresponding elements of $\mathbb{Q}[G]$ and $\operatorname{End}_{\mathbb{Q}}(A)$ by the same letter. Now let $H \subseteq G$ be any subgroup. Recall Eq. (2.24) defining the idempotent p_H of $\mathbb{Q}[G]$, and consider the induced abelian subvariety

$$A_H := \operatorname{Im}(p_H) = \operatorname{Im}(np_H)$$

where n is any positive integer such that $np_H \in \operatorname{End}(A)$. In general there is no group action on A_H. However, there is a natural algebra action, given as follows:

Consider the subalgebra of $\mathbb{Q}[G]$ given by

$$\mathcal{H}_H := \mathbb{Q}[H \backslash G / H] = p_H \mathbb{Q}[G] p_H$$

It is called the *Hecke algebra for* H *in* G. If $\mathbb{Q}[G]$ denotes as usual the algebra of $\mathbb{Q}$-valued functions on G, the Hecke algebra $\mathcal{H}_H$ is the subalgebra of functions which are constant on all double cosets HgH. Restricting the homomorphism $\rho : \mathbb{Q}[G] \to \operatorname{End}_{\mathbb{Q}}(A)$ to $\mathcal{H}_H$, we get an algebra homomorphism

$$\widetilde{\rho} : \mathcal{H}_H \to \operatorname{End}_{\mathbb{Q}}(A_H).$$

For the sake of completeness, we outline in this section the main results on the action of $\mathcal{H}_H$ on the abelian subvareity A_H and deduce corresponding decompositions of the abelian subvariety A_H. We do not give proofs, since we do not need the results. For full proofs we refer to [8] and for a minor point to [5].

As in Sect. 2.8, let $W_1, \ldots, W_r$ denote the irreducible rational representations of G. To each W_i we associated in (2.23) a central idempotent e_{W_i} of $\mathbb{Q}[G]$ such that

$$1 = e_{W_1} + \cdots + e_{W_r}.$$

Let ρ_H denote the representation of G induced by the trivial representation of H. Then we have according to [8, equation (2.3)]

$$\rho_H = \sum_{i=1}^{r} a_i W_i \quad \text{with} \quad a_i = \frac{1}{s_i} \dim_{\mathbb{C}}(V_i^H) \tag{2.35}$$

where V_i is a complex irreducible representation of G, Galois associated to W_i, and s_i the Schur index of V_i. Renumbering if necessary, let $\{W_1, \ldots, W_u\}$ be the set of all W_i such that $a_i \neq 0$. As shown in [10, Chapter 11], there is a bijection from this set to the set $\{\widetilde{W}_1, \ldots, \widetilde{W}_u\}$ of all irreducible rational representations of the algebra $\mathcal{H}_H$. An analogous statement is valid for the complex Hecke algebra $\mathbb{C}[H\backslash G/H]$.

Let $\widetilde{V}_i$ denote the representation of $\mathbb{C}[H\backslash G/H]$ associated to the complex irreducible representation V_i of above. Then $\widetilde{V}_i$ is Galois associated to $\widetilde{W}_i$, and according to Corollary 2.8.2 and [10, Theorem 11.25], the dimension of $\widetilde{W}_i$ is given by

$$\dim_{\mathbb{Q}}(\widetilde{W}_i) = \dim_{\mathbb{Q}}(W_i^H) = [L_i : \mathbb{Q}] \dim_{\mathbb{C}}(V_i^H) \tag{2.36}$$

where L_i denotes the field of definition of the representation V_i.

For $i = 1, \ldots, u$, recall the central idempotents of $\mathcal{H}_H$ given by

$$f_{H,\widetilde{W}_i} := p_H e_{W_i} = e_{W_i} p_H.$$

Then p_H decomposes as

$$p_H = \sum_{i=1}^{u} f_{H,\widetilde{W}_i}. \tag{2.37}$$

Defining for $i = 1, \ldots, u$ the abelian subvarieties

$$A_{H,\widetilde{W}_i} := \mathrm{Im}(f_{H,\widetilde{W}_i}),$$

we obtain the following isogeny decomposition of A_H, given by the addition map

$$\mu : A_{H,\widetilde{W}_1} \times \cdots \times A_{H,\widetilde{W}_u} \to A_H. \tag{2.38}$$

It is uniquely determined by H and the action of G and is called the *isotypical decomposition of A_H*.

In order to decompose A_H further, we need some results of [8, page 2631]. Consider $i \in \{1, \ldots, u\}$. Then there exist primitive idempotents $p_1, \ldots, p_{a_i}$ in the simple algebra $\mathcal{H}_H f_{H,\widetilde{W}_i}$ with $a_i = \frac{1}{s_i} \dim V_i^H$ as above which are H-invariant under multiplication from both sides satisfying

$$f_{H,\widetilde{W}_i} = p_1 + \cdots + p_{a_i}.$$

Since all left ideals $\mathcal{H}_H p_j$ are isomorphic for $j = 1, \ldots, a_i$, the abelian subvarieties

$$B_j := A_H^{p_j} \subset A_H$$

are isogenous to each other. Thus we obtain the following decomposition of A_H up to isogeny, called the *Hecke algebra decomposition* with respect to the action of the Hecke algebra $\mathcal{H}_H$.

Proposition 2.10.1 *Let* $\{\widetilde{W}_1, \ldots, \widetilde{W}_u\}$ *denote the irreducible rational representations of the Hecke algebra* $\mathcal{H}_H$ *and* $a_i = \frac{1}{s_i} \dim V_i^H$ *where* V_i *is Galois associated to* W_i *for* $i = 1, \ldots, u$. *Then there are abelian subvarieties* B_i *of* A_H *and an* $\mathcal{H}_H$-*equivariant isogeny*

$$B_1^{a_1} \times \cdots \times B_u^{a_u} \to A_H,$$

where $\mathcal{H}_H$ *acts on* $B_i^{a_i}$ *via the representation* $\widetilde{W}_i$ *(with appropriate multiplicity).*

Chapter 3
Prym Varieties

In this chapter we give the definition of Prym varieties and derive the properties which we need in the subsequent chapters. For any finite cover of smooth projective curves $f : \widetilde{C} \to C$, the *Prym variety* $P(f) = P(\widetilde{C}/C)$ is defined to be the complementary abelian subvariety of f^*J in $\widetilde{J}$ with respect to the canonical polarization of $\widetilde{J}$. Here J and $\widetilde{J}$ denote the Jacobian varieties of C and $\widetilde{C}$, respectively.

So our notion of Prym variety is more general than the original notion introduced by Mumford in [27]. We call his more special Prym varieties *principally polarized Prym varieties*. Hence for us Prym varieties are not necessarily principally polarized.

In Sect. 3.1 we recall the usual definitions for finite covers of curves including an existence theorem for Galois covers. Moreover, we prove, following [33], a result computing for a Galois cover $f : \widetilde{C} \to C$ with group G and any subgroup H the genus of the curve $C_H = \widetilde{C}/H$ in terms of the signature of the cover f and the groups G and H.

Section 3.2 contains the definition of Prym varieties together with first properties. In particular we derive the type of the induced polarization in the cases we need and compute the degree of the isogeny $f^*J \times P(f) \to \widetilde{J}$ as well as of the canonical map $P(k) \times P(h) \to P(f)$ for any composition of covers $f = h \circ k$.

In Sect. 3.3 we give explicit formulas for the two-division points of the Prym variety $P(f)$ for any double cover $f : \widetilde{C} \to C$. These will be needed at several points in the subsequent chapters.

In some cases in later chapters, the factors of the Jacobian $\widetilde{J}$ are not Prym varieties, but Prym varieties $P(k_1, k_2)$ for a pair of covers $k_i : \widetilde{C} \to C_i$, $i = 1, 2$. Section 3.4 contains their definition together with some of their properties.

Finally, in Sect. 3.5 we consider Galois covers $f : \widetilde{C} \to C$ with group G. In particular we study the isotypical and the group algebra decomposition of $\widetilde{J}$ as well as the induced decompositions of JC_H for a subgroup $H \subset G$. Our main result is Corollary 3.5.10 as already outlined in the introduction.

© The Author(s), under exclusive license to Springer Nature Switzerland AG 2022   39
H. Lange, R. E. Rodríguez, *Decomposition of Jacobians by Prym Varieties*, Lecture
Notes in Mathematics 2310, https://doi.org/10.1007/978-3-031-10145-8_3

3.1   Finite Covers of Curves

3.1.1   Definitions and Elementary Results

By a *curve* we always understand a complex smooth connected projective curve. If C is a curve, its genus is by definition the number $g = g_C = h^0(C, \omega_C)$, where ω_C denotes the canonical line bundle on C. A *cover* $f : \widetilde{C} \to C$ is by definition a surjective holomorphic map of curves. The *multiplicity of* f at a point $p \in \widetilde{C}$, denoted $\mathrm{mult}_p(f)$, is the unique positive integer m such that there are local coordinates near p and $f(p)$ with f of the form $z \mapsto z^m$. A point $p \in \widetilde{C}$ is called a *ramification point of* f if $\mathrm{mult}_p(f) > 1$. In this case the number

$$r_p(f) := \mathrm{mult}_p(f) - 1$$

is called the *ramification index of* f at p. If p is a ramification point of f, the point $f(p) \in C$ is called a *branch point of* f. We denote always by $R(f)$ the set of ramification points of f and by $B(f)$ the set of branch points of f. These are always finite sets. Hence the number

$$r(f) := \sum_{p \in \widetilde{C}} r_p(f)$$

is an integer ≥ 0. It is called the *ramification degree* of the cover f.

For any point $q \in C \setminus B(f)$, the number of points $|f^{-1}(q)|$ is independent of q and called the *degree of the cover* f. For $q \in B(f)$ consider its fiber $f^{-1}(q) = \{p_1, \ldots, p_r\} \subset \widetilde{C}$. The *cycle structure of* f at q is the r-tuple $(n_1, \ldots, n_r)$ where $n_i = \mathrm{mult}_{p_i}(f)$ for $i = 1, \ldots, r$. The *Hurwitz formula* for f is the following relation between the genera $g_{\widetilde{C}}$ and g_C and the degree d of f:

$$g_{\widetilde{C}} = d(g_C - 1) + 1 + \frac{1}{2} r(f) \qquad (3.1)$$

For any curve $\widetilde{C}$, we denote by $\mathrm{Aut}(\widetilde{C})$ its group of automorphisms; that is, holomorphic bijective maps $\widetilde{C} \to \widetilde{C}$. Now let $\widetilde{C}$ be a curve and G a finite group with $G \subset \mathrm{Aut}(\widetilde{C})$. The quotient $C := \widetilde{C}/G$ admits a unique structure of a curve such that the quotient map $f : \widetilde{C} \to C$ is a cover of degree $d = |G|$. We call it a *Galois cover* with group G. Equivalently, G is the Galois group of the function field extension $\mathbb{C}(\widetilde{C})/\mathbb{C}(C)$.

Let $B(f) = \{q_1, \ldots q_s\}$ be the branch locus of f. For any $q_i \in B(f)$, choose a preimage $p_i \in R(f)$. The multiplicity $\mathrm{mult}_{p_i}(f)$ is equal to the order of the stabilizer subgroup $G_{p_i} := \mathrm{Stab}_{p_i}(G) \subset G$ and hence is independent of the choice of p_i in the fiber of q_i. In fact, the stabilizers of different preimages are conjugate subgroups in G. To be more precise, if $p \in \widetilde{C}$ with non-trivial stabilizer G_p, then the points in its orbit, denoted by $O^G(p)$, have stabilizers running through the complete conjugacy class of G_p in G. For a point $q \in O^G(p)$, we have $G_q = G_p$ if and only if q is an element of the orbit $O^{N_G(G_p)}(p)$ of p under the normalizer of G_p in G.

3.1.2 The Signature of a Galois Cover

Let $f : \widetilde{C} \to C$ be a Galois cover with group G. For i in $\{1, \ldots, s = |B(f)|\}$, let $p_i \in f^{-1}(q_i)$, and let $n_i = |G_i|$ denote the order of its stabilizer $G_i := G_{p_i}$. The *signature* of the Galois cover $f : \widetilde{C} \to C$ is by definition the tuple $(g; n_1, \ldots, n_s)$, where g is the genus of C. Note that as orders of subgroups of G, the numbers n_i are always divisors of $|G|$ and moreover ≥ 2 as multiplicities of ramification points.

For a Galois cover $f : \widetilde{C} \to C$ of signature $(g; n_1, \ldots n_s)$, the Hurwitz formula can also be written as

$$g_{\widetilde{C}} = |G|(g - 1) + 1 + \frac{|G|}{2} \sum_{i=1}^{s} (1 - \frac{1}{n_i}). \tag{3.2}$$

In order to recall an existence result for Galois covers for any finite group G, we need some preliminaries.

Assume g and s are non-negative integers and $n_1, \ldots, n_s$ are integers greater or equal to two. A $(2g + s)$-tuple

$$(a_1, \ldots, a_g, b_1, \ldots, b_g, c_1, \ldots, c_s)$$

of elements of G is called a *generating vector of type* $(g; n_1, \ldots, n_s)$ if the following conditions are satisfied:

(a) The elements $a_1, \ldots, a_g, b_1, \ldots, b_g, c_1, \ldots c_s$ generate the group G.
(b) c_i is of order n_i for $i = 1, \ldots, s$.
(c) $\prod_{i=1}^{g} [a_i, b_i] \prod_{j=1}^{s} c_j = 1$, where $[a_i, b_i] = a_i b_i a_i^{-1} b_i^{-1}$.

The following result is a consequence of the Uniformization Theorem:

Theorem 3.1.1 *Given a finite group G, there is a curve $\widetilde{C}$ of genus $g_{\widetilde{C}}$ on which G acts with signature $(g; n_1, \ldots, n_s)$ if and only if the following two conditions hold:*

(i) *The Hurwitz formula* (3.2).
(ii) *G admits a generating vector $(a_1, \ldots, a_g, b_1, \ldots, b_g, c_1, \ldots, c_s)$ of type $(g; n_1, \ldots, n_s)$.*

As an immediate consequence, we get

Corollary 3.1.2 *Given a finite group G and a curve C, there is always a Galois cover $f : \widetilde{C} \to C$ with group G.*

Proof It is easy to see that for a sufficiently high number s, there is always a generating vector with this number of elements c_i.   □

Let G be any finite group. The following example for a Galois cover with group G is particularly easy. We need it later when we give an example of such a cover for the groups in Chaps. 4 to 6.

Example 3.1.3 Let $c_1, \ldots c_{s-1}$ be a set of generators of G, and let $c_s :=$ $(\prod_{i=1}^{s-1} s_i)^{-1}$. If c_i is of order $n_i \geq 2$ for all i, then $(c_1, \ldots, c_s)$ is a generating vector of type $(0; n_1, \ldots, n_s)$. Finally, let $g_{\widetilde{C}} = -|G| + 1 + \frac{|G|}{2} \sum_{i=1}^{s} (1 - \frac{1}{n_i})$.

If $g_{\widetilde{C}}$ is a non-negative integer, then according to Theorem 3.1.1, there is a Galois cover $f : \widetilde{C} \to \mathbb{P}^1$ branched over s pairwise different points of $\mathbb{P}^1$.

More generally, let C be a curve of genus g. Then let $a_1, \ldots, a_g, b_1, \ldots, b_g,$ $c_1, \ldots, c_{s-1}$ be a set of generators of G with $s \geq 1$. Defining

$$c_s := \left(\prod_{i=1}^{g} [a_i, b_i] \prod_{i=1}^{s-1} c_i \right)^{-1},$$

then $(a_1, \ldots, a_g, b_1, \ldots, b_g, c_1, \ldots, c_s)$ is a generating vector for a Galois cover $f : \widetilde{C} \to C$ with group G.

3.1.3 The Geometric Signature of a Galois Cover

The geometric signature is a refined version of the notion of signature of a Galois cover. In Example 3.1.5 we see that there are Galois covers with the same signature, but different geometric signatures.

Let $f : \widetilde{C} \to C$ be a Galois cover with group G and branch locus $B(f) =$ $\{q_1, \ldots, q_s\}$. For every $q_i \in B(f)$, choose a point $p_i \in f^{-1}(q_i)$, and let $G_i := G_{p_i}$. The *geometric signature* of the Galois cover $f : \widetilde{C} \to C$ is by definition the tuple $(g; G_1, \ldots, G_s)$. To be more precise, instead of the subgroups G_i, one should have written the conjugacy class of subgroups of which G_i is a member, but with a slight abuse of notation, instead of considering the conjugacy class of G_j, we just write G_j.

The geometric signature of f determines the signature of $f : \widetilde{C} \to C$: in fact, let $n_i := |G_{p_i}|$ for all i. Then the signature of f with geometric signature $(g; G_1, \ldots, G_s)$ is $(g; n_1, \ldots, n_s)$.

The following theorem is a refinement of Theorem 3.1.1. It is a criterion for the existence of a Galois cover with a given geometric signature.

Theorem 3.1.4 *Given a finite group G, there is a curve $\widetilde{C}$ of genus $g_{\widetilde{C}}$ on which G acts with geometric signature $(g; G_1, \ldots, G_s)$ if and only if the following three conditions hold:*

(i) *The Hurwitz formula* (3.2).

(ii) *G admits a generating vector $(a_1, \ldots, a_g, b_1, \ldots, b_g, c_1, \ldots, c_s)$ of type $(g; n_1, \ldots, n_s)$ where n_i is the order of G_i.*

(iii) *The group generated by the element c_i of the generating vector is conjugate to G_i for $i = 1, \ldots, s$.*

We omit the proof of the theorem here, but refer to [33, Theorem 4.1], since we only use it in the following example.

Example 3.1.5 Let G denote the Klein group of order 4, generated by a_1 and a_2, with $a_3 = a_1 a_2$, and $G_j = \langle a_j \rangle$ the corresponding subgroups of G for $j = 1, 2, 3$. Since G is abelian, each of these subgroups forms a whole conjugacy class of subgroups.

Let s be an integer greater or equal to four. Then there are curves $\widetilde{C}$ on which G acts with signature $(0; 2, \ldots, 2)$, with s numbers 2, since generating vectors for G can be written down; if s is even, $(a_1, a_1, a_2, a_2, \ldots, a_2, a_2)$ is an example, and if s is odd, $(a_1, a_2, a_3, a_1, a_1, \ldots, a_1, a_1)$ works. All such curves $\widetilde{C}$ have genus $s - 3$ according to (3.2).

However, there are different geometric signatures associated to the same signature: consider non-negative integers m_1, m_2, and m_3 such that $m_1 + m_2 + m_3 = s$ and $m_i + m_j$ is even and greater or equal to two for all $i \neq j$, and choose the generating vector as

$$(\overbrace{a_1, a_1, \ldots, a_1}^{m_1}, \overbrace{a_2, a_2, \ldots, a_2}^{m_2}, \overbrace{a_3, a_3, \ldots, a_3}^{m_3}).$$

Then the signature is $(0; \overbrace{2, \ldots, 2}^{s})$ and the geometric signature is

$$(0; \overbrace{G_1, G_1, \ldots, G_1}^{m_1}, \overbrace{G_2, G_2, \ldots, G_2}^{m_2}, \overbrace{G_3, G_3, \ldots, G_3}^{m_3}).$$

According to Theorem 3.1.4, there exist covers with this geometric signature. Given s, the curves $\widetilde{C}$ all have the same genus, but they can be essentially different from each other, since the genera $g_{\widetilde{C}/G_j}$ of the quotients $\widetilde{C}/G_j$ are given by

$$g_{\widetilde{C}/G_j} = \frac{1}{2}(m_i + m_k) - 1$$

for $\{i, j, k\} = \{1, 2, 3\}$.

The following theorem, due to Rojas [33, Corollary 3.4], will be very helpful.

Theorem 3.1.6 *Let $f : \widetilde{C} \to C$ be a Galois cover with group G and geometric signature $(g; G_1, \ldots, G_s)$. Then for any subgroup $H \subset G$, the genus of the quotient $C_H = \widetilde{C}/H$ is given by*

$$g_{C_H} = [G : H](g - 1) + 1 + \frac{1}{2} \sum_{j=1}^{s} ([G : H] - |H\backslash G/G_j|).$$

Proof Consider the decomposition of f given by H:

$$f : \tilde{C} \xrightarrow{h} C_H \xrightarrow{k} C.$$

We prove the theorem by computing the ramification degree $r(k)$ appearing in the Hurwitz formula

$$g_{C_H} = [G : H](g - 1) + 1 + \frac{r(k)}{2}.$$

Fixing $j \in \{1, \ldots, s\}$, the subgroup G_j is the stabilizer of a point $p \in \tilde{C}$. It suffices to compute the contribution to $r(k)$ of the points in the fibre $f^{-1}(f(p))$.

First we observe that for any point q in $f^{-1}(f(p)) = O_p^G$, the orbit of p under G, its stabilizer G_q for the total cover f satisfies $G_q = g_q G_p g_q^{-1}$ where $q = g_q(p)$ for some $g_q \in G$. Also, the stabilizer of q for the H-action corresponding to the cover $h : \tilde{C} \to C_H$ is given by $G_q \cap H$, with $|h^{-1}(h(q))| = \frac{|H|}{|H \cap G_q|}$.

Now consider the decomposition of G into its double cosets with respect to H and G_j

$$G = H x_1 G_j \sqcup H x_2 G_j \sqcup \ldots \sqcup H x_r G_j.$$

Then g_q belongs to a unique double coset $H x_i G_j$, hence $q = g_q(p) \in H x_i G_j(p)$, and

$$H x_i G_j(p) = H x_i(p) = O_{x_i(p)}^H = O_q^H = h^{-1}(h(q)).$$

Calling $q_i := x_i(p)$ in the fiber of f over p, we see that the fiber $k^{-1}(f(p))$ consists of the $|H\backslash G/G_j|$ points $\{h(q_i)) : 1 \leq i \leq r\}$ in C_H, and the multiplicity of k at each $h(q_i)$ is $\frac{|G_{q_i}|}{|H \cap G_{q_i}|}$.

Therefore the contribution of the points in the fibre $f^{-1}(f(p))$ to $r(k)$ is given by

$$\sum_{i=1}^{r} \left(\frac{|G_{q_i}|}{|H \cap G_{q_i}|} - 1 \right) = \sum_{i=1}^{r} \left(\frac{|x_i G_j x_i^{-1}|}{|H \cap x_i G_j x_i^{-1}|} \right) - r$$

$$= \sum_{i=1}^{r} \left(\frac{|G_j|}{|x_i^{-1} H x_i \cap G_j|} \right) - |H\backslash G/G_j|$$

$$= [G : H] - |H\backslash G/G_j|,$$

where the last equality is a well-known property of double cosets. Thus

$$r(k) = \sum_{j=1}^{s}([G:H] - |H\backslash G/G_j|).$$

□

Corollary 3.1.7 *Let the assumptions be as in the previous theorem, and consider subgroups $H \subset N \subset G$. Then we have for the ramification degree $r(h)$ of the corresponding cover $h : C_H \to C_N$,*

$$r(h) = \sum_{i=1}^{s}(\deg h \cdot |N\backslash G/G_i| - |H\backslash G/G_i|).$$

Proof Applying the Hurwitz formula to the cover h and Theorem 3.1.6 to the curves C_H and C_N, we get

$$r(h) = 2g(C_H) - 2 - 2[N:H](g(C_N) - 1)$$

$$= 2\left([G:H](g-1) + 1 + \frac{1}{2}\sum_{j=1}^{s}([G:H] - |H\backslash G/G_j|)\right) - 2$$

$$-2[N:H]\left([G:N](g-1) + \frac{1}{2}\sum_{j=1}^{s}([G:N] - |H\backslash G/G_j|)\right)$$

$$= \sum_{i=1}^{s}([N:H]|N\backslash G/G_i| - |H\backslash G/G_i|).$$

This gives the assertion, since $\deg h = [N:H]$.

□

3.2 Prym Varieties of Covers of Curves

In this section we define the Prym variety of a finite cover of curves, recall some basic properties from [3, Section 12], and prove some more special results.

3.2.1 Definition of Prym Varieties

For any curve C, its Jacobian is defined as

$$JC := H^0(\omega_C)^* / H_1(C, \mathbb{Z})$$

where ω_C denotes the canonical bundle of C and $H_1(C, \mathbb{Z})$ is embedded in the dual vector space of $H^0(\omega_C)$ via integration. The intersection form on $H_1(C, \mathbb{Z})$ defines a principal polarization $\Theta = \Theta_C$ in a canonical way. Hence, if C is of genus g, then (JC, Θ) is a principally polarized abelian variety of dimension g.

Any divisor D of degree 0 on C can be written as a finite sum $D = \sum_{\nu=1}^{N}(p_\nu - q_\nu)$ with points $p_\nu, q_\nu \in C$. Then the map

$$D \mapsto \left\{ \omega \mapsto \sum_{\nu=1}^{N} \int_{q_\nu}^{p_\nu} \omega \right\}$$

induces a well-defined map $\mathrm{Pic}^0(C) \to JC$ of the group of line bundles of degree 0 on C into the Jacobian of C. It is a canonical isomorphism

$$a : \mathrm{Pic}^0(C) \xrightarrow{\sim} JC$$

called the *Abel-Jacobi isomorphism* of C. Often we identify $\mathrm{Pic}^0(C) = JC$ via a.

Now let $f : \widetilde{C} \to C$ be a finite cover of curves of degree d, and let $g_C = g$ and $g_{\widetilde{C}} = \widetilde{g}$. In order to avoid trivialities, in this section we assume $g \geq 1$ and $d \geq 2$. Denote by $(\widetilde{J}, \widetilde{\Theta})$ and (J, Θ) the corresponding principally polarized Jacobians. There are two homomorphisms relating the Jacobians JC and $J\widetilde{C}$: if we consider the Jacobians as the groups of line bundles of degree 0, the pullback of line bundles gives a homomorphism

$$f^* : J \to \widetilde{J}$$

and the norm map $\mathrm{Nm}_f : \mathcal{O}_{\widetilde{C}}(\sum r_\nu p_\nu) \mapsto \mathcal{O}_C(\sum r_\nu f(p_\nu))$ gives a homomorphism, also called *norm map*

$$\mathrm{Nm}_f : \widetilde{J} \to J.$$

The definitions immediately give

$$\mathrm{Nm}_f \circ f^* = d_J$$

where d_J denotes multiplication by d on J. Moreover, identifying $\widetilde{J}$ and J with their dual abelian varieties via the canonical principal polarizations, we have according to [3, page 331, equation (2)]

$$\widehat{f^*} = \mathrm{Nm}_f. \tag{3.3}$$

Denoting $Y = \operatorname{Im} f^*$, the map f^* factorizes into an isogeny $j : J \to Y$ and the canonical embedding $\iota_Y : Y \hookrightarrow \tilde{J}$. With $\phi_Y := \phi_{\iota_Y^* \tilde{\Theta}}$, the following diagram commutes

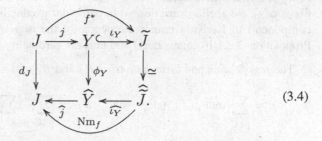

$$(3.4)$$

Here we identify J with its dual $\hat{J}$ and $\tilde{J}$ with its dual $\hat{\tilde{J}}$ via their canonical principal polarizations.

From [3, Lemma 12.3.1 and Proposition 12.3.2], we have

Proposition 3.2.1

(a) *The principal polarizations* $\tilde{\Theta}$ *and* Θ *are related by*

$$(f^*)^* \tilde{\Theta} \equiv d\Theta.$$

(b) *The norm map* Nm_f *and the norm-endomorphism* N_Y *of* Y *in* $\tilde{J}$ *are related by*

$$f^* \operatorname{Nm}_f = \frac{d}{e(Y)} N_Y$$

where $e(Y)$ *denotes the exponent of the restricted polarization* $\Theta|_Y$.

Note that [3, Proposition 11.4.3] gives the following criterion for the map f^* to be injective.

Proposition 3.2.2 *The homomorphism* $f^* : J \to \tilde{J}$ *is not injective if and only if* f *factorizes via a cyclic étale cover* h *of degree* ≥ 2:

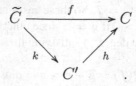

Furthermore, the proof of [3, Proposition 11.4.3] gives

Proposition 3.2.3 *In case* h *is étale cyclic, the kernel of* h^* *is of degree* $\deg h$ *and is generated by the line bundle defining* h. *Hence, if* k^* *is injective, then* $\operatorname{Ker} f^*$ *is cyclic of order* $\deg h$.

Proposition 3.2.4 *For any cover* $f : \widetilde{C} \to C$, *the number of components of* KerNm_f *equals the cardinality of* $\mathrm{Ker} f^*$.

Proof Consider the diagram (3.4). From Proposition 2.2.1(a) we know that the fibres of $\widehat{i_Y}$ are abelian varieties, so in particular irreducible. Hence the number of components of KerNm_f equals the degree of the isogeny $\widehat{j}$, which according to Proposition 2.2.1(b) equals deg j and thus the cardinality of $\mathrm{Ker} f^*$.   □

The *ramification* and *branch divisors* R_f and B_f of $f : \widetilde{C} \to C$ are defined by

$$R_f := \sum_{p \in \widetilde{C}} (\mathrm{mult}_p(f) - 1)p \quad \text{and} \quad B_f := \sum_{q \in C} [\sum_{p \in f^{-1}(q)} (\mathrm{mult}_p(f) - 1)]q$$

where $\mathrm{mult}_p(f)$ denotes the multiplicity of f at the point $p \in \widetilde{C}$. According to (3.1), the Hurwitz formula relates the genus $\widetilde{g}$ of $\widetilde{C}$ and the genus g of C by $\widetilde{g} = d(g - 1) + 1 + \frac{1}{2}\deg(R_f)$. In particular, the degree of the ramification divisor R_f is always even.

Let $f : \widetilde{C} \to C$ and $Y = \mathrm{Im} f^*$ be as in diagram (3.4). The *Prym variety* of f is by definition the complementary abelian subvariety of Y in $J\widetilde{C}$ with respect to the canonical polarization $\widetilde{\Theta}$

$$P(f) = P(\widetilde{C}/C) := Y^c = \mathrm{Im}(f^*)^c.$$

Another description of $P(f)$ is given by

$$P(f) = \mathrm{Ker}(\mathrm{Nm}_f)^0. \tag{3.5}$$

This follows from Eq. (3.3) and Proposition 3.2.1(b) or from Corollary 2.5.6.

According to Corollary 2.5.2 and the Hurwitz formula, we have

$$\dim P(f) = \widetilde{g} - g = (d - 1)(g - 1) + \frac{1}{2}\deg(R_f). \tag{3.6}$$

Remark 3.2.5 Observe that we assumed the conditions $d \geq 2, g \geq 1$. If $d = 1$, then $P(f) = 0$. If $g = 0$, then $P(f) = \widetilde{J}$. So these cases are not interesting. Moreover we may assume that $\deg(R_f) > 0$ if $g = 1$, since otherwise again $P(f) = 0$.

For most values of d, g and $\deg(R_f)$ satisfying these conditions, the inequality

$$\dim P(f) = \widetilde{g} - g > g = \dim JC$$

holds. The exceptions are as follows.

(a) f is étale of degree 2, in which case any $g \geq 2$ will work, and $\dim P(f) = g - 1 < g = \dim JC$.

(b) f is of degree 2 and ramified in exactly 2 points, in which case any $g \geq 1$ will work, and dim $P(f) = g = \dim JC$.
(c) f is étale of degree 3 and $g(C) = 2$, in which case dim $P(f) = 2 = \dim JC$.
(d) f of degree $d \geq 2$ with $g(\widetilde{C}) = 2$ and $g(C) = 1$.

3.2.2 Polarizations of Prym Varieties

The Prym variety $P(f)$ is a polarized abelian variety with respect to the restriction of the canonical polarization $\widetilde{\Theta}$

$$\Theta_{P(f)} := \widetilde{\Theta}|_{P(f)}.$$

According to Corollary 2.4.2 (b), the polarization $\Theta_{P(f)}$ is a multiple of a principal polarization if and only if it is of type $(d, \ldots, d)$. Generally this principal polarization is denoted by Ξ. Originally Mumford (in [27]) denoted the polarized abelian varieties $(P(f), \Xi)$ as Prym varieties. Here we use a more general notation. We call the Mumford's varieties *principally polarized Prym varieties*.

The theorem [3, Theorem 12.3.3] tries to determine all covers leading to ppav's. Unfortunately one case is forgotten there. This case (case c) in Theorem 3.2.6) was investigated in [22]. The following theorem gives the correct result. The proof is just an application of Proposition 2.6.3, Corollary 2.4.2, and the Hurwitz formula.

Theorem 3.2.6 *Let $f : \widetilde{C} \to C$ be a cover of degree $d \geq 2$ and $g(C) = g \geq 1$ with f ramified if $g = 1$. The cover f defines a principally polarized abelian variety if and only if f is of one of the following four cases:*

(a) *f is étale of degree 2.*
(b) *f is of degree 2 and ramified in exactly 2 points.*
(c) *f is non-cyclic étale of degree 3 and $g(C) = 2$.*
(d) *$\deg f \geq 2$, $g(\widetilde{C}) = 2$ and $g(C) = 1$.*

Note that $J[d]$ is the kernel of the polarization $d\Theta_J$, the d-fold of the canonical polarization on J. We get from Proposition 3.2.1(a) and Proposition 2.4.1

Proposition 3.2.7 $\operatorname{Ker} f^* \subset J[d]$ *is isotropic with respect to the Weil form $e^{d\Theta}$ of the polarization $d\Theta$ on J.*

The type of $\Theta_{P(f)}$ can be fairly complicated. We know the type of this polarization by Theorem 3.2.6 in the exceptional cases of Remark 3.2.5. We work out only the following two other cases.

Proposition 3.2.8 *Let $f : \widetilde{C} \to C$ be a cover of degree $d \geq 2$ and $g(C) = g \geq 1$ with f ramified if $g = 1$; furthermore, assume f is not one of the exceptional cases in Remark 3.2.5.*

Then $\dim P(f) = \widetilde{g} - g > g = \dim J(C)$ *and*

(a) *If* f *does not factorize via an étale cyclic cover of degree* ≥ 2, *then* $\Theta_{P(f)}$ *is of type* $(1, \ldots, 1, d, \ldots, d)$ *with g numbers d and* $\widetilde{g} - 2g$ *numbers* 1
(b) *If* f *factorizes as* $f = hk$ *with* $h : C' \to C$ *cyclic étale of degree* $d_1 < d$ *and k does not factorize via a cyclic étale cover of degree* ≥ 2, *then* $\Theta_{P(f)}$ *is of type* $(1, \ldots, 1, \frac{d}{d_1}, d, \ldots, d)$ *with g − 1 numbers d.*

Proof (a) is an immediate consequence of Propositions 3.2.1 (a), 3.2.2, and 2.6.3.

Proof of (b): According to Proposition 3.2.1(a), $(f^*)^*\widetilde{\Theta}$ is of type $(d, \ldots, d)$. Let $Y = \mathrm{Im} f^* \subset \widetilde{J}$. The map $j : J \to Y$ is an isogeny of degree d_1. Hence by Propositions 3.2.7 and 2.4.1, $\widetilde{\Theta}|_Y$ is of type $(\frac{d}{d_1}, d, \ldots d)$. The assertion follows from Proposition 2.6.3, since $P(f) = Y^c$. □

3.2.3 The Degrees of the Decomposition Isogeny

Let again $f : \widetilde{C} \to C$ be a cover of degree d. We want to compute the degree of the isogeny $P(f) \times f^* J \to \widetilde{J}$.

Recall that for any subgroup K of $J[d]$, $K^\perp$ denotes the orthogonal complement in $J[d]$ with respect to the Weil form $e^{d\Theta}$.

Proposition 3.2.9 *The homomorphism* $f^* : J \to \widetilde{J}$ *induces an isomorphism*

$$(\mathrm{Ker} f^*)^\perp / \mathrm{Ker} f^* \to K(\Theta_{f^*J}) = K(\Theta_{P(f)}) = P(f) \cap f^* J \subseteq P[d]$$

and

$$|P(f) \cap f^* J| = \frac{|J[d]|}{|\mathrm{Ker} f^*|^2}.$$

Proof It follows from Proposition 2.4.1 with descent theory that f^* induces an isomorphism $(\mathrm{Ker} f^*)^\perp / \mathrm{Ker} f^* \to K(\Theta_{f^*J})$. Hence

$$(\mathrm{Ker} f^*)^\perp = (f^*)^{-1}(K(\Theta_{f^*J})).$$

Equation $K(\Theta_{f^*J}) = K(\Theta_{P(f)}) = P(f) \cap f^* J$ is a special case of Corollary 2.6.2. In order to show that $P(f) \cap f^* J \subseteq J[d]$, note that $x \in P(f) \cap f^* J$ if and only if $x \in P(f)$ and $x = f^*(x')$ for some $x' \in J$. This implies $0 = \mathrm{Nm}_f(x) = \mathrm{Nm}_f(f^*(x')) = dx'$ and hence $dx = 0$.

To show the last assertion, note that the above isomorphism implies the first equation of

$$|P(f) \cap f^* J| = \frac{|(\mathrm{Ker} f^*)^\perp|}{|\mathrm{Ker} f^*|} = \frac{|J[d]|}{|\mathrm{Ker} f^*|^2}.$$

The last equation follows from the non-degenerateness of the Weil form $e^{d\Theta}$ on $J[d]$.   □

The next result follows inmediately.

Corollary 3.2.10 *The addition map*

$$\alpha : P(f) \times f^*J \to \tilde{J}$$

is an isogeny of degree

$$\deg \alpha = \frac{|J[d]|}{|\mathrm{Ker}\, f^*|^2}.$$

Clearly the map

$$\beta : P(f) \times J \to \tilde{J}, \qquad \beta(x, y) = x + f^*(y)$$

is an isogeny. Let $\pi_2 : P(f) \times J \to J$ denote the canonical projection. We have

$$\mathrm{Ker}\,\beta = \{(-f^*(x), x) \mid x \in J[d] \text{ and } f^*(x) \in P(f)\}. \tag{3.7}$$

This follows from the fact that $x \mapsto (-f^*(x), x)$ is a section of $\mathrm{Ker}\,\beta \to \pi_2(\mathrm{Ker}\,\beta)$.

Corollary 3.2.11 $\pi_2(\mathrm{Ker}\,\beta) = (\mathrm{Ker}\, f^*)^{\perp}$ *and β is an isogeny of degree*

$$\deg \beta = |(\mathrm{Ker}\, f^*)^{\perp}| = \frac{|J[d]|}{|\mathrm{Ker}\, f^*|}.$$

Proof By (3.7) and Proposition 3.2.9, we have

$$\pi_2(\mathrm{Ker}\,\alpha) = \{x \in J \mid f^*(x) \in P(f)\} = (f^*)^{-1}(f^*J \cap P(f)) = (\mathrm{Ker}\, f^*)^{\perp}.$$

Therefore $|\mathrm{Ker}\,\beta| = |(\mathrm{Ker}\, f^*)^{\perp}|$ and Proposition 3.2.9 gives also the second assertion.   □

3.2.4   Degrees of Isogenies Arising from a Decomposition of $f : \tilde{C} \to C$

Consider the following diagram of covers of smooth projective curves

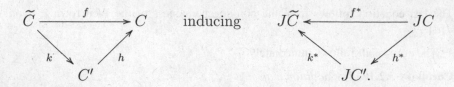

We want to study the relation of the Prym variety $P(f)$ with $P(h)$ and $P(k)$.
   Note first that

$$\operatorname{Ker} f^* = h^{*-1}(\operatorname{Im} h^* \cap \operatorname{Ker} k^*). \tag{3.8}$$

For the sake of abbreviation, we denote $d_f := \deg f$ and similarly d_k for k and d_h for h.
   Our main result is the following theorem.

Theorem 3.2.12

(a): *The map*

$$\psi : P(k) \times P(h) \to P(f), \qquad (x, y) \mapsto x + k^* y$$

is an isogeny of degree

$$\deg \psi = |P(h)[d_k]|\frac{|\operatorname{Im} h^* \cap \operatorname{Ker} k^*|}{|\operatorname{Ker} k^*|}.$$

(b) *The map*

$$\Psi : P(k) \times P(h) \times JC \to J\widetilde{C}, \qquad (x, y, z) \mapsto x + k^* y + f^* z$$

is an isogeny of degree

$$\deg \Psi = \frac{|JC'[d_k]|}{|\operatorname{Ker} k^*|} \cdot \frac{|JC[d_h]|}{|\operatorname{Ker} h^*|} = \frac{|P(h)[d_k]|}{|\operatorname{Ker} k^*|} \cdot \frac{|JC[d_f]|}{|\operatorname{Ker} h^*|}.$$

   Note first the following lemma.

Lemma 3.2.13

$$\operatorname{Ker} \psi \subset P(k)[d_k] \times k^*(P(h)[d_k]).$$

Proof Let $(x, k^* y) \in \operatorname{Ker} \psi$; that is, $x + k^*(y) = 0$. Applying the norm of k, we get

$$0 = \operatorname{Nm}_k(x) + \operatorname{Nm}_k(k^* y) = d_k y.$$

So $y \in P(h)[d_k]$ and thus $\operatorname{Ker}\psi \subset \{(-k^*y, k^*y) \mid y \in P(h)[d_k]\}$. Hence also k^*y and x are d_k-division points, which give the assertion. $\square$

Proof of the theorem By Corollary 3.2.11 we have isogenies

$$\alpha : P(k) \times JC' \to J\widetilde{C}, \quad (x, y) \mapsto x + k^*y \quad \text{of degree} \quad \deg\alpha = \frac{|JC'[d_k]|}{|\operatorname{Ker}k^*|},$$

$$\beta : P(h) \times JC \to JC', \quad (y, z) \mapsto y + h^*z \quad \text{of degree} \quad \deg\beta = \frac{|JC[d_h]|}{|\operatorname{Ker}h^*|},$$

and

$$\gamma : P(f) \times JC \to J\widetilde{C}, \quad (x, z) \mapsto x + f^*z \quad \text{of degree} \quad \deg\gamma = \frac{|JC[d_f]|}{|\operatorname{Ker}f^*|}.$$

Using (3.5) one checks that $P(k) \subset P(f)$ and also $k^*P(h) \subset P(f)$. This implies

$$\operatorname{Im}\psi \subset P(f).$$

Hence we have the commutative diagram

$$
\begin{array}{ccc}
P(k) \times P(h) \times JC & \xrightarrow{\;1_{P(k)} \times \beta\;} & P(k) \times JC' \\[2pt]
{\scriptstyle \psi \times 1_{JC}}\big\downarrow & \searrow{\scriptstyle \Psi} & \big\downarrow{\scriptstyle \alpha} \\[2pt]
P(f) \times JC & \xrightarrow[\;\gamma\;]{} & J\widetilde{C}.
\end{array}
$$

This implies that ψ and Ψ are isogenies and that

$$|\operatorname{Ker}\Psi| = |\operatorname{Ker}\psi| \cdot |\operatorname{Ker}\gamma| = |\operatorname{Ker}\alpha| \cdot |\operatorname{Ker}\beta|.$$

This gives the first equation of (b), namely,

$$|\operatorname{Ker}\Psi| = |\operatorname{Ker}\psi| \cdot |\frac{|JC[d_f]|}{|\operatorname{Ker}f^*|} = \frac{|JC'[d_k]|}{|\operatorname{Ker}k^*|} \cdot \frac{|JC[d_h]|}{|\operatorname{Ker}h^*|}.$$

The last equality implies

$$|\operatorname{Ker}\psi| = \frac{|JC'[d_k]| \cdot |JC[d_h]|}{JC[d_f]} \cdot \frac{|\operatorname{Ker}f^*|}{|\operatorname{Ker}k^*| \cdot |\operatorname{Ker}h^*|}.$$

Now note that, if g, g' and $\widetilde{g}$ denote the genera of C, C' and $\widetilde{C}$, respectively,

$$\frac{|JC'[d_k]| \cdot |JC[d_h]|}{JC[d_f]} = \frac{d_k^{2g'} \cdot d_h^{2g}}{d_f^{2g}} = d_k^{2 \dim P(h)} = |P(h)[d_k]|.$$

From (3.8) we get

$$|\mathrm{Ker} f^*| = |\mathrm{Ker} h^*| \cdot |\mathrm{Im} h^* \cap \mathrm{Ker} k^*|.$$

Together this implies (a) and the second equation of (b). □

Corollary 3.2.14
(a): *The addition map*

$$\alpha_0 : P(k) \times k^* P(h) \to P(f),$$

is an isogeny of degree

$$\deg \alpha_0 = |P(h)[d_k]| \frac{|\mathrm{Im} h^* \cap \mathrm{Ker} k^*|}{|\mathrm{Ker} k^*| \cdot |\mathrm{Ker}_{|P(h)}^*|}.$$

(b) *The addition map*

$$\alpha : P(k) \times k^* P(h) \times f^* JC \to J\widetilde{C}$$

is an isogeny of degree

$$\deg \alpha = \frac{|P(h)[d_k]|}{|\mathrm{Ker} k^*| \cdot |\mathrm{Ker}_{|P(h)}^*|} \cdot \frac{|JC[d_f]|}{|\mathrm{Ker} h^*| \cdot |\mathrm{Ker} f^*|}.$$

Proof For the proof of (a), consider the following commutative diagram

$$P(k) \times P(h)$$

with maps $1_{P(k)} \times k_{|P(h)}^*$ and ψ

$$P(k) \times k^* P(h) \xrightarrow{\quad \alpha_0 \quad} P(f)$$

which gives

$$\deg \alpha_0 = \frac{\deg \psi}{|\mathrm{Ker} k_{|P(h)}^*|} = |P(h)[d_k]| \frac{|\mathrm{Im} h^* \cap \mathrm{Ker} k^*|}{|\mathrm{Ker} k^*| \cdot |\mathrm{Ker}_{|P(h)}^*|}.$$

Similarly, for (b) consider the commutative diagram

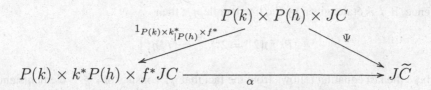

giving

$$\deg \alpha = \frac{\deg \Psi}{|\operatorname{Ker} k^*_{|P(h)}| \cdot |\operatorname{Ker} f^*|} = \frac{|P(h)[d_k]|}{|\operatorname{Ker} k^*| \cdot |\operatorname{Ker} k^*_{|P(h)}|} \cdot \frac{|JC[d_f]|}{|\operatorname{Ker} h^*| \cdot |\operatorname{Ker} f^*|}.$$

$\square$

3.3   Two-Division Points of Prym Varieties of Double Covers

Let $f : \widetilde{C} \to C$ be a double cover; that is, $C = \widetilde{C}/\langle \tau \rangle$ with an involution τ on $\widetilde{C}$. As always, let $\widetilde{J} = J\widetilde{C}$ and $J = JC$. In this subsection we want to describe explicitly the group of two-division points of the Prym variety $P(f)$.

Proposition 3.3.1

(a) *Let $f : \widetilde{C} \to C$ be an étale double cover given by the line bundle $\eta_f \in JC[2]$. Then the two-division points of the Prym variety $P(f)$ are*

$$P(f)[2] = f^* J \cap P(f) = f^*(\eta_f^\perp).$$

(b) *Let $f : \widetilde{C} \to C$ be a ramified double cover. Then f^* induces an isomorphism followed by an injection*

$$J[2] \simeq f^* J \cap P(f) \hookrightarrow P(f)[2].$$

Proof By Proposition 3.2.9, f^* induces an isomorphism

$$(\operatorname{Ker} f^*)^\perp / \operatorname{Ker} f^* \simeq f^* J \cap P(f) \subset P(f)[2], \tag{3.9}$$

where $\perp$ denotes the orthogonal complement in $J[2]$ with respect to the Weil form on $J[2]$, and moreover,

$$|(\operatorname{Ker} f^*)^\perp| = \frac{|J[2]|}{|\operatorname{Ker} f^*|}.$$

Hence, if f is étale, given by the line bundle η_f, then

$$|P(f)[2]| = 2^{2g-2} = |f^*\eta_f^{\perp}|$$

where the last equation follows from the fact that $\eta_f^{\perp}$ consists of 2^{2g-1} elements and contains η_f. So (3.9) implies (a).

If f is ramified, by Proposition 3.2.3, $\operatorname{Ker} f^* = 0$ and hence (b) is a direct consequence of (3.9). $\square$

Now let f be ramified with ramification degree $r > 0$. The Hurwitz formula implies that r is even and, if C is of genus g, then

$$\dim P(f) = g - 1 + \frac{r}{2} \quad \text{and hence} \quad |P(f)[2]| = 2^{2g-2+r}.$$

The cover f being ramified implies $|\operatorname{Ker} f^*| = 0$; hence by Proposition 3.3.1(b) we have

$$f^*J[2] \subset P(f)[2] \quad \text{and} \quad \frac{|P(f)[2]|}{|f^*J[2]|} = 2^{r-2}. \tag{3.10}$$

If $r = 2$, this implies that $P(f)[2] = f^*J[2]$.

Remark 3.3.2 This gives again a proof of Theorem 3.2.6 (b) saying that $P(f)$ is principally polarized if $r = 2$.

Let $p_1, \ldots, p_r$ denote the ramification points of f with corresponding branch points $q_i = f(p_i)$ for $i = 1, \ldots, r$. For $i = 2, \ldots, r$, choose

$$m_i \in \operatorname{Pic}^0(C) \quad \text{such that} \quad m_i^2 = \mathcal{O}_C(q_1 - q_i)$$

and consider the line bundle

$$\mathcal{F}_i := \mathcal{O}_{\widetilde{C}}(p_i - p_1) \otimes f^*(m_i) \quad \text{for} \quad i = 2, \ldots, r. \tag{3.11}$$

Since $\operatorname{Nm}_f \mathcal{F}_i = \mathcal{O}_C(q_i - q_1) \otimes m_i^2 = \mathcal{O}_C$ and $\operatorname{Ker} \operatorname{Nm}_f = P(f)$ by Propositions 3.2.2 and 3.2.4, we conclude that

$$\mathcal{F}_i \in P(f)[2] \quad \text{for all} \quad i = 2, \ldots, r.$$

With these notations we have

Proposition 3.3.3 *Let $f : \widetilde{C} \to C$ be a ramified double cover of ramification degree r (necessarily even and ≥ 2). For any $\ell \in P(f)[2]$, there exists $m \in J[2]$ and unique $v_i \in \{0, 1\}, i = 2, \ldots r - 1$ such that*

$$\ell = f^*(m) \otimes \mathcal{F}_2^{v_2} \otimes \cdots \otimes \mathcal{F}_{r-1}^{v_{r-1}}.$$

In other words,

$$P(f)[2] = f^*J[2] \bigoplus_{i=2}^{r-1} \mathcal{F}_i \mathbb{Z}/2\mathbb{Z}.$$

Proof Suppose first $r = 2$. According to (3.10) we have $P(f)[2] = f^*J[2]$ which is the assertion in this case.

So suppose $r \geq 4$. Let $\mathcal{D} \in \text{Pic}^{r/2}(C)$ denote the line bundle defining the double cover f. This means

$$\mathcal{D}^2 = \mathcal{O}_C(q_1 + \cdots + q_r) \quad \text{and} \quad f^*(\mathcal{D}) = \mathcal{O}_{\widetilde{C}}(p_1 + \cdots + p_r).$$

Consider the Abel-Jacobi map

$$a : \widetilde{C}^{(\widetilde{g})} \longrightarrow \widetilde{J}, \quad D \mapsto \mathcal{O}_{\widetilde{C}}(D - \widetilde{g} \cdot p_1).$$

Given $\ell \in P(f)[2]$, the involution σ acts on the linear system $a^{-1}(\ell)$. Hence there is a $D \in a^{-1}(\ell)$ which is fixed under the action of σ. This implies that D must be of the form

$$D = f^*E + \sum_{i=1}^{s} p_{l_i} \quad \text{with an effective } E \in \text{Div}(C) \text{ and } s \leq r.$$

It follows that

$$\ell = \mathcal{O}_{\widetilde{C}}\left(\sum_{i=1}^{s}(p_{l_i} - p_1)\right) \otimes f^*(n) \quad \text{for some } n \in \text{Pic}^0(C).$$

But $\ell^2 = \mathcal{O}_{\widetilde{C}}$ and hence $n \in J[2]$, since f^* is injective. This implies

$$\ell = \mathcal{F}_{l_1} \otimes \cdots \otimes \mathcal{F}_{l_s} \otimes f^*(m)$$

with $m = n \otimes m_{l_1}^{-1} \cdots \otimes m_{l_s}^{-1} \in J[2]$.

Since the square of any line bundle $\mathcal{F}_i$ is a pullback under f, for the existence of the decomposition, we only have to show that, up to a pullback under f, $\mathcal{F}_r$ can be expressed by the other $\mathcal{F}_i$. This and the uniqueness of the decomposition follow from the following lemma. $\qquad\square$

Lemma 3.3.4 *Let the notations be as in Proposition 3.3.3. Then the following conditions are equivalent:*

(i) $\mathcal{F}_{i_1} \otimes \cdots \otimes \mathcal{F}_{i_k} \in f^*J[2]$ *with* $2 \leq i_1 < \cdots < i_k \leq r$.
(ii) $\{i_1, \ldots, i_k\} = \{2, \ldots, r\}$.

Proof (i) $\Rightarrow$ (ii): We may label the points in such a way that $\{i_1, \ldots, i_k\} = \{2, \ldots, k+1\}$. Then we have to show that if $\mathcal{F}_2 \otimes \cdots \otimes \mathcal{F}_{k+1} = f^*(n)$ with $n \in J[2]$, then $k + 1 = r$.

The assumption $\mathcal{F}_2 \otimes \cdots, \otimes \mathcal{F}_{k+1} = f^*(n)$ with $n \in J[2]$ is equivalent to

$$\mathcal{O}_{\widetilde{C}}\left(\sum_{i=2}^{k+1} p_i - k p_1\right) = f^*(n \otimes m_2^{-1} \otimes \cdots \otimes m_{k+1}^{-1}).$$

If $k + 1$ is even, define

$$\mathcal{F} := n \otimes m_2^{-1} \otimes \cdots \otimes m_{k+1}^{-1} \otimes \mathcal{O}_C\left(\frac{k+1}{2} q_1\right).$$

Then by the choice of the m_j,

$$\mathcal{F}^2 = m_2^{-2} \otimes \cdots \otimes m_k^{-2} \otimes \mathcal{O}_C((k+1)q_1) = \mathcal{O}_C(\sum_{i=1}^{k+1} q_i)$$

and

$$f^*(\mathcal{F}) = f^*(n \otimes m_2^{-1} \otimes \cdots \otimes m_{k+1}^{-1}) \otimes \mathcal{O}_{\widetilde{C}}((k+1)p_1)$$

$$= \mathcal{O}_{\widetilde{C}}\left(\sum_{i=2}^{k+1} p_i - k p_1\right) \otimes \mathcal{O}_{\widetilde{C}}((k+1)p_1) = \mathcal{O}_{\widetilde{C}}\left(\sum_{i=1}^{k+1} p_i\right).$$

This is a contradiction if $k + 1 < r$, since then the cover f would be ramified only at $k + 1$ points.

If $k + 1$ would be odd, we would define

$$\mathcal{F} := n \otimes m_2^{-1} \otimes \cdots \otimes m_{k+1}^{-1} \otimes \mathcal{O}_C\left(\frac{k}{2} q_1\right)$$

and have that $\mathcal{F}^2 = \mathcal{O}_C(\sum_{i=2}^{k+1} q_i)$ and $f^*(\mathcal{F}) = \mathcal{O}_{\widetilde{C}}(\sum_{i=2}^{k+1} p_i)$ which are impossible, since $k + 1 < r$ in any case, r being an even number.

(ii) $\Rightarrow$ (i): Let $\mathcal{D} \in \mathrm{Pic}^{r/2}(C)$ be as in the proof of Proposition 3.3.3, and assume $\{i_1, \ldots, i_k\} = \{2, \ldots, r\}$. Then

$$\mathcal{F}_2 \otimes \cdots \otimes \mathcal{F}_r = \mathcal{O}_{\widetilde{C}}(p_2 + \cdots + p_r - (r-1)p_1) \otimes f^*(m_2 \otimes \cdots \otimes m_r)$$

$$= f^*\left(\mathcal{D}(-\frac{r}{2} q_1) \otimes m_2 \cdots \otimes m_r\right).$$

Since

$$\left(\mathcal{D}(-\frac{r}{2}q_1) \otimes m_2 \cdots \otimes m_r \right)^2 = \mathcal{O}_C(q_1 + \cdots + q_r) \otimes \mathcal{O}_C(-rq_1) \otimes_{i=2}^r \mathcal{O}_C(q_1 - q_r) = \mathcal{O}_C,$$

this implies $\mathcal{F}_2 \otimes \cdots \otimes \mathcal{F}_r \in f^* J[2]$. $\qquad\qquad \square$

The following corollary gives a different set of generators of the group $P(f)[2]/f^* J[2]$, which we will need in later sections. For $i = 1, \ldots, r/2$ choose elements

$$n_i \in \mathrm{Pic}^0(C) \quad \text{such that} \quad n_i^2 = \mathcal{O}_C(q_{2i-1} - q_{2i})$$

and for $j = 1, \ldots r/2 - 1$ elements

$$n_i' \in \mathrm{Pic}^0(C) \quad \text{such that} \quad n_i'^2 = \mathcal{O}_C(q_{2i} - q_{2i+1})$$

and consider the line bundles

$$\mathcal{G}_i := \mathcal{O}_{\widetilde{C}}(p_{2i} - p_{2i-1}) \otimes f^*(n_i) \quad \text{for} \quad i = 1, \ldots r/2$$

and

$$\mathcal{H}_i := \mathcal{O}_{\widetilde{C}}(p_{2i+1} - p_{2i}) \otimes f^*(n_i') \quad \text{for} \quad i = 1 \ldots, r/2 - 1.$$

Corollary 3.3.5 *With these notations we have*

$$P(f)[2] = f^* J[2] \bigoplus_{i=1}^{r/2-1} \mathcal{G}_i \mathbb{Z}/2\mathbb{Z} \bigoplus_{j=1}^{r/2-1} \mathcal{H}_j \mathbb{Z}/2\mathbb{Z}.$$

Proof It suffices to show that modulo $f^* J[2]$ the $\mathcal{F}_i$, $i = 2, \ldots, r - 1$ can be expressed by the $r - 2$ line bundles $\mathcal{G}_i$ and $\mathcal{H}_j$. But modulo $f^* J[2]$ we have for $i = 1, \ldots, r/2 - 1$.

$$\mathcal{G}_1 \otimes \mathcal{H}_1 \otimes \cdots \otimes \mathcal{G}_{i-1} \otimes \mathcal{H}_{i-1} \otimes \mathcal{G}_i = \mathcal{F}_{2i}$$

and

$$\mathcal{G}_1 \otimes \mathcal{H}_1 \otimes \cdots \otimes \mathcal{G}_{i-1} \otimes \mathcal{H}_{i-1} \otimes \mathcal{G}_i \otimes \mathcal{H}_i = \mathcal{F}_{2i+1}.$$

$\qquad\qquad \square$

Another type of generators of $\oplus_{i=2}^{r-1} \mathcal{F}_i \mathbb{Z}/2\mathbb{Z}$ is given as follows: Let the notation be as for Proposition 3.3.3. For $i = 1, \ldots r - 1$ choose $n_i \in \mathrm{Pic}^0(C)$ such that

$$n_i^2 = \mathcal{O}_C(f(p_i) - f(p_{i+1}))$$

and define for $i = 1, \ldots, r - 1$,

$$\mathcal{G}_i := \mathcal{O}_{\widetilde{C}}(p_i - p_{i+1}) \otimes f^*(n_i).$$

Corollary 3.3.6 *With these notations we have*

$$P(f)[2] = f^* J[2] \bigoplus_{i=1}^{s-1} \mathcal{G}_i \mathbb{Z}/2\mathbb{Z} \bigoplus_{i=s+1}^{r-1} \mathcal{G}_i \mathbb{Z}/2\mathbb{Z}.$$

Proof Clearly the $\mathcal{G}_i$ are contained in $P(f)[2]$. One checks in the same way as for Lemma 3.3.4 that $\mathcal{G}_1, \ldots \mathcal{G}_{s-1}, \mathcal{G}_{s+1}, \ldots, \mathcal{G}_{r-1}$ are linear independent. This implies that modulo $f^* J[2]$,

$$\bigoplus_{i=2}^{r-1} \mathcal{F}_i \mathbb{Z}/2\mathbb{Z} = \bigoplus_{i=1}^{s-1} \mathcal{G}_i \mathbb{Z}/2\mathbb{Z} \bigoplus_{i=s+1}^{r-1} \mathcal{G}_i \mathbb{Z}/2\mathbb{Z}.$$

Hence the assertion follows from Proposition 3.3.3.   □

Remark 3.3.7 The description of the two-division points of Prym varieties of double covers generalizes easily to the description of the n-division points of Prym varieties of cyclic covers of degree $n \geq 2$. For $n = 3$ this is done (with proof) in Sect. 4.2.5.

3.4 Prym Varieties of Pairs of Covers

Consider the following commutative diagram of finite morphisms of smooth projective curves:

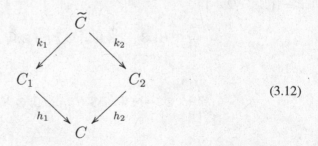

$$(3.12)$$

In this section we want to introduce and study the Prym variety of the pair (k_1, k_2) following [20] and [21].

Lemma 3.4.1 *Suppose both h_1 and h_2 do not factorize via the same morphism $C' \to C$ of degree ≥ 2. Then $k_1^* P(h_1)$ is an abelian subvariety of the Prym subvariety $P(k_2)$.*

Proof Suppose first that both k_1 and k_2 do not factorize via a morphism $k : \widetilde{C} \to C'$ of degree ≥ 2. The universal property of the fibre product over C gives the diagram

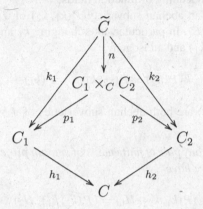

where n is the normalization map. According to [16, Proposition 6.5.8] we have for any line bundle $\ell \in \mathrm{Pic}(C_1)$

$$\mathrm{Nm}_{p_2} p_1^*(\ell) = h_2^* \mathrm{Nm}_{h_1}(\ell).$$

According to [16, Section 6], the norm map is also defined for the normalization map, and we have

$$\mathrm{Nm}_{k_2} k_1^*(\ell) = \mathrm{Nm}_{p_2} \mathrm{Nm}_n n^* p_1^*(\ell) = \mathrm{Nm}_{p_2} p_1^*(\ell)$$

where the last equation follows, since n is of degree 1. Combining both equations, we get for any $\ell \in P(h_1)$

$$\mathrm{Nm}_{k_2} k_1^*(\ell) = h_2^* \mathrm{Nm}_{h_1}(\ell) = 0,$$

which gives the assertion in the special case.

In the general case, suppose k_i factorizes as $k_i = k_i' k$ with some morphism $k : \widetilde{C} \to C'$ for $i = 1$ and 2 such that the pair (k_1', k_2') satisfies the above assumptions. Then $k_1'^* P(h_1)$ is an abelian subvariety of $P(k_2')$. But then, by Theorem 3.2.12(a), $k_1^* P(h_1)$ is an abelian subvariety of $P(k_2)$.   □

The canonical principal polarization of $\widetilde{J}$ induces a polarization on $P(k_2)$. Hence according to Corollary 2.5.2, the complementary abelian subvariety of $k_1^* P(h_1)$ in $P(k_2)$ is well defined. We denote it by $P(k_1, k_2)$ and call it the *Prym variety of the pair $P(k_1, k_2)$*. The addition map induces an isogeny

$$k_1^* P(h_1) \times P(k_1, k_2) \to P(k_2).$$

Note that $P(k_1, k_2)$ only depends on $k_1 : \widetilde{C} \to C_1$ and $k_2 : \widetilde{C} \to C_2$. In fact, the curve C and the covers $h_i : C_i \to C$ are uniquely determined: C is the smooth projective curve corresponding to the function field $\mathbb{C}(C_1) \cap \mathbb{C}(C_2)$ and the maps h_i to the corresponding embedding of function fields.

In this way, we get an abelian subvariety $P(k_1, k_2)$ of $\widetilde{J}$ for any pair of finite curve maps $k_i : \widetilde{C} \to C_i$. In particular, interchanging k_1 and k_2 we get the Prym variety of the pair (k_2, k_1) and an isogeny

$$k_2^* P(h_2) \times P(k_2, k_1) \to P(k_1).$$

$P(k_1, k_2)$ and $P(k_2, k_1)$ are both abelian subvarieties of $\widetilde{J}$ with induced polarizations, say H_1 and H_2.

Proposition 3.4.2 *For any pair of finite maps of smooth projective curves $k_1 : \widetilde{C} \to C_1$ and $k_2 : \widetilde{C} \to C_2$, we have*

$$(P(k_1, k_2), H_1) = (P(k_2, k_1), H_2)$$

as polarized abelian subvarieties of $\widetilde{J}$.

Proof It suffices to show that $P(k_1, k_2) = P(k_2, k_1)$, since both polarizations are induced by the canonical polarization on $\widetilde{J}$. The addition maps induce isogenies

$$k_2^* h_2^* J \times k_2^* P(h_2) \times k_1^* P(h_1) \times P(k_1, k_2) \longrightarrow k_2^* J C_2 \times P(k_2) \longrightarrow \widetilde{J}$$

and

$$k_1^* h_1^* J \times k_1^* P(h_1) \times k_2^* P(h_2) \times P(k_2, k_1) \longrightarrow k_1^* J C_1 \times P(k_1) \longrightarrow \widetilde{J}$$

where all abelian varieties are subvarieties of $\widetilde{J}$. Obviously we have $k_1^* h_1^* J = k_2^* h_2^* J$. Hence, if Z denotes the image of $k_2^* h_2^* J \times k_2^* P(h_2) \times k_1^* P(h_1)$ in $\widetilde{J}$, the addition map gives isogenies

$$Z \times P(k_1, k_2) \to \widetilde{J} \quad \text{and} \quad Z \times P(k_2, k_1) \to \widetilde{J}.$$

Now the corresponding decompositions of the tangent spaces are orthogonal with respect to the hermitian form associated to the canonical polarization of $\widetilde{J}$. This implies that on the one hand $P(k_1, k_2)$ and on the other hand $P(k_2, k_1)$ are the complement of the abelian subvariety Z in $\widetilde{J}$. Since the complement is uniquely determined, this implies $P(k_1, k_2) = P(k_2, k_1)$. $\square$

Corollary 3.4.3 *With the notations of above, we have*

$$P(k_1, k_2) \subseteq [P(k_1) \cap P(k_2)]^0.$$

Proof By definition $P(k_1, k_2) \subseteq P(k_2)$ and by Proposition 3.4.2, we have $P(k_1, k_2) \subseteq P(k_1)$. Since $P(k_1, k_2)$ is connected, this implies the assertion. $\square$

In all cases we checked, there holds equality in Corollary 3.4.3. We do not know whether this holds always.

We will compute the dimension of $P(k_1, k_2)$ only in the most important case, namely, when the hypotheses of Lemma 3.4.1 are satisfied and $\widetilde{C}$ is the normalization of $C_1 \times_C C_2$; that is, the function fields satisfy $\mathbb{C}(C_1)\mathbb{C}(C_2) = \mathbb{C}(\widetilde{C})$ and $\mathbb{C}(C_1) \cap \mathbb{C}(C_2) = \mathbb{C}(C)$.

Let $d_1 := \deg k_1 = \deg h_2$ and $d_2 := \deg k_2 = \deg h_1$. Moreover, for any cover f of smooth projective curves, let $r(f)$ denote the ramification degree of f. Then we have

Proposition 3.4.4

$$\dim P(k_1, k_2) = (d_1 - 1)(d_2 - 1)(g(C) - 1) + \frac{1}{2}[r(k_1) + (d_1 - 1)r(h_1) - r(h_2)].$$

Proof

$$\dim P(k_1, k_2) = \dim P(k_2) - \dim P(h_1) = g(\widetilde{C}) - g(C_2) - g(C_1) + g(C).$$

So the proof is an application of the Hurwitz formula which we omit. $\square$

We compute the type of the polarization on $P(k_1, k_2)$ only in a special case. Recall that for any polarization L on an abelian variety A, the group $K(L)$ denotes the kernel of the induced isogeny $\phi_L : A \to \widehat{A}$. Recall furthermore that the polarization L is of type $(d_1, \ldots, d_g)$ if and only if $K(L) \simeq (\mathbb{Z}/d_1\mathbb{Z} \times \cdots \times \mathbb{Z}/d_g\mathbb{Z})^2$. Hence, in order to compute the type of a polarization L, it suffices to determine the group $K(L)$.

Suppose we are given a diagram (3.12) with the additional properties:

(i) $C = \mathbb{P}^1$.
(ii) Both h_1 and h_2 do not factorize via the same morphism $C' \to \mathbb{P}^1$ of degree ≥ 2.
(iii) $\widetilde{C}$ is the normalization of $C_1 \times_{\mathbb{P}^1} C_2$.
(iv) $d_1 := \deg k_1 \geq 2$ and $d_2 := \deg k_2 \geq 2$ are prime to each other.

Let Θ denote the canonical polarization of $\widetilde{J}$ and denote

$$L := \Theta|_{P(k_2)},$$

$$L_{k_1^* P(h_1)} := L|_{k_1^* P(h_1)} = \Theta|_{k_1^* P(h_1)}, \quad \text{and}$$

$$L_{P(k_1, k_2)} := L|_{P(k_1, k_2)} = \Theta|_{P(k_1, k_2)}.$$

Then we have,

Theorem 3.4.5 *Given a diagram* (3.12) *with the additional properties* (i) *to* (iv). *Then*

$$K(L_{P(k_1,k_2)}) \simeq K(L) \oplus K(L_{k_1^* P(h_1)}).$$

Proof By definition $P(k_2)$ is the complementary abelian subvariety of $k_2^* J C_2$ in $\tilde{J}$. According to Propositions 3.2.2 and 3.2.1, $\Theta|_{k_2^* J C_2}$ is of type $(d_{21}, \ldots d_{2s}, d_2, \ldots, d_2)$ where d_{2i} divides d_2 for $i = 1, \ldots, s$ and $0 \leq s <$ dim J_{C_2}. Since dim $P(k_2) \geq$ dim $J C_2$, Proposition 2.6.3 implies that L is of type $(1, \ldots, 1, d_{21}, \ldots d_{2s}, d_2, \ldots, d_2)$. In particular, $K(L)$ is of exponent d_2.

On the other hand, $P(h_1) = J C_1$. By the same reason, $L_{k_1^* J C_1} = \Theta|_{k_1^* P(h_1)}$ is of type $(d_{11}, \ldots, d_{1t}, d_1, \ldots, d_1)$ where d_{1i} divides d_1 for $i = 1, \ldots, t <$ dim $J C_1$. In particular, $L_{k_1^* P(h_1)}$ is of exponent d_1.

Now $P(k_1, k_2)$ is by definition the complementary abelian subvariety of $k_1^* P(h_1)$ in $P(k_2)$. Hence, according to Lemma 2.6.9, there is an exact sequence

$$0 \to k_1^* P(h_1) \cap P(k_1, k_2) \to K(L_{k_1^* P(h_1)}) \to K(L) \cap k_1^* P(h_1) \to 0$$

The assumption $\gcd(d_1, d_2) = 1$ implies by the above reasoning that

$$\gcd(|K(L)|, |K(k_1^* P(h_1))|) = 1.$$

Hence $K(L) \cap k_1^* P(h_1) = 0$ and thus

$$K(L_{k_1^* P(h_1)}) = k_1^* P(h_1) \cap P(k_1, k_2).$$

But then Lemma 2.6.7 yields

$$|K(L_{k_1^* P(h_1)})| \cdot |K(L_{P(k_1,k_2)})| = |K(L_{k_1^* P(h_1)})|^2 \cdot |K(L)|$$

Hence, dividing by $|K(L_{k_1^* P(h_1)})|$ gives $|K(L_{P(k_1,k_2)})| = |K(L_{k_1^* P(h_1)})| \cdot |K(L)|$. Since d_1 and d_2 are prime to each other, this gives the assertion. □

3.5 Galois Covers of Curves

In this section we study Galois covers $f : \tilde{C} \to C$ with finite group G. In particular we consider the isotypical and the group algebra decomposition of $\tilde{J}$ as well as the induced decompositions of $J C_H$ for a subgroup $H \subset G$.

3.5.1 Jacobians and Pryms of Intermediate Covers

Let $f : \widetilde{C} \to C$ be a Galois cover; that is, there is a subgroup $G \subseteq \mathrm{Aut}(\widetilde{C})$ such that $C = \widetilde{C}/G$ and f is the canonical quotient map. For any $\sigma \in G$, we denote the induced automorphism on the Jacobian $\widetilde{J} := J\widetilde{C}$ by the same symbol σ and by $\langle \sigma \rangle$ the group generated by σ. Consider the homomorphism Nm_G defined as

$$\mathrm{Nm}_G : \widetilde{J} \to \widetilde{J}, \qquad x \mapsto \sum_{\sigma \in G} \sigma(x).$$

Nm_G and Nm_f are related as follows.

Proposition 3.5.1 *For any Galois cover* $f : \widetilde{C} \to C$ *with group* G, *we have*

$$\mathrm{Nm}_G = f^* \circ \mathrm{Nm}_f.$$

Proof This follows immediately from the definitions. To be more precise, every $x \in \widetilde{J}$ is of the form $x = \mathcal{O}_{\widetilde{C}}(\sum_{i=1}^{r}(p_i - q_i))$ for some $r \geq 1$ and points $p_i, q_i \in \widetilde{C}$. Hence

$$f^*\mathrm{Nm}_f(x) = f^*\mathcal{O}_C\left(\sum_{i=1}^{r}(f(p_i) - f(q_i)) \right) = \mathcal{O}_{\widetilde{C}}\left(\sum_{i=1}^{r}(f^{-1}f(p_i) - f^{-1}f(q_i)) \right)$$

$$= \mathcal{O}_{\widetilde{C}}\left(\sum_{i=1}^{r}\sum_{\sigma \in G}(\sigma(p_i) - \sigma(q_i)) \right) = \mathrm{Nm}_G(x).$$

$\square$

As an immediate consequence, we get

Corollary 3.5.2 *For any Galois cover* $f : \widetilde{C} \to C$ *with group* G, *we have*

(a) $\mathrm{Im} f^* = \{x \in \widetilde{J} \mid x = \sum_{\sigma \in G} \sigma(y) \text{ for some } y \in \widetilde{J}\}$
(b) $P(f) = (\mathrm{Ker}\,\mathrm{Nm}_G)^0$

For any subset $S \subseteq \widetilde{J}$ and any subgroup $H \subseteq G$, we denote

$$S^H := \{x \in S \mid \sigma(x) = x \text{ for all } \sigma \in H\}$$

the set of fixed points of H in S. For any element $\sigma \in G$, we write S^σ instead of $S^{\langle \sigma \rangle}$.

Corollary 3.5.3 *Let* $f : \widetilde{C} \to C$ *be a Galois cover with group* G *of degree* d, *and let* $P(f)$ *denote the corresponding Prym variety. Then*

(a) $(\widetilde{J}^G)^0 = f^*J$.
(b) $\widetilde{J}^G = f^*J + P(f)_0$ *with* $P(f)_0 = P(f) \cap \widetilde{J}^G \subseteq P(f)[d]$.

Proof The subvarieties f^*J and $P(f)$ are complementary abelian subvarieties in $\tilde{J}$ with respect to the canonical polarization. Hence, using Corollary 3.5.2(a), we get

$$\tilde{J}^G = (f^*J)^G + P(f)^G = f^*J + P(f)^G.$$

So Corollary 3.5.2 implies assertion (b). The other assertions also follow from Corollary 3.5.2.   □

Corollary 3.5.4 *Let* $f : \tilde{C} \to C$ *be a double cover given by the involution* $\sigma :$ $\tilde{C} \to \tilde{C}$. *Then we have*

$$\tilde{J}^\sigma = f^*J + P(f)[2].$$

Compare this with Proposition 3.3.3.

Proof From Corollary 3.5.3 we know that

$$\tilde{J}^\sigma = f^*J + P(f)_0 \qquad \text{with} \qquad P(f)_0 = P(f) \cap \tilde{J}^\sigma \subseteq P(f)[2].$$

Since σ is an involution, we clearly have $P(f)[2] \subset \tilde{J}^\sigma$ which gives the assertion.
                                                                      □

For any subgroup $H \subset G$, we denote

$$C_H := \tilde{C}/H.$$

In particular, $C = C_G$. So we have the following commutative diagram

$$\tilde{C} \xrightarrow{\quad f \quad} C = C_G$$

with maps $k : \tilde{C} \to C_H$ and $h : C_H \to C = C_G$.

The following proposition generalizes Proposition 3.5.1.

Proposition 3.5.5 *Let* $\{\sigma_1, \sigma_2, \ldots, \sigma_r\}$ *be a set of representatives for* G/H. *Then*

$$f^*\mathrm{Nm}_h(y) = \sum_{j=1}^r \sigma_j k^*(y) \quad \text{for all} \quad y \in JC_H.$$

Proof Every $y \in JC_H$ is of the form $y = \mathcal{O}_{C_H}\left(\sum_{i=1}^s (\overline{p}_i - \overline{q}_i)\right)$ with points $\overline{p}_i, \overline{q}_i \in C_H$ and some $s \geq 1$. Choose $p_i \in k^{-1}(\overline{p}_i)$ and $q_i \in k^{-1}(\overline{q}_i)$. Then

$$f^* \mathrm{Nm}_h(y) = f^* \mathcal{O}_C \left(\sum_{i=1}^{s} (h(\overline{p_i}) - h(\overline{q_i})) \right) = \mathcal{O}_{\widetilde{C}} \left(\sum_{i=1}^{s} \sum_{\sigma \in G} (\widetilde{\sigma}(p_i) - \sigma(q_i)) \right)$$

$$= \mathcal{O}_{\widetilde{C}} \left(\sum_{i=1}^{s} \sum_{j=1}^{r} \sigma_j \sum_{\tau \in H} (\tau(p_i) - \tau(q_i)) \right)$$

$$= \mathcal{O}_{\widetilde{C}} \left(\sum_{j=1}^{r} \sigma_j \sum_{\tau \in H} \sum_{i=1}^{s} (\tau(p_i) - \tau(q_i)) \right) = \sum_{j=1}^{r} \sigma_j k^*(y).$$

$\square$

Proposition 3.5.6 *With the assumptions of Proposition 3.5.5, we have*

$$k^* P(h) = \left\{ x \in \widetilde{J}^H \mid \sum_{j=1}^{r} \sigma_j(x) = 0 \right\}^0 .$$

Proof Consider the abelian subvariety $A := \{ x \in \widetilde{J}^H \mid \sum_{j=1}^{r} \sigma_j(x) = 0 \}^0$ of $\widetilde{J}$.
We first claim $k^* P(h) \subseteq A$. To see this, let $z \in P(h)$. Then

$$\mathrm{Nm}_f(k^* z) = \mathrm{Nm}_h \mathrm{Nm}_k k^*(z) = |H| \mathrm{Nm}_h(z) = 0.$$

Hence

$$k^* P(h) \subseteq (\mathrm{Ker} \mathrm{Nm}_f)^0 = (\mathrm{Ker} \mathrm{Nm}_G)^0$$

which gives

$$|H| \sum_{j=1}^{r} \sigma_j h^*(z) = \sum_{g \in G} h^*(z) = 0.$$

Since clearly $k^* P(h) \subseteq \widetilde{J}^H$, this implies $k^* P(h) \subseteq A$.
Conversely, let $x \in A$. Then

$$\mathrm{Nm}_G(x) = \sum_{j=1}^{r} \sigma_j \sum_{\tau \in H} \tau(x) = |H| \sum_{j=1}^{r} \sigma_j(x) = 0$$

which gives $A \subseteq (\mathrm{Ker} \mathrm{Nm}_G)^0 = (\mathrm{Ker} \mathrm{Nm}_f)^0$. Hence $\mathrm{Nm}_k(A) \subseteq (\mathrm{Ker} \mathrm{Nm}_h)^0 = P(h)$ and thus

$$A = |H| A = \sum_{\tau \in H} \tau(A) = k^* \mathrm{Nm}_k(A) \subseteq k^* P(h).$$

$\square$

3.5.2   Isotypical and Group Algebra Decompositions of Intermediate Covers

Let $f : \widetilde{C} \to C$ be a Galois cover with group G and let $H \subseteq N \subseteq G$ be subgroups. In this subsection we want to discuss the isogeny decompositions of the Jacobian $JC_H = J(\widetilde{C}/H)$ and the Prym variety $P(C_H/C_N)$ determined by an appropriate combination of rational representations of G with the representations ρ_H and ρ_N of G, induced by the trivial representations of H and N, respectively.

As above, denote $\widetilde{J} := J\widetilde{C}$ and $J := JC$. Assume that the rational irreducible representations $W_1, \dots, W_r$ of G are labelled in such a way that W_1 is the trivial representation. Hence, if $e_{W_1}, \dots, e_{W_r}$ are the corresponding symmetric idempotents of $\mathbb{Q}[G]$, the isotypical decomposition of $\widetilde{J}$ is

$$\widetilde{J} \sim J \times \widetilde{J}^{e_{W_2}} \times \cdots \times \widetilde{J}^{e_{W_r}} \tag{3.13}$$

and according to Theorem 2.9.2 and Eq. (2.30), the group algebra decomposition of $\widetilde{J}$ is

$$\widetilde{J} \sim J \times B_2^{\frac{\dim V_2}{s_i}} \times \cdots \times B_r^{\frac{\dim V_r}{s_r}}, \tag{3.14}$$

where $B_i := \mathrm{Im} q_i$ with q_i a primitive idempotent in $\mathbb{Q}[G]e_{W_i}$ and s_i the Schur index of V_i.

Now for any subgroup $H \subset G$, consider the idempotents

$$p_H = \frac{1}{|H|} \sum_{h \in H} h \qquad \text{and} \qquad f_H^j = p_H e_{W_j}$$

as in Sect. 2.8. With these notations we have

Proposition 3.5.7 *Let $H \subseteq G$ be a subgroup and $\pi_H : \widetilde{C} \to C_H(= \widetilde{C}/H)$ the corresponding cover. Then we have*

$$\mathrm{Im}\, p_H = \pi_H^*(JC_H) \qquad and$$

$$\mathrm{Im}(f_H^j) \sim B_j^{\frac{\dim V_j^H}{s_j}} \qquad for\ all\ j = 1, \dots, r.$$

Proof First, $\mathrm{Im}(p_H)$ is the connected component containing 0 of the subvariety $\widetilde{J}^H$ of $\widetilde{J}$ fixed by H and therefore is equal to $\pi_H^*(JC_H)$ which is of course isogenous to JC_H.

For the last assertion, note first that according to the definition of f_H^j, its image is contained in $\text{Im}(p_H)$ and according to Proposition 2.8.5 and Sect. 2.9 we have

$$\text{Im}(f_H^j) \sim B_j^{\frac{\dim V_j^H}{s_j}} .$$

□

Corollary 3.5.8 *Let the hypotheses be as in Proposition 3.5.7.*

(i) *Up to isogeny there is a decomposition:*

$$JC_H \sim J \times B_2^{\frac{\dim V_2^H}{s_2}} \times \cdots \times B_r^{\frac{\dim V_r^H}{s_r}} .$$

(ii) *For $j = 2, \ldots, r$ the abelian subvariety $A_H^j := \text{Im}(f_H^j)$ is uniquely determined by W_j and H, and the canonical map*

$$\mu : J \times A_H^2 \times \cdots \times A_H^r \to JC_H$$

is an isogeny.

Proof Since p_H and e_{W_j} are uniquely determined by H and W_j, so is A_H^j. The definition of f_H^j and the equation $\sum e_{W_j} = 1$ give

$$\sum_{j=1}^r f_H^j = \sum_{j=1}^r p_H e_{W_j} = p_H$$

Since the f_H^j are orthogonal to each other, this together with Proposition 3.5.7 gives both assertions, choosing isogenies $B_j^{\frac{\dim V_j^H}{s_j}} \to A_H^j$. □

Note that in general there is no group action on JC_H. However, as we saw in Sect. 2.10, the Hecke algebra $\mathcal{H}_H = \mathbb{Q}[H \backslash G / H] = p_H \mathbb{Q}[G] p_H$ for H in G acts on JC_H by restricting $\rho : \mathbb{Q}[G] \to \text{End}_{\mathbb{Q}}(\tilde{J})$ to $\tilde{\rho} : \mathcal{H}_H \to \text{End}_Q(JC_H)$. Here we use bijective correspondence between the rational irreducible representations W of G such that $\langle \rho_H, W \rangle \neq 0$ and the rational irreducible representations $\tilde{W}$ of $\mathbb{Q}[H \backslash G / H]$. As follows from Proposition 2.8.6, the central idempotents of $\mathbb{Q}[H \backslash G / H]$ are given precisely by those f_H^j which are non-zero or equivalently by those with $V_j^H \neq 0$.

Hence omitting the factors 0 in Corollary 3.5.8(i) and the corresponding factors in (ii), the isogenies in the Corollary coincide with the Hecke algebra decomposition of JC_H of Proposition 2.10.1 and the isotypical decomposition of JC_H given in Eq. (2.38).

In a similar way, one can express the following corollary concerning Prym varieties of intermediate covers in terms of representations of Hecke algebras, for which we refer to [8], but prove it directly.

Corollary 3.5.9 *Let the hypotheses be as in Proposition 3.5.7. For any subgroups* $H \subset N \subset G$, *we have, for the Prym variety* $P(C_H/C_N)$,

(i) *For* $j = 2, \ldots, r$, *let* $t_j := \dfrac{\dim V_j^H}{s_j} - \dfrac{\dim V_j^N}{s_j}$. *Then an associated group algebra decomposition of* $P(C_H/C_N)$ *is*

$$P(C_H/C_N) \sim B_2^{t_2} \times \cdots \times B_r^{t_r}.$$

(ii) *Let* $A_H^j = \operatorname{Im}(f_H^j)$ *and* $A_N^j = \operatorname{Im}(f_N^j)$. *For* $j = 2, \ldots, r$, *we have* $A_N^j \subset A_H^j$. *Let* $P_{H,N}^j$ *be the complementary abelian subvariety of* A_N^j *in* A_H^j *with respect to the induced polarization. Then the associated isotypical decomposition of* $P(C_H/C_N)$ *is given by the addition map*

$$\mu : P_{H,N}^2 \times \cdots \times P_{H,N}^r \to P(C_H/C_N).$$

Proof
(i): Corollary 3.5.8(i) gives associated algebra decompositions of JC_H and JC_N. On the other hand, we have an isogeny

$$JC_N \times P(C_H/C_N) \sim JC_H.$$

This gives

$$P(C_H/C_N) \times J \times B_2^{\frac{\dim V_2^N}{s_2}} \times \cdots \times B_r^{\frac{\dim V_r^N}{s_r}} \sim J \times B_2^{\frac{\dim V_2^H}{s_2}} \times \cdots \times B_r^{\frac{\dim V_r^H}{s_r}}$$

This gives (i), since up to isogeny one can cancel factors.
(ii): follows from Corollary 3.5.8 for H and N together with the fact that the addition map $\mu : JC_N \times P(C_H/C_N) \to JC_H$ is an isogeny. $\square$

The following result is a special case of Corollary 3.5.9. It was quoted in the introduction as Theorem. It is in fact the most important result of this section, in the sense that it will be applied most often in the subsequent chapters.

Corollary 3.5.10 *Let the hypotheses be as above with the additional assumption that there are subgroups* $H \subset N \subset G$ *such that*

$$\rho_H \simeq \rho_N \oplus W_j$$

for one of the irreducible rational representation W_j *of* G, $j \in \{2, \ldots, r\}$, *and let* B_j *be the image of a primitive idempotent associated by* W_j *as in (3.14). Then* B_j *is an abelian subvariety of* A_{W_j}, *and there is an isogeny*

$$P(C_H/C_N) \sim B_j.$$

Let

$$n_j := \frac{\dim V_j}{s_j}.$$

Then there is an isogeny

$$(P(C_H/C_N))^{n_j} \sim A_{W_j}$$

of the n_j-th power of the Prym variety $P(C_H/C_N)$ onto the isotypical component A_{W_j} associated to W_j.

Proof Let t_i be as defined in Corollary 3.5.9. According to Lemma 2.8.1,

$$\rho_H = \rho_N \bigoplus_{i=2}^{r} t_i W_i.$$

But by hypothesis we have

$$\rho_H = \rho_N \oplus W_j.$$

Therefore $t_j = 1$ and $t_i = 0$ for $i \neq j$. $\qquad \square$

3.5.3 Decomposition of the Tangent Space of the Prym Variety Associated to a Pair of Subgroups

Let G be a finite group acting on a curve $\widetilde{C}$. For any subgroup $H \subseteq G$, let $\pi_H : \widetilde{C} \to C_H := \widetilde{C}/H$ denote the canonical projection onto the corresponding quotient curve. For any pair of subgroups $H \subset N \subset G$, let $\pi_{H,N} : C_H \to C_N$ be the canonical projection. We want to use the group action of G in order to decompose the Prym variety

$$P(\pi_{H,N}) = P(C_H/C_N).$$

Let

$$T_0 \widetilde{J} = V = n_1 V_1 \oplus \cdots \oplus n_r V_r \tag{3.15}$$

denote the isotypical decomposition of the tangent space of $\widetilde{J}$ at 0. According to Sect. 2.9.2, it is orthogonal with respect to the canonical bilinear character form $\langle \cdot, \cdot \rangle$.

Proposition 3.5.11 *Assume that there is an integer s such that the decomposition (3.15) satisfies for the subgroups $H \subseteq N \subseteq G$ that*

$$\rho_H = \rho_N \oplus \bigoplus_{i=1}^{s} (\dim V_i^H) V_i.$$

Then we have for the differential of the map $\pi_H : \widetilde{C} \to C_H$ and the tangent space of the Prym variety

$$d\pi_H^*(T_0 P(C_H/C_N)) = n_1 V_1^H \oplus \cdots \oplus n_s V_s^H.$$

with n_i as given in Eq. (3.15).

Proof According to Lemma 2.9.4, we have a G-equivariant decomposition

$$(T_0 \widetilde{J})^H = (T_0 \widetilde{J})^N \oplus n_1 V_1^H \oplus \cdots \oplus n_s V_s^H,$$

orthogonal with respect to the inner product $\langle \cdot, \cdot \rangle$.

For every subgroup F of G, we have the equality

$$(T_0 \widetilde{J})^F = d\pi_F^*(T_0 J C_F),$$

since $\pi_F^* : J C_F \to \widetilde{J}^F$ is an isogeny. Hence the above decomposition may be written as

$$d\pi_H^*(T_0 J C_H) = d\pi_N^*(T_0 J C_N) \oplus n_1 V_1^H \oplus \cdots \oplus n_s V_s^H.$$

But $d\pi_H^*(T_0 P(C_H/C_N))$ is the complement of $d\pi_N^*(T_0 J C_N)$ inside $d\pi_H^*(T_0 J C_H)$ with respect to the restriction of the canonical polarization of $\widetilde{J}$ to $\pi_H^*(J C_H)$. So the uniqueness of the complement and the preceding decomposition give the assertion. $\square$

For the following corollary, we need some special notation. Suppose there are subgroups H, N_1 and N_2 of G with

$$H \subset N_1 \cap N_2 \subset G$$

such that the decomposition (3.15) of $T_0 \widetilde{J}$ can be written in the form

$$T_0 \widetilde{J} = \bigoplus_{i=1}^{s} n_i V_i \oplus \bigoplus_{j=1}^{t_1} a_j^1 U_j^1 \oplus \bigoplus_{k=1}^{t_2} a_k^2 U_k^2 \oplus \bigoplus_{\ell=1}^{u} b_\ell U_\ell,$$

that is, we have that the set of occurring $V_i, U_j^1, U_k^2, U_\ell$ equals just the set of all V_i, and the coefficients a_j^1, a_k^2, b_ℓ coincide with the corresponding coefficients n_i. Then we have

Corollary 3.5.12 *Suppose in addition that we have for $k = 1$ and 2,*

$$\rho_H - \rho_{N_k} = \bigoplus_{i=1}^{s} (\dim V_i^H) V_i \oplus \bigoplus_{j=1}^{t_k} (\dim (U_j^k)^H) U_j^k.$$

Then we have the following decomposition:

$$d\pi_H^*[T_0(P(C_H/C_{N_1})) \cap T_0(P(C_H/C_{N_2}))] = n_1 V_1^H \oplus \cdots \oplus n_s V_s^H.$$

Proof Proposition 3.5.11 gives for $k = 1$ and 2,

$$d\pi_H^*\left[T_0 P(C_H/C_{N_k})\right] = \bigoplus_{i=1}^{s} n_i V_i^H \oplus \bigoplus_{j=1}^{t_k} a_j^k (U_j^k)^H.$$

Since the right-hand sides intersect exactly in $\bigoplus_{i=1}^{s} n_j V_j^H$, this implies the assertion.   □

Remark 3.5.13 Corollary 3.5.12 generalizes immediately to any finite number of subgroups N_i containing H such that the representations $\rho_H - \rho_{N_i}$ have a common piece of the form $\bigoplus_{j=1}^{s} (\dim V_j^H) V_j$ and an obvious generalization of the assumption of the corollary. We do not need this.

We will apply only the following corollary.

Corollary 3.5.14 *Let $f : \widetilde{C} \to C$ be a Galois cover with group G, and let H, N_1 and N_2 be subgroups of G with*

$$H \subset N_1 \cap N_2$$

such that there are rational representations W, W_1 and W_2 with W irreducible, not in W_1 nor in W_2, and satisfying that W_1 and W_2 do not admit a common irreducible representation. If then

$$\rho_H = \rho_{N_1} \oplus W \oplus W_1 = \rho_{N_2} \oplus W \oplus W_2.$$

Then there is an abelian subvariety B_W associated to W as in (3.14) such that

$$B_W = [P(C_H/C_{N_1}) \cap P(C_H/C_{N_2})]^0.$$

Proof The isotypical decomposition of the tangent space $T_0\tilde{J}$ is induced by the isotypical decomposition of the Jacobian $\tilde{J}$. This implies that the analytic isotypical decomposition induces and is given by the corresponding decomposition of the rational representation of G. The assumptions of the corollary are just the translation of a special case of the assumptions of Corollary 3.5.12. This gives

$$d\pi_H^*(T_0(P(C_H/C_{N_1})) \cap T_0(P(C_H/C_{N_2})) = T_0 B_W$$

which implies the assertion.   □

3.5.4   The Dimension of an Isotypical Component

Let G be a finite group acting on a smooth projective curve $\tilde{C}$. The action induces an action on the Jacobian $\tilde{J}$ of $\tilde{C}$. Let W denote an irreducible rational representation with corresponding isotypical component A_W as well as a group algebra component B_W. In this subsection we will give a formula for their dimensions. Note that, although B_W is not uniquely determined, its dimension is.

Let $f : \tilde{C} \to C$ be the corresponding Galois cover with group G and geometric signature $(g; G_1, \ldots, G_s)$. Let $W_1, \ldots, W_r$ be a complete set of irreducible rational representations of G with W_1 the trivial representation. For any W_i, $i = 2, \ldots, r$, choose an associated irreducible complex representation V_i. The following theorem is due to Rojas [33].

Theorem 3.5.15 *The multiplicity h_i of W_i in the rational representation $\rho_r = \sum_{j=1}^r h_j W_j$ of $End_{\mathbb{Q}}(\tilde{J})$ is given by*

$$h_i = \frac{1}{s_i}\left(2(g-1)\dim V_i + \sum_{j=1}^s \left(\dim V_i - \dim V_i^{G_j}\right)\right)$$

for all $i \geq 2$, where s_i is the Schur index of V_i and $h_1 = 2g$.

Proof Assume

$$\rho_r = \sum_{j=1}^r h_j W_j$$

and consider a full set of representatives $\{H_1 = \{1_G\}, H_2, \ldots, H_r\}$ of the conjugacy classes of cyclic subgroups of G, chosen such that the order of H_j is non-decreasing. Note that we use here the well-known fact that the number of irreducible rational representations of a finite group G equals the number of conjugacy classes of cyclic subgroups of G (see [9, Corollary 39.5]).

Then for each subgroup H_k we have

$$\dim(\rho_r^{H_k}) = \sum_{j=1}^{r} h_j \dim(W_j)^{H_k}$$

and it is known that $\dim(\rho_r^{H_k}) = 2g_{C_{H_k}}$ (see [13, Section V.2.2]). Hence we obtain an $r \times r$ linear system of equations for the unknowns $h_1, \ldots h_r$ given by

$$2g_{C_{H_k}} = \sum_{j=1}^{r} h_j \dim(W_j^{H_k}) \ , \quad j = 1, \ldots, r. \tag{3.16}$$

Now consider the matrix

$$A = (a_{j,k}) \quad \text{given by} \quad a_{j,k} := \dim(W_j^{H_k}).$$

Next we show that this matrix is invertible, from where the system (3.16) will have a unique solution, and then verify that the proposed values for the h_j do satisfy it.

Recall that the rational conjugacy class of an element g of a group G, denoted by $\overline{g}$, consists of all elements of G that generate a subgroup conjugate to the subgroup generated by g; hence the number of rational conjugacy classes in G equals the number of conjugacy classes of cyclic subgroups, in our case r. Choose representatives $g_1 = 1_G, g_2, \ldots, g_r$ for these rational conjugacy classes such that each g_i is a generator of H_i, and also recall that the rational character table for G consists of r rows and r columns, where the j-th row contains the values $(\chi_{W_j}(g_1), \ldots, \chi_{W_j}(g_r))$ and that this matrix is invertible.

Observing that

$$a_{j,k} = \dim(W_j^{H_k}) = \langle W_j, \rho_{H_k} \rangle = \frac{1}{|H_k|} \sum_{h \in H_k} \chi_{W_j}(h) = \frac{1}{|H_k|} \sum_{t=1}^{r} \chi_{W_j}(g_t) \, |H_k \cap \overline{g_t}|,$$

and noting that $|H_k \cap \overline{g_t}| = 0$ for $t > k$, we see that the j-th row of A is obtained by applying elementary row transformations to the rational character table of G that do not involve the rows below the j-th one. Thus A is invertible.

All that is left to verify is that the h_i given by

$$h_i = \frac{1}{s_i} \left(2(g-1) \dim V_i + \sum_{j=1}^{s} \left(\dim V_i - \dim V_i^{G_j} \right) \right)$$

for $i \geq 2$ and $h_1 = 2g$ satisfy (3.16).

Rewriting

$$h_1 = 2 + \frac{1}{s_1}\left(2(g-1)\dim V_1 + \sum_{j=1}^{s}\left(\dim V_1 - \dim V_1^{G_j}\right)\right),$$

fixing $H = H_k$ for $1 \le k \le r$, using Corollary2.8.2, and replacing the expressions for h_j into the right hand side of (3.16), we obtain

$$\sum_{j=1}^{r} h_j \dim(W_j^H) = \sum_{j=1}^{r} h_j [L_j : \mathbb{Q}] \dim(V_j^H)$$

$$= 2 + \left(2(g-1)\dim V_1 + \sum_{j=1}^{r}\left(\dim V_1 - \dim V_1^{G_j}\right)\right)$$

$$+ \sum_{j=2}^{r}\left(2(g-1)\dim V_j + \sum_{t=1}^{s}\left(\dim V_j - \dim V_j^{G_t}\right)\right)[K_j : \mathbb{Q}]\dim(V_j^H)$$

$$= 2\left(1 + (g-1)\sum_{V \in \mathrm{Irr}_{\mathbb{C}}(G)} \dim(V)\dim(V^H)+\right.$$

$$\left. + \frac{1}{2}\sum_{V \in \mathrm{Irr}_{\mathbb{C}}(G)}\left(\sum_{t=1}^{s}\left(\dim(V) - \dim(V^{G_t})\right)\dim(V^H)\right)\right)$$

where the last equality follows since $\dim V_j = \dim V_j^\gamma$ and $\dim V_j^H = \dim(V_j^\gamma)^H$ for all γ in $\mathrm{Gal}(K_j/\mathbb{Q})$.

Now

$$\sum_{V \in \mathrm{Irr}_{\mathbb{C}}(G)} \dim(V)\dim(V^H) = \frac{1}{|H|}\sum_{V \in \mathrm{Irr}_{\mathbb{C}}(G)} \dim(V)\sum_{h \in H}\chi_V(h)$$

$$= \frac{1}{|H|}\sum_{h \in H}\left(\sum_{V \in \mathrm{Irr}_{\mathbb{C}}(G)}\dim(V)\chi_V\right)(h)$$

$$= \frac{1}{|H|}\sum_{h \in H}\chi_{\rho_{1_G}}(h)$$

$$= \frac{|G|}{|H|} = [G : H]$$

and similarly, one checks for each $1 \le t \le s$ using [34, Proposition 22],

$$\sum_{V \in \mathrm{Irr}_{\mathbb{C}}(G)} \dim(V^{G_t})\dim(V^H) = |H \backslash G / G_t|.$$

So finally, applying Theorem 3.1.6, we obtain

$$\sum_{j=1}^{r} h_j \dim(W_j)^H = 2\left(1 + (g-1)[G:H] + \frac{1}{2}\sum_{t=1}^{s}([G:H] - |H\backslash G/G_t|)\right)$$

$$= 2g_{C_H}$$

as required. $\square$

Corollary 3.5.16 *With the notation of Theorem 3.5.16, we obtain*

$$\rho_r = 2\chi_0 + 2(g-1+s)\rho_{1_G} - \sum_{t=1}^{s}\rho_{G_t}.$$

Proof Recall from Lemma 2.8.1 that

$$\rho_H = \bigoplus_{j=1}^{r} \frac{1}{s_j}\dim(V_j^H)\,W_j$$

for each subgroup H of G.
  Therefore

$$2\chi_0 + 2(g-1+s)\rho_{1_G} - \sum_{t=1}^{s}\rho_{G_t}$$

$$= 2\chi_0 + 2(g-1+s)\left(\chi_0 + \bigoplus_{j=2}^{r}\frac{1}{s_j}\dim(V_j)\,W_j\right)$$

$$- \sum_{t=1}^{s}\left(\chi_0 + \bigoplus_{j=2}^{r}\frac{1}{s_j}\dim(V_j^{G_t})\,W_j\right)$$

$$= 2g\,\chi_0 + \bigoplus_{j=2}^{r}\frac{1}{s_j}\left((2(g-1)+s)\dim(V_j) - \sum_{t=1}^{s}\dim(V_j^{G_t})\right)W_j$$

$$= \sum_{j=1}^{r} h_j W_j = \rho_r.$$

$\square$

  The following corollary can be conveniently applied by using the group theoretical computer program Magma.

Corollary 3.5.17 *For any* $i > 0$, *the dimensions of any group algebra component* B_{W_i} *and the isotypical component* A_i *are*

$$\dim B_{W_i} = [L_i : \mathbb{Q}] \left(\dim V_i (g - 1) + \frac{1}{2} \sum_{j=1}^{s} \left(\dim V_i - \dim V_i^{G_j} \right) \right)$$

and

$$\dim A_{W_i} = \dim V_i \cdot [K_i : \mathbb{Q}] \left(\dim V_i (g - 1) + \frac{1}{2} \sum_{j=1}^{s} \left(\dim V_i - \dim V_i^{G_j} \right) \right)$$

where L_i *denotes the field of definition of* V_i *and* K_i *its character field.*

Proof According to Proposition 2.9.3, we have

$$\dim B_i = \frac{1}{2} h_i s_i [L_i : \mathbb{Q}].$$

Inserting the formula of Theorem 3.5.15 gives the first equation. The second equation follows from this one and Eq. (2.30). $\qquad\square$

Chapter 4
Covers of Degree 2 and 3

In this and the next chapter, we will study the decomposition of covers of degree $d = 2, 3$, and 4. Our point of view is the following: Let $f' : C' \to C$ be a cover of degree d of smooth projective curves, and let $f : \widetilde{C} \to C$ be its Galois closure. Using the methods of Sect. 3.5, we compute the isotypical and group algebra decomposition of $\widetilde{J} := J\widetilde{C}$, which induces a decomposition of JC'.

Another point of view is to start with the easy cases and gradually come to the more complicated groups. Hence in this chapter we start with the easiest case that is the case of a double cover $f : \widetilde{C} \to C$, which is always Galois with group $G = \mathbb{Z}/2\mathbb{Z}$. Here the isotypical and the group algebra decomposition is given by the addition map $f^*J \times P(f) \to \widetilde{J}$. Moreover, the degree of this isogeny is given.

Section 4.2 deals with covers $f' : C' \to C$ of degree 3. Here we have to consider two cases: either f' is Galois and thus cyclic of degree 3 or non-Galois, in which case its Galois closure $f : \widetilde{C} \to C$ has Galois group $\mathcal{S}_3$, the group of all permutations of degree 3. The Galois case is very similar to the degree-two case. For the sake of completeness, we give the result in Sect. 4.2.1.

The rest of the section deals with the non-Galois case. The results in this case were first given in [31] with a slightly different proof. In Sects. 4.2.2 and 4.2.3, we work out the possible ramifications and genera of the covers involved and recall the irreducible rational representations of $\mathcal{S}_3$. In Sect. 4.2.4 we compute the isotypical and the group algebra decompositions of $\widetilde{J}$. Finally in the last section, we compute the degrees of the corresponding isogenies.

4.1 Covers of Degree 2

By what we have seen in Chap. 3, the case of covers of degree 2 is easy and well known. We include it here for the sake of completeness and in order to outline the way of reasoning in the more complicated cases.

© The Author(s), under exclusive license to Springer Nature Switzerland AG 2022
H. Lange, R. E. Rodríguez, *Decomposition of Jacobians by Prym Varieties*, Lecture
Notes in Mathematics 2310, https://doi.org/10.1007/978-3-031-10145-8_4

Any cover of degree 2 is cyclic, so $C' = \widetilde{C}$, and we start with a double cover $f : \widetilde{C} \to C$. We always assume that the genus of C is g. According to the Hurwitz formula, f is branched at an even number 2β of points of C, which is equal to the ramification degree $r(f) = 2\beta$, and we get for the genus of $\widetilde{C}$:

$$g_{\widetilde{C}} = 2g - 1 + \beta.$$

This implies that the Prym variety $P(f) = P(\widetilde{C}/C)$ is of dimension

$$\dim P(f) = g - 1 + \beta.$$

Of course, if $g = 0$, then $\beta > 0$. If $g = 1$ and $\beta = 0$, then $P(f) = 0$, so this case is uninteresting.

The Prym variety $P(f)$ of f is by definition the complement of f^*J in $\widetilde{J}$ with respect to the canonical principal polarization Θ of $\widetilde{J}$. As every polarization on $\widetilde{J}$, the polarization Θ induces a polarization on every abelian subvariety $A \subset \widetilde{J}$, which we denote by Θ_A and we get for the types of the abelian subvarieties f^*J and $P(f)$:

Proposition 4.1.1 *In this case we have the following:*

(i) *The polarization Θ_{f^*J} is of type $(1, 2, \ldots, 2)$, if f is étale, and type $(2, \ldots, 2)$ otherwise.*

(ii) *The polarization $\Theta_{P(f)}$ is of type $(1, \ldots, 1, 2, \ldots, 2)$ with $g - 1$ numbers 2, if f is étale, and with g numbers 2 otherwise.*

Proof (i) follows from Propositions 3.2.1 and 3.2.3. Using this, (ii) is a special case of Proposition 3.2.8. $\qquad\qquad\qquad\qquad\qquad\qquad\qquad\qquad\qquad\qquad\qquad\qquad\qquad\quad$ $\square$

In order to compute the isotypical and group algebra decompositions of $\widetilde{J}$, we have to determine the irreducible rational representations of the Galois group $G \simeq \mathbb{Z}/2\mathbb{Z}$ of f. Since G is cyclic of order 2, there are only two such representations, the trivial representation χ_0 and the alternating representation χ_1. It is clear that G acts on f^*J by χ_0 and on $P(f)$ by χ_1. This implies

Proposition 4.1.2 *Both the isotypical and group algebra decompositions of $\widetilde{J}$ are given by the addition map*

$$\alpha : f^*J \times P(f) \longrightarrow \widetilde{J}.$$

As a special case of Corollary 3.2.10, we obtain the following proposition.

Proposition 4.1.3 *The addition map $\alpha : f^*J \times P(f) \to \widetilde{J}$ is an isogeny of degree*

$$\deg \alpha = \begin{cases} 2^{2g-2} & \text{if } f \text{ is étale}; \\ 2^{2g} & \text{otherwise}. \end{cases}$$

4.2 Covers of Degree 3

Let $f' : C' \to C$ be a cover of degree 3 with Galois closure $f : \widetilde{C} \to C$. The Galois group G of f is a subgroup of the symmetric group $\mathcal{S}_3$ of degree 3, which is of order 6. Hence there are only two possibilities: G is either cyclic of order 3 or is $\mathcal{S}_3$ itself.

4.2.1 Cyclic Covers of Degree 3

Let $f' : C' \to C$ be cyclic of degree 3, i.e., $f' = f : \widetilde{C} \to C$. Then f is either étale or totally ramified. Hence the ramification degree $r(f)$ and the number of branch points $b(f)$ are related by

$$r(f) = 2b(f)$$

and the Hurwitz formula reads

$$g_{\widetilde{C}} = 3g - 2 + b(f)$$

where as always $g = g_C$, and the Prym variety $P(f)$ is of dimension

$$\dim P(f) = 2g - 2 + b(f).$$

The same proof as for Proposition 4.1.1 gives for the types of the induced polarizations.

Proposition 4.2.1 *In this case we have the following:*

(i) *The polarization $\Theta_{f_* J}$ is of type $(1, 3, \ldots, 3)$ if f is étale and type $(3, \ldots, 3)$ otherwise.*

(ii) *The polarization $\Theta_{P(f)}$ is of type $(1, \ldots, 1, 3, \ldots, 3)$ with $g - 1$ numbers 3 if f is étale and g numbers 3 otherwise.*

In order to compute the irreducible rational representations of G, let σ be a generator of G. Recall that the characters of G are given by

$$\chi_k(\sigma) = \xi_3^k \quad \text{for} \quad k = 0, 1, 2 \quad \text{where} \quad \xi_3 := e^{2\pi i/3}.$$

Clearly the trivial representation χ_0 of G is defined over the rationals, i.e., $\chi_0 = W_0 \otimes \mathbb{C}$, where W_0 denotes the trivial rational representation. On the other hand, there is an irreducible rational representation W_1 of degree 2 such that

$$\chi_1 \oplus \chi_2 = W_1 \otimes \mathbb{C}.$$

Obviously W_0 and W_1 are exactly all irreducible rational representations of G.

Proposition 4.2.2 *The isotypical component of $\widetilde{J}$ corresponding to W_0 (respectively, W_1) is f^*J (respectively, $P(f)$), and both the isotypical and group algebra decompositions are given by the addition map*

$$\alpha : f^*J \times P(f) \to \widetilde{J}.$$

Proof The assertion on W_0 is obvious. In order to prove that $P(f)$ is the isotypical component corresponding to W_1, we apply Corollary 3.5.10 to the subgroups $H = \{1\}$ and $N = G$: Since

$$\rho_{\{1\}} = W_0 \oplus W_1 = \rho_G \oplus W_1,$$

Corollary 3.5.10 gives that $P(f) = P(\widetilde{C}/C) = P(C_{\{1\}}/C_G)$ is the isotypical component corresponding to W_1.

It remains to show that $P(f)$ is also a group algebra component of $\widetilde{J}$. This follows from Proposition 3.5.7, since the Schur index of the character χ_1 is 1.   □

As in the case of a double cover, we obtain as a special case of Corollary 3.2.10 the following proposition.

Proposition 4.2.3 *The addition map $\alpha : f^*J \times P(f) \to \widetilde{J}$ is an isogeny of degree*

$$\deg \alpha = \begin{cases} 3^{2g-2} & \text{if } f \text{ is étale;} \\ 3^{2g} & \text{otherwise.} \end{cases}$$

4.2.2   Non-cyclic Covers of Degree 3: The Galois Closure

Let $f' : C' \to C$ be a cover of degree 3, which is not cyclic. Then its Galois closure $f : \widetilde{C} \to C$ is of degree 6 with Galois group $G := S_3$. So G is of the form

$$G = \langle \sigma, \tau \mid \sigma^3 = \tau^2 = (\sigma\tau)^2 = 1 \rangle.$$

The subgroups $\langle \sigma^i \tau \rangle$ for $i = 0, 1$, and 2 (of order 2) are conjugate to each other. So it suffices to consider the following diagram of subcovers

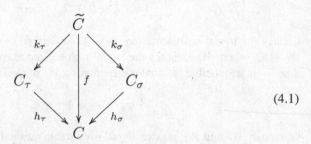

(4.1)

with $\deg k_\tau = \deg h_\sigma = 2$ and $\deg k_\sigma = \deg h_\tau = 3$. We may assume that $C' = C_\tau$. So h_τ is non-cyclic.

Lemma 4.2.4 *Let $c \in C$ be a branch point of f. There are exactly two possibilities for the ramification of the curves of the diagram:*

(i) h_τ *is totally ramified over c. Then h_σ and k_τ are unramified over c and $h_\tau^{-1}(c)$, respectively, and k_σ is totally ramified over each point of $h_\sigma^{-1}(c)$.*

(ii) h_τ *is simply ramified over c. Then h_σ is ramified over c, k_τ is unramified over the ramification point of h_τ and ramified over the remaining point of $h_\tau^{-1}(c)$, and k_σ is unramified over $h_\sigma^{-1}(c)$.*

Proof Since f is non-cyclic of degree $6 = 2 \cdot 3$, there are two possibilities for f over c: $f^{-1}(c)$ consists of either two points $\{P_1, P_2\}$, both fixed by σ and permuted by any involution, or of *three* points $\{Q_1, Q_2, Q_3\}$, each fixed by exactly one of the *three* involutions $\sigma^{-j}\tau\sigma^j$, $1 \leq j \leq 3$, in $\mathcal{S}_3$, and permuted by σ. To see this, note that if one involution would admit more than one fixed point in the same fiber, then there would be two involutions with the same fixed point, but any two involutions generate the group $\mathcal{S}_3$, so the stabilizer group of this point would not be cyclic. We label these fixed points in such a way that Q_j is fixed by $\sigma^{-j}\tau\sigma^j$.

In the first case, h_σ and k_τ are necessarily unramified over c, and each point of $h_\tau^{-1}(c)$ and k_σ is totally ramified over each point of $h_\sigma^{-1}(c)$. Hence we are in case (i).

In the second case, observe that τ acts on each fiber $\{Q_1, Q_2, Q_3\}$ by fixing exactly one point, say Q_3, and exchanging the other *two* points: $\tau(Q_j) = Q_{3-j}$ for $j = 1, 2$. Therefore k_τ ramifies at Q_3 and is unramified over the ramification point of $h_\tau^{-1}(c)$. Then h_σ is ramified over c and k_σ is unramified over the point $h_\sigma^{-1}(c)$. ☐

Here are pictures of the two local ramification types (i) and (ii) of diagram (4.1).

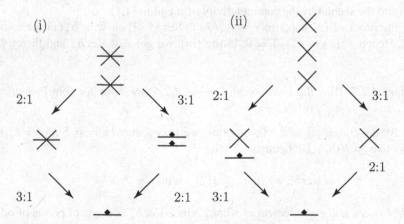

Denote by α (respectively, β) the number of branch points of f of type (i) (respectively, of type (ii)). In terms of fixed points, this means

$$\alpha := \frac{1}{2}|\operatorname{Fix}(\sigma)| \quad \text{and} \quad \beta := |\operatorname{Fix}(\tau)|.$$

Note that β is even, since it coincides with the number of ramification points of the double cover h_σ. Moreover, $\beta \geq 2$ if $C = \mathbb{P}^1$, since $\mathbb{P}^1$ does not admit connected étale double covers. The Hurwitz formula then gives

Lemma 4.2.5 *Given a diagram* (4.1) *with* $\mathcal{S}_3 \subset \operatorname{Aut}(\widetilde{C})$, α *and* β *as above and with* $g = g_C$ *the genus of* C. *Then we have for the genera of the other curves*

$$g_{C_\tau} = 3g - 2 + \alpha + \frac{\beta}{2}, \quad g_{C_\sigma} = 2g - 1 + \frac{\beta}{2}, \quad g_{\widetilde{C}} = 6g - 5 + 2\alpha + \frac{3}{2}\beta.$$

Lemma 4.2.6 *Given a diagram* (4.1) *with* $\mathcal{S}_3 \subset \operatorname{Aut}(\widetilde{C})$. *Then*

(i) $h_\tau^* : J \to JC_\tau$ *is injective.*
(ii) k_τ^* *is injective if and only* h_σ^* *is injective.*
(iii) $\operatorname{Ker} h_\sigma^* = \operatorname{Ker} f^*$.
(iv) $\operatorname{Ker} k_\tau^* = h_\tau^*(\operatorname{Ker} h_\sigma^*)$.
(v) $h_\sigma^*(J) \cap \operatorname{Ker} k_\sigma^* = \{0\}$.

Note that Lemma 4.2.6 is valid for τ replaced by any involution $\sigma^i \tau$, since all involutions in $\mathcal{S}_3$ are conjugate.

Proof (i) and (ii) follow from Proposition 3.2.2, (i) since h_τ is non-cyclic and (ii) since k_τ is ramified if and only if h_σ is ramified.

As for (iii), clearly $\operatorname{Ker} h_\sigma^* \subseteq \operatorname{Ker} f^*$. So let $x \in \operatorname{Ker} f^*$. Then $h_\tau^*(x) \subset \operatorname{Ker} k_\tau^* \subseteq JC_\tau[2]$. By (i), h_τ is injective. Hence $x \in J[2]$. On the other hand, $h_\sigma^*(x) \in \operatorname{Ker} k_\sigma^* \subseteq JC_\sigma[3]$. Since $x \in J[2]$, we conclude $h_\sigma^*(x) = 0$; that is, $x \in \operatorname{Ker} h_\sigma^*$.

For (iv) just note that $\operatorname{Ker} h_\sigma^* = \operatorname{Ker} f^* = (h_\tau^*)^{-1}(\operatorname{Ker} k_\tau^*)$ with the first equality by (iii) and the second by the commutativity of diagram (4.1).

For the proof of (v), suppose $y \in h_\sigma^*(J) \cap \operatorname{Ker} k_\sigma^*$. Then $y = h_\sigma^*(x)$ for some $x \in J$. Hence $f^*(x) = k_\sigma^*(y) = 0$. Using (iii), we get $x \in \operatorname{Ker} h_\sigma^*$ and therefore $y = 0$. $\qquad\square$

Corollary 4.2.7 *With the notation of Lemma 4.2.6, we have for any involution* $\sigma^i \tau \in \mathcal{S}_3$: $k_{\sigma^i \tau}^*|_{P(h_{\sigma^i \tau})}$ *is injective.*

Proof By the conjugacy of the involutions, we may assume $i = 0$. Suppose k_τ^* is not injective on $P(h_\tau)$. By Lemma 4.2.6(ii),

$$\operatorname{Ker} h_\sigma^* = \{0, \eta\} \subseteq J[2] \quad \text{with} \quad \eta \neq 0.$$

Then $h_\tau^*(\eta)$ is a non-zero element of $\operatorname{Ker} k_\tau^*$, since $\operatorname{Ker} h_\tau^*$ consists of points of odd order. But $h_\tau^*(J) \cap P(h_\tau)$ also consists of points of odd order. Hence $h_\tau^*(\eta) \notin P(h_\tau)$. $\qquad\square$

4.2.3 The Irreducible Rational Representations of S_3

There are two complex representations of S_3 of degree 1, namely, the trivial representation χ_0 and the alternating representation χ_1 defined by

$$\chi_1(\sigma^j) = 1 \quad \text{and} \quad \chi_1(\sigma^j \tau) = -1 \quad \text{for} \quad j = 0, 1, 2.$$

There is only one more irreducible complex representations (see [34]), which is of degree two and given by

$$V_1(\sigma^j) = \begin{pmatrix} \xi_3^j & 0 \\ 0 & \xi_3^{-j} \end{pmatrix} \quad \text{and} \quad V_1(\sigma^j \tau) = \begin{pmatrix} 0 & \xi_3^{-j} \\ \xi_3^j & 0 \end{pmatrix},$$

for all j, where $\xi_3 = e^{2\pi i/3}$. All three representations are defined over the rationals. This is clear for χ_0 and χ_1, but also

$$V_1 = W_1 \otimes \mathbb{C}$$

with an irreducible rational representation W_1. This follows from the fact that the character field of V_1 is $\mathbb{Q}(\xi_3 + \xi_3^{-1}) = \mathbb{Q}$ and its Schur index is 1 [34].

It follows that the rational group algebra of S_3 decomposes as follows:

$$\mathbb{Q}[S_3] = \chi_0 \oplus \chi_1 \oplus 2W_1. \tag{4.2}$$

4.2.4 Curves with an S_3-action: Decomposition of $\tilde{J}$

Given a diagram of curves (4.1) with $S_3 \subseteq \mathrm{Aut}(\tilde{C})$, we want to determine the isotypical and group algebra decompositions of $\tilde{J} = J\tilde{C}$.

Let $A_{\chi_0}, A_{\chi_1}, A_{W_1}$ denote the isotypical components of $\tilde{J}$ corresponding to the indicated irreducible rational representations. Then the isotypical decomposition of $\tilde{J}$ is given by the addition map

$$\mu : A_{\chi_0} \times A_{\chi_1} \times A_{W_1} \to \tilde{J}.$$

The following proposition identifies the isotypical components by the maps of diagram (4.1).

Proposition 4.2.8

(i) $A_{\chi_0} = f^* J$.
(ii) $A_{\chi_1} = k_\sigma^* P(h_\sigma)$.
(iii) $A_{W_1} = P(k_\sigma)$.

Proof

(i) is clear, since S_3 acts on f^*J and on no other abelian subvariety of $\tilde{J}$ by the trivial representation χ_0.

(ii): We apply Corollary 3.5.10 to the pair of subgroups $H = \langle \sigma \rangle$ and $N = G$ of $G = S_3$. Since

$$\rho_H = \chi_0 + \chi_1 = \rho_N + \chi_1,$$

Corollary 3.5.10 implies

$$A_{\chi_1} = B_{\chi_1} \sim P(C_H/C_N) = P(h_\sigma).$$

This gives (ii), since $k_\sigma^* P(h_\sigma)$ is the corresponding abelian subvariety of $\tilde{J}$.

(iii): We apply Corollary 3.5.9 to the pair of subgroups $H = 1$ and $N = \langle \sigma \rangle$ of S_3. Since

$$\rho_H = \chi_0 + \chi_1 + 2W_1 = \rho_N + \chi_1 + W_1$$

If B_{W_1} and similarly B_{χ_1} are defined as in Sect. 3.5.2, Corollary 3.5.9 implies

$$P(k_\sigma) = P(C_H/C_N) \sim B_{\chi_1}^{t_{\chi_1}} \times B_{W_1}^{t_{W_1}}$$

with

$$t_{\chi_1} = \dim \chi_1^H - \dim \chi_1^N = 1 - 1 = 0$$

and

$$t_{W_1} = \dim W_1^H - \dim W_1^N = 2 - 0 = 2$$

Note that all Schur indices are 1. On the other hand, we have, according to Corollary 3.5.8,

$$A_{W_1} \sim B_{W_1}^{\dim W_1} = B_{W_1}^2.$$

Together this implies (iii), since $P(k_\sigma)$ is the abelian subvariety corresponding to $B_{W_1}^2$.   □

Remark 4.2.9 One can also see directly that S_3 acts on $k_\sigma^* P(h_\sigma)$ by the alternating representation χ_1. Let τ be any transposition of S_3. Since $\langle \sigma \rangle$ is a normal subgroup of S_3, τ induces an involution τ' on C_σ such that

$$\tau^* \circ k_\sigma^* = k_\sigma^* \circ \tau'^*.$$

By Corollary 3.5.2(b) we have $P(h_\sigma) = \text{Ker}(1 + \tau')^0$. It follows that $\tau'(y) = -y$ for any $y \in P(h_\sigma)$. Hence

$$\tau^*(k_\sigma^*(y)) = -k_\sigma^*(y)$$

for any $y \in P(h_\sigma)$ and any involution $\tau \in S_3$. We also have $k_\sigma^* P(h_\sigma) \subset k_\sigma^* J C_\sigma \subset \widetilde{J}^{\langle \sigma^* \rangle}$; that is, σ acts trivially on $k_\sigma^* P(h_\sigma)$. So S_3 acts on $k_\sigma^* P(h_\sigma)$ by the representation χ_1. This gives $k_\sigma^* P(h_\sigma) \subseteq A_{\chi_1}$. We get even equality here computing dimensions using, for example, Corollary 3.5.17 and the Hurwitz formula.

The following corollary is an immediate consequence of Proposition 4.2.8.

Corollary 4.2.10 *With the notations of diagram* (4.1), *the addition map*

$$\mu : P(k_\sigma) \times k_\sigma^* P(h_\sigma) \times f^* J \longrightarrow \widetilde{J}$$

is the isotypical decomposition of $\widetilde{J}$ with respect to the action of S_3.

According to Theorem 2.9.2 and Eq. (4.2), the group algebra decompositions of $\widetilde{J}$ are of the form

$$\widetilde{J} \sim f^* J \times k_\sigma^* P(h_\sigma) \times B_{W_1}^2$$

with an abelian subvariety B_{W_1} of $A_{W_1} = P(k_\sigma)$. In the next section, we want to compute the degree of such an isogeny. For this we need to know more about a particular B_{W_1}. The following lemma gives a particular abelian subvariety B_{W_1} for a group algebra decomposition of $\widetilde{J}$. Let

$$B_{W_1} := P(h_\tau). \tag{4.3}$$

Then we have the following lemma.

Lemma 4.2.11 *The map*

$$\varphi : P(h_\tau) \times P(h_\tau) \to P(k_\sigma), \quad (z_1, z_2) \mapsto k_\tau^*(z_1) + \sigma k_\tau^*(z_2)$$

is an isogeny with kernel

$$K = \{(z, -z) \in P(h_\tau)[3] \times P(h_\tau)[3] \mid k_\tau^*(z) \text{ is } S_3 - invariant\}.$$

Proof Denote

$$B := k_\tau^* P(h_\tau).$$

According to Corollary 4.2.7, the map $k_\tau^* | P(h_\tau) : P(h_\tau) \to B$ is injective. Hence it suffices to show that the composed map

$$\alpha : B \times B \overset{1 \times \sigma}{\to} B \times \sigma(B) \overset{\mu}{\to} P(k_\sigma),$$

where μ denotes the addition map, is an isogeny with kernel

$$K' := \{(x, -x) \in B[3] \times B[3] \mid x \text{ is } S_3 - invariant\}.$$

Note first that by Corollary 3.5.2,

$$P(k_\sigma) = \{z \in \widetilde{J} \mid \sum_{i=0}^{2} \sigma^i z = 0\}^0$$

and

$$B = \{x \in \widetilde{J}^\tau \mid \sum_{i=0}^{2} \sigma^i x = 0\}^0.$$

Hence both B and $\sigma(B)$ are contained in $P(k_\sigma)$ and thus $\mu(B \times \sigma(B)) \subseteq P(k_\sigma)$. From Lemma 4.2.5 one checks that

$$2 \dim B = \dim P(k_\sigma).$$

Hence it suffices to show the assertion on the kernel of $\alpha : B \times B \to P(k_\sigma)$.
  Suppose $(x_1, x_2) \in \operatorname{Ker} \alpha$, that is,

$$x_1 + \sigma(x_2) = 0.$$

Using the fact that both x_1 and x_2 are fixed under τ, we get

$$x_1 + \sigma^2(x_2) = x_1 + \tau \sigma(x_2) = \tau(x_1 + \sigma(x_2)) = 0.$$

Combining the two equations, we get

$$\sigma^2(x_1) = -\sigma^2(\sigma^2(x_2)) = -\sigma(x_2) = x_1.$$

In other words, x_1 is invariant under the action of $\langle \sigma^2 \rangle = \langle \sigma \rangle$. But then $B \subseteq P(k_\tau)$ implies

$$0 = (1 + \sigma + \sigma^2)(x_1) = 3x_1;$$

that is, $x_1 \in B[3]$. Moreover, since x_1 is invariant under S_3 and $x_2 = -\sigma^2(x_1) = -x_1$, we get $\operatorname{Ker} \alpha \subset K'$. That conversely $K' \subseteq \operatorname{Ker} \alpha$ is obvious.   □

  From Lemma 4.2.11 we get the following explicit group algebra decomposition of $\widetilde{J}$ with respect to the action of S_3.

Corollary 4.2.12 *Let the notations be as in diagram (4.1). Then with the notations of Lemma 4.2.11, the map*

$$P(h_\tau)^2 \times k_\sigma^* P(h_\sigma) \times f^* J \to \widetilde{J}, \qquad (x_1, x_2, y, z) \mapsto \alpha(x_1, x_2) + y + z$$

is a group algebra decomposition of $\widetilde{J}$ with respect to the action of $\mathcal{S}_3$.

According to Corollary 3.5.8, we obtain from this the induced decomposition of the degree 3 cover $h_\tau : C_\tau \to C$. It does not give anything new. We only outline it for the sake of completeness.

Corollary 4.2.13 *The decomposition of the Jacobian JC_τ, induced from the decomposition of $\widetilde{J}$, is the usual decomposition $JC_\tau \sim J \times P(h_\tau)$.*

Proof Corollary 3.5.8 gives with Proposition 3.5.7

$$JC_\tau \sim J \times B_{\chi_1}^{\dim \chi_1^\tau} \times B_{W_1}^{\dim V_1^\tau} = J \times B_{W_1},$$

since all Schur indices are 1 and $\dim \chi_1^\tau = 0$ and $\dim V_1^\tau = 1$. The assertion follows with Eq. (4.3). □

Example 4.2.14 To give an easy example of a Galois cover with group $\mathcal{S}_3$, we use Example 3.1.3. Let $(c_1, c_2, c_3) = (\sigma, \tau, \tau\sigma^2)$. Then (c_1, c_2, c_3) is a generating vector of type $(0; 3, 2, 2)$ giving an $\mathcal{S}_3$-cover $\widetilde{C} \to \mathbb{P}^1$ of genus $g_{\widetilde{C}} = 0$. To get a cover of positive genus, just take moreover $(c_4, c_5, c_6) = (c_1, c_2, c_3)$. It is also easy to give an $\mathcal{S}_3$-cover of a curve of positive genus. We leave this to the reader.

4.2.5　The Degree of the Isogeny ψ

In Chap. 5 we need the degree of the natural map

$$\psi : P(h_\tau)^2 \times P(h_\sigma) \times J \to \widetilde{J}.$$

Before we compute it in Theorem 4.2.16, we need the degree of the map $\varphi : P(h_\tau) \times P(h_\tau) \to P(k_\sigma)$ of Lemma 4.2.11 for the proof of which we follow [31].

Lemma 4.2.15

$$\deg \varphi = \begin{cases} 3^{2g} & \text{if } \alpha = 0; \\ 3^{2g+\alpha-1} & \text{if } \alpha > 0. \end{cases}$$

Proof According to Lemma 4.2.11, the degree of φ equals the cardinality of the set K, which implies

$$\deg \varphi = |L| \qquad \text{with} \qquad L := \{z \in P(h_\tau)[3] \mid \sigma k_\tau^*(z) = k_\tau^*(z)\}.$$

Note that we clearly have

$$h_\tau^*(J[3]) \subseteq L. \tag{4.4}$$

We distinguish two cases:

Case (i): $\alpha = 0$. According to Lemma 4.2.4, this means that the cyclic degree 3 cover $k_\sigma : \widetilde{C} \to C_\sigma$ is étale.
We have to show that $L \subseteq h_\tau^*(J[3])$. Since k_σ is étale cyclic, we know that $\widetilde{J}^\sigma = \mathrm{Ker}(1-\sigma) = k_\sigma^* JC_\sigma$. In particular $\mathrm{Ker}(1-\sigma)$ is connected. This implies

$$\mathrm{Ker}(1-\sigma) = \mathrm{Im}(1+\sigma+\sigma^2).$$

Now according to Lemma 4.2.7, the map $k_\tau^*|_{P(h_\tau)}$ is injective. Hence k_τ^* induces an isomorphism

$$k_\tau^*|_L : L \xrightarrow{\simeq} k_\tau^* L = k_\tau^*(P(h_\tau)[3]) \cap \mathrm{Ker}(1-\sigma) = k_\tau^*(P(h_\tau)[3]) \cap \mathrm{Im}(1+\sigma+\sigma^2).$$

It follows that for any $z \in L$, there is a $w \in \widetilde{J}$ such that $k_\tau^*(z) = w + \sigma w + \sigma^2 w$. Let $x = \mathrm{Nm}_f(w)$. We claim that $x \in J[3]$ with $z = h_\tau^*(-x)$. But this implies $L \subseteq h_\tau^*(J[3])$ which was to be shown.
For the proof of the assertion, note that

$$3x = 3\,\mathrm{Nm}_f(w) = \mathrm{Nm}_f(k_\tau^* z) = \mathrm{Nm}_{h_\tau}(\mathrm{Nm}_{k_\tau}(k_\tau^* z)) = \mathrm{Nm}_{h_\tau}(2z) = 0.$$

Therefore $k_\tau^*(x) \in P(h_\tau)[3]$. But then $z + h_\tau(x) \in P(h_\tau)[3]$. However, $k_\tau^*(z + h_\tau^*(x)) = 3k_\tau^*(z) = 0$. So $z + h_\tau^*(x) \in \mathrm{Ker}\, k_\tau^* \subset JC_\tau[2]$ which gives $z = h_\tau^*(-x)$ as claimed.

Case (ii): $\alpha > 0$. Let $\{q_1, \ldots, q_\alpha\} \subseteq C_\tau$ denote the set of total ramification points of h_τ, and let for $i = 1, \ldots, \alpha$, $k_\tau^*(q_i) = \{p_i, p_i'\} \subseteq \widetilde{C}$. Then we have

$$\tau p_i = p_i', \ \tau p_i' = p_i \qquad \text{and} \qquad \sigma p_i = p_i, \ \sigma p_i' = p_i'.$$

If $\alpha \geq 2$, choose for every j, $2 \leq j \leq \alpha$,

$$m_j \in \mathrm{Pic}^0(J) \quad \text{such that} \quad m_j^3 = \mathcal{O}_C(h_\tau(q_1) - h_\tau(q_j))$$

and define

$$\mathcal{F}_j := \mathcal{O}_{C_\tau}(q_j - q_1) \otimes h_\tau^*(m_j) \in \mathrm{Pic}^0(C_\tau).$$

We claim that $\mathcal{F}_j \in L$ for $2 \leq j \leq \alpha$.

To see this, note that $\mathrm{Nm}_{h_\tau} \mathcal{F}_j = \mathcal{O}_C(h_\tau(q_j) - h_\tau(q_1)) \otimes m_j^3 = \mathcal{O}_C$ and $\mathrm{Ker}\,\mathrm{Nm}_{h_\tau} = P(h_\tau)$ by Propositions 3.2.2 and 3.2.4. We conclude that $\mathcal{F}_j \in P(h_\tau)[3]$. Since clearly $\sigma^* k_\tau^* \mathcal{F}_j = k_\tau^* \mathcal{F}_j$, this implies $\mathcal{F}_j \in L$.

Assertion Any element $z \in L$ can be written in a unique way as

$$z = \mathcal{F}_2^{b_2} \otimes \cdots \otimes \mathcal{F}_\alpha^{b_\alpha} \otimes h_\tau^* \mathcal{M} \tag{4.5}$$

with $0 \le b_j \le 2$ and $\mathcal{M} \in J[3]$.

PROOF of the assertion. **(a)**: *Existence of a presentation* (4.5): Choose an integer n with $2n \ge g_{\widetilde{C}}$. Then the Abel-Jacobi map

$$\widetilde{C}^{(2n)} \to \mathrm{Pic}^0(\widetilde{C}), \qquad D \mapsto \mathcal{O}_{\widetilde{C}}(D - n(p_1 + p_1'))$$

is surjective. Clearly $\widetilde{C}^{(2n)}$ is a projective bundle over $\mathrm{Pic}^0(\widetilde{C})$. We denote by $|\mathcal{G}|$ its fibre over an element $\mathcal{G} \in \mathrm{Pic}^0(\widetilde{C})$.

Now let $z \in L$. Then $k_\tau^*(z) \in \mathrm{Pic}^0(\widetilde{C})$ with $\sigma k_\tau^*(z) = k_\tau^*(z)$. It follows that σ acts as a projective transformation on $|k_\tau^*(z)|$. Any such action has fixed points. So we may choose a divisor $D \in |k_\tau^*(z)|$ which is σ-invariant. Then D is of the form

$$D = \sum_{i=1}^{\alpha} (n_i p_i + n_i' p_i') + \sum_{j=1}^{s} (1 + \sigma + \sigma^2) w_j,$$

with non-negative integers n_i, n_i' and s points $w_j \in \widetilde{C}$. We also know that $\mathcal{O}_{\widetilde{C}}(D - n(p_1 + p_1')) = k_\tau^*(z)$ and that $\tau k_\tau^*(z) = k_\tau^*(z)$ and $z^3 = \mathcal{O}_{C_\tau}$. Hence we obtain

$$k_\tau^*(z^{-1}) = k_\tau^*(z^2) = k_\tau^*(z) + \tau k_\tau^*(z)$$
$$= \mathcal{O}_{\widetilde{C}}(D - n(p_1 + p_1') + \tau D - n(p_1 + p_1'))$$
$$= \mathcal{O}_{\widetilde{C}}\left(\sum_{i=1}^{\alpha} t_i(p_i + p_i') + \sum_{j=1}^{s} \mathcal{S}_3(w_j) - 2n(p_1 + p_1')\right)$$

with $t_j = m_j + n_j$ and where $\mathcal{S}_3(w_j)$ denotes the $\mathcal{S}_3$-orbit of w_j.

Applying Nm_{k_τ} we obtain

$$z = z^{-2} = \mathrm{Nm}_{k_\tau} k_\tau^*(z^{-1})$$
$$= \mathcal{O}_{C_\tau}\left(\sum_{i=1}^{\alpha} 2t_i q_i + \sum_{j=1}^{s} 2h_\tau^*(f(w_j)) - 4nq_1\right)$$

$$= \mathcal{O}_{C_\tau} \left(\sum_{i=2}^{\alpha} 2t_i(q_i - q_1) + \sum_{j=1}^{s} [2h_\tau^*(f(w_j)) - 6q_1] \right).$$

The last equation comes from the fact that the degree of the line bundle is zero. Thus z can be written in the form

$$z = \mathcal{O}_{C_\tau} \left(\sum_{i=2}^{\alpha} b_i(q_i - q_1) \right) + h_\tau^*(\mathcal{L})$$

with $0 \le b_i \le 2$ and $\mathcal{L} \in \mathrm{Pic}^0(J)$.

Since h_τ^* is injective, this proves the Lemma in the case $\alpha = 1$.

Assume $\alpha \ge 2$. The we have also the equalities

$$z \otimes \mathcal{F}_2^{-b_2} \otimes \cdots \otimes \mathcal{F}_\alpha^{-b_\alpha} = h_\tau^*(\mathcal{L} \otimes m_2^{-b_2} \otimes \cdots \otimes m_\alpha^{-b_\alpha})$$

and

$$\left(\mathcal{L} \otimes m_2^{-b_2} \otimes \cdots \otimes m_\alpha^{-b_\alpha} \right)^3 = \mathcal{O}_C$$

which completes the existence proof in the case $\alpha \ge 2$.

(**b**): *Uniqueness for* $\alpha \ge 2$: We have to prove the uniqueness of the coefficients b_i modulo $h_\tau^*(J[3])$.

So, assume that there is a non-trivial combination of the $\mathcal{F}_j$ with coefficients in $\mathbb{Z}/3\mathbb{Z}$ which is trivial modulo $h_\tau^*(J[3])$. Reordering the indices if necessary, we may assume that there are integers r and s with $2 \le s \le r \le \alpha$ and an $m \in \mathrm{Pic}^0(C)$ such that

$$\mathcal{O}_{C_\tau}(q_s + \cdots + q_r + 2q_{r+1} + \cdots + 2q_\alpha - (r - s + 1 + 2(\alpha - r))q_1 =$$

$$= h_\tau^*(m \otimes m_s^{-1} \otimes \cdots \otimes m_r^{-1} \otimes m_{r+1}^{-2} \otimes \cdots \otimes m_\alpha^{-2}).$$

Write $r - s + 1 + 2(\alpha - r) = 3n - \epsilon$ with $0 \le \epsilon \le 2$. Then the equality may be written as

$$\mathcal{O}_{C_\tau}(q_s + \cdots + q_r + 2q_{r+1} + \cdots + 2q_\alpha + \epsilon q_1) =$$

$$= h_\tau^*(m \otimes m_s^{-1} \otimes \cdots \otimes m_r^{-1} \otimes m_{r+1}^{-2} \otimes \cdots \otimes m_\alpha^{-2}) \otimes \mathcal{O}_C(nh_\tau(q_1)) = h_\tau^*(\mathcal{D})$$

with

$$\mathcal{D} = m \otimes m_s^{-1} \otimes \cdots \otimes m_r^{-1} \otimes m_{r+1}^{-2} \otimes \cdots \otimes m_\alpha^{-2} \otimes \mathcal{O}_C(nh_\tau(q_1)) \in \mathrm{Pic}^0(C)$$

and with

$$\mathcal{D}^3 = \mathcal{O}_C(h_\tau(q_s) + \cdots + h_\tau(q_r) + 2h_\tau(q_{s+1}) + \cdots + 2h_\tau(q_\alpha) + \epsilon h_\tau(q_1))$$

by the definition of the m_i. But this contradicts the fact that h_τ is non-Galois. This completes the proof of the assertion.

So we proved $L = h_\tau^*(J[3])$ if $\alpha = 0$ and $\alpha = 0$ and $L = h_\tau^*(J[3]) \oplus_{i=2}^\alpha \mathcal{F}_i \mathbb{Z}/3\mathbb{Z}$ if $\alpha > 1$. Since h_τ^* is injective in any case and $\deg \varphi = |L|$, this completes the proof of Lemma 4.2.15.

$\square$

Theorem 4.2.16 *Let the notations be as in diagram* (4.1). *Then the map*

$$\psi : P(h_\tau)^2 \times P(h_\sigma) \times J \to \tilde{J}, \qquad (x_1, x_2, u, v) \mapsto k_\tau^*(x_1) + \sigma k_\tau^*(x_2) + k_\sigma^*(u) + f^*(v)$$

is an isogeny of degree

$$\deg \psi = \begin{cases} 2^{2g-1} 3^{6g-3+\alpha} & \text{if } \beta = 0; \\ 2^{2g} 3^{6g-3+\alpha+\beta} & \text{if } \beta > 0. \end{cases}$$

Proof The map ψ fits into the following diagram

$$
\begin{array}{c}
P(h_\tau)^2 \times P(h_\sigma) \times J \\
{\scriptstyle \varphi \times k_\sigma^*|_{P(h_\sigma)} \times f^*} \Big\downarrow \qquad \qquad \searrow {\scriptstyle \psi} \\
P(k_\sigma) \times k_\sigma^* P(h_\sigma) \times f^* J \xrightarrow[\mu]{} J
\end{array}
$$

where μ denotes the addition map. Hence we have

$$\deg \psi = \deg \varphi \cdot \deg k_\sigma^*|_{P(h_\sigma)} \cdot \deg f^* \cdot \deg \mu. \qquad (4.6)$$

Corollary 3.2.14 applied to the factorization $f = k_\sigma \circ h_\sigma$ gives, using Lemma 4.2.5,

$$
\begin{aligned}
\deg \mu &= \frac{|P(h_\sigma)[3]|}{|\ker k_\sigma^*| \cdot |\mathrm{Ker}\, k_\sigma^*|_{P(h_\sigma)}|} \cdot \frac{|J[6]|}{|\mathrm{Ker}\, h_\sigma^*| \cdot |\mathrm{Ker}\, f^*|} \\
&= \frac{2^{2g} 3^{4g-2+\beta}}{|\ker k_\sigma^*| \cdot |\mathrm{Ker}\, k_\sigma^*|_{P(h_\sigma)}| \cdot |\mathrm{Ker}\, h_\sigma^*| \cdot |\mathrm{Ker}\, f^*|}
\end{aligned}
$$

and hence

$$\deg \psi = \deg \varphi \cdot \frac{2^{2g}3^{4g-2+\beta}}{|\ker k_\sigma^*| \cdot |\operatorname{Ker} h_\sigma^*|}.$$

Hence inserting Lemma 4.2.15 and

$$|\operatorname{Ker} k_\sigma^*| = \begin{cases} 3 \text{ if } \alpha = 0 \\ 1 \text{ if } \alpha > 0 \end{cases} \quad \text{and} \quad |\operatorname{Ker} h_\sigma^*| = \begin{cases} 2 \text{ if } \beta = 0 \\ 1 \text{ if } \beta > 0 \end{cases}$$

completes the proof of the proposition.   □

Chapter 5
Covers of Degree 4

In this chapter we study covers $f' : C' \to C$ of degree 4. It is the main chapter of the book. As mentioned already in the introduction, most of the results of this chapter appeared already in the unpublished paper [32]. Our contribution consists only in some improvements of the proofs. In particular we use more the representation theory of the corresponding groups.

There are five possibilities for f': Either it is a Galois cover. Then it is either cyclic of degree 4 or given by the action of the Klein group of order 4. In both cases we write $f : \widetilde{C} \to C$ instead of $f' : C' \to C$. Or f' is non-Galois. If $f : \widetilde{C} \to C$ is its Galois closure, then the corresponding Galois group G is either the dihedral group D_4 of order 8, the alternating group $\mathcal{A}_4$ of order 12, or the symmetric group $\mathcal{S}_4$ of order 24.

In each case we compute the isotypical decomposition as well as a group algebra decomposition of the Jacobian $\widetilde{J} = J\widetilde{C}$ with respect to the Galois group G. In all cases the factors are given either by Jacobians or Prym varieties of subcovers. In most cases this is an application of Corollary 3.5.10 (the theorem of the introduction). We will also compute the degree of the map giving the isotypical decomposition.

As an application we get several isogenies between isogeny factors. In a sense some of these isogenies may be considered as generalizations or analogues of some classical n-gonal constructions. In particular we obtain a new proof of the bigonal construction as well as direct generalizations of the trigonal construction in the $\mathcal{A}_4$ and $\mathcal{S}_4$ cases.

In Sect. 5.1 we consider the easy case of cyclic covers of degree 4 and in the second section Galois covers $f : \widetilde{C} \to C$ with group the Klein group of order 4. Here the main difficulty is the computation of the degree of an isogeny between the product of two Prym varieties of double covers of C and the Prym variety of the opposite double cover of a subcover. This will be applied later in several sections.

Sections 5.3 and 5.4 deal with Galois covers with the dihedral group of order 8. We obtain two isogenies between Prym varieties of double covers of the diagram

H. Lange, R. E. Rodríguez, *Decomposition of Jacobians by Prym Varieties*, Lecture
Notes in Mathematics 2310, https://doi.org/10.1007/978-3-031-10145-8_5

of subcovers, one considering $\widetilde{C}$ with the action of a subgroup of D_4 isomorphic to the Klein group K_4 and the other considering an intermediate cover with action of the quotient group isomorphic to K_4. It turns out that the latter isogeny is just the bigonal construction of [11] (see Theorem 5.4.4). We use it to give a new proof of Pantazis' Theorem (see Theorem 5.4.12). The advantage of our approach is that we can give precise conditions on the covers.

In Sects. 5.5 and 5.6, we study Galois covers with group the alternating group $\mathcal{A}_4$ of degree 4. After proving the usual decompositions of the Jacobian $\widetilde{J}$, we compute the degree of an isogeny (Theorem 5.5.6) which turns out to be a generalization of the trigonal construction. We make this precise by giving a proof of the usual trigonal construction for these covers in Theorem 5.6.2.

Finally, Sects. 5.7. and 5.8 deal with Galois covers with group the symmetric group $\mathcal{S}_4$ of degree 4. In Sect. 5.8 we derive the decompositions of the Jacobian $\widetilde{J}$ as well as prove some isogenies between Prym varieties of subcovers arising by considering the curve $\widetilde{C}$ with an action of a subgroup or a quotient group using results of former sections. One of them is the generalization of the trigonal construction in the $\mathcal{A}_4$ case mentioned above. Section 5.8 gives another generalization of the trigonal construction, not coming from an action of a subgroup which turns out to be more difficult.

5.1 Cyclic Covers of Degree 4

Let $f : \widetilde{C} \to C$ be a cyclic cover of smooth projective curves with Galois group

$$G = \langle \sigma \mid \sigma^4 = 1 \rangle \subset \mathrm{Aut}(\widetilde{C}).$$

Let C_{σ^2} denote the quotient of $\widetilde{C}$ by the subgroup $\langle \sigma^2 \rangle$ of order 2. So we have the following diagram of covers

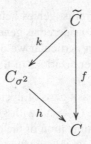

where as always we denote $C_{\sigma^2} := \widetilde{C}/\langle \sigma^2 \rangle$. Let α be the number of fixed points of σ and β be the number of fixed points of σ^2, which are not fixed by σ. If as usual g denotes the genus of C, the genera of C_{σ^2} and $\widetilde{C}$ are

$$g_{C_{\sigma^2}} = 2g - 1 + \frac{\alpha}{2} \quad \text{and} \quad g_{\widetilde{C}} = 4g - 3 + \frac{3}{2}\alpha + \frac{\beta}{2}. \tag{5.1}$$

So the Prym varieties are of dimension

$$\dim P(h) = g - 1 + \frac{\alpha}{2}, \quad \dim P(k) = 2g - 2 + \alpha + \frac{\beta}{2}, \quad \dim P(f) = 3g - 3 + \frac{3}{2}\alpha + \frac{\beta}{2}.$$

Remark 5.1.1 The equations (5.1) impose the conditions

$$\alpha \equiv \beta \equiv 0 \mod 2$$

on the numbers α and β. Moreover, if $g = 0$, we must have $\alpha \geq 2$ in order that the covers are connected.

Recall that Θ denotes the canonical polarization of $\widetilde{J}$ with induced polarizations $\Theta_{P(f)}$ on $P(f)$ and $\Theta_{P(k)}$ on $P(k)$. Similarly let $\Theta_{P(h)}$ denote the polarization on $P(h)$ induced by the canonical polarization of JC_{σ^2}. It is easy to apply Proposition 3.2.8 in order to compute the types of these polarizations.

In order to compute the rational irreducible representations of G, recall that the characters of G are given by

$$\chi_k(\sigma) = \xi_4^k \quad \text{for} \quad k = 0, 1, 2, 3 \quad \text{where} \quad \xi_4 := e^{2\pi i/4}.$$

Clearly the trivial representation χ_0 and the representation χ_2 of G are defined over the rationals. On the other hand, there is an irreducible rational representation W of degree 2 such that

$$\chi_1 \oplus \chi_3 = W \otimes \mathbb{C}.$$

Proposition 5.1.2 *The isotypical component corresponding to χ_0 (respectively, χ_2) is f^*J (respectively, $k^*P(h)$) and the isotypical component corresponding to the representation W is $P(k)$.*

Both the isotypical and group algebra decompositions $\widetilde{J}$ are given by the addition map

$$\mu : f^*J \times k^*P(h) \times P(k) \to \widetilde{J}.$$

Proof The assertion on χ_0 is obvious. The action of G on $P(h)$ is given by the the representation $\sigma \mapsto -1$ on $P(h)$ which implies the assertion on the isotypical component corresponding to χ_2. In order to prove that $P(k)$ is the isotypical component corresponding to W, we apply Corollary 3.5.10 to the subgroups $H = \{1\}$ and $N = \langle \sigma^2 \rangle$: Since

$$\rho_{\{1\}} = \mathbb{Q}[G] \doteq \chi_1 \oplus \chi_2 \oplus W = \rho_{\langle\chi_2\rangle} \oplus W,$$

Corollary 3.5.10 gives that $P(k) = P(\widetilde{C}/C_{\sigma^2})$ is the isotypical component of $\widetilde{C}$ corresponding to W. Hence the addition map gives the isotypical decomposition of $\widetilde{J}$.

It remains to show that $P(k)$ is also a group algebra component of $\widetilde{J}$. This follows from Proposition 3.5.7, since all complex irreducible representations are of dimension 1 and of Schur index 1.   □

Proposition 5.1.3 *Let* $f : \widetilde{C} \to C$ *be a cyclic cover of degree* 4.

(i) *The map*

$$\psi : P(k) \times P(h) \to P(f), \qquad (x, y) \mapsto x + k^* y$$

is an isogeny of degree

$$\deg \psi = 2^{2g-1+\alpha};$$

(ii) *The map*

$$\Psi : P(k) \times P(h) \times J \to \widetilde{J}, \qquad (x, y, z) \mapsto x + k^* y + f^* z$$

is an isogeny of degree

$$\deg \Psi = \begin{cases} 2^{6g-4} & \alpha = \beta = 0, \\ 2^{6g-3} & \text{if } \alpha = 0, \ \beta > 0, \\ 2^{6g-2+\alpha} & \alpha > 0. \end{cases}$$

Proof This follows directly from Theorem 3.2.12 using Eq. (5.1), the ramification of h and k as well as $|\operatorname{Im} h^* \cap \operatorname{Ker} k^*| = |\operatorname{Ker} k^*|$.   □

As a consequence we get the degree of the addition map in the case of the isotypical decomposition using that, if f is étale, given by the 4-division point η, we have according to Proposition 3.3.1

$$|\operatorname{Ker} k^*|_{P(h)}| = \begin{cases} 2 \text{ if } h^*(\eta) \in P(k), \\ 1 \text{ otherwise.} \end{cases} \tag{5.2}$$

Corollary 5.1.4 *Let* $f : \widetilde{C} \to C$ *be a cyclic cover of degree* 4.

(i) *The addition map* $v : P(k) \times k^* P(h) \to P(f)$ *is an isogeny of degree*

$$\deg v = \frac{2^{2g-1+\alpha}}{|\operatorname{Ker} k^*|_{P(h)}|}.$$

(ii) *The addition map* $\mu : P(k) \times k^* P(h) \times J \to \widetilde{J}$ *of the isotypical decomposition is of degree*

$$\deg \mu = \begin{cases} \dfrac{2^{6g-6}}{|\operatorname{Ker} k^*|_{P(h)}|} & if \quad \alpha = \beta = 0 \\ 2^{6g-4} & if \quad \alpha = 0, \beta > 0 \\ 2^{6g-2+\alpha} & if \quad \alpha > 0 \end{cases}$$

where $|\operatorname{Ker} k^*|_{P(h)}|$ *is given by* (5.2).

Proof This follows from Proposition 5.1.3 using the diagram

$$P(k) \times k^* P(h) \times J \to \widetilde{J} \xrightarrow{\quad\quad\mu\quad\quad} \widetilde{J}$$

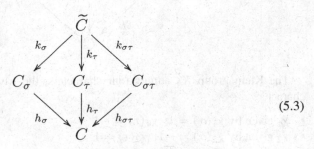

and $|\operatorname{Ker} f^*| = \begin{cases} 4 \ if \ \alpha = \beta = 0 \\ 1 \ otherwise. \end{cases}$ $\qquad\qquad\square$

5.2 The Klein Group of Order 4

5.2.1 Decompositions of $\widetilde{J}$

Let $f : \widetilde{C} \to C$ be a Galois cover with Galois group the Klein group of order 4,

$$K_4 := \langle \sigma, \tau | \sigma^2 = \tau^2 = (\sigma\tau)^2 = 1 \rangle.$$

As before we denote for any $\kappa \in K_4$ by C_κ the quotient of $\widetilde{C}$ by the group $\langle \kappa \rangle$. Then we have the following commutative diagram

(5.3)

where all maps are double covers. According to the Hurwitz formula, the ramification degree of any double cover is even. This implies that the number of fixed points of any involution of $\widetilde{C}$ is even. We denote

$$2s := \text{number of fixed points of } \sigma;$$

$$2t := \text{number of fixed points of } \tau;$$

$$2r := \text{number of fixed points of } \sigma\tau.$$

Proposition 5.2.1 *If g is the genus of C, then the signature of the Galois cover $f : \widetilde{C} \to C$ is $(g; 2, \dots, 2)$, where the number of numbers 2 is $r + s + t$.*
We have for the genera of the curves in diagram (5.3)

$$g_{\widetilde{C}} = 4g - 3 + s + t + r,$$

$$g_{C_\sigma} = 2g - 1 + \frac{t+r}{2}, \quad g_{C_\tau} = 2g - 1 + \frac{s+r}{2}, \quad g_{C_{\sigma\tau}} = 2g - 1 + \frac{s+t}{2}.$$

Proof Since the stabilizer group of any point is a cyclic subgroup of G, any point in the ramification locus $R(f)$ is of multiplicity 2. If $p \in \widetilde{C}$ is a fixed point of some involution, say σ, then h_σ is unramified over $f(p)$, and k_σ is ramified in both points p and $\tau(p)$. Moreover, σ induces an involution on the other two intermediate covers C_τ and $C_{\sigma\tau}$, which in turn implies that k_τ and $k_{\sigma\tau}$ are unramified over the corresponding ramification point. Since an analogous statement is valid for the involutions τ and $\sigma\tau$, it follows that the signature of f is $(g; 2, \dots 2)$ with $r + s + t$ numbers 2.

The formulas for the genera follow from the Hurwitz formula.   □

In particular the ramification degrees of the maps in diagram (5.3) are as follows:

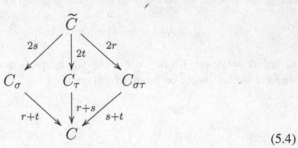

$$(5.4)$$

The Klein group K_4 admits four characters, the trivial character χ_0 and the characters

- χ_1 given by $\chi_1(\sigma) = 1$, $\chi_1(\tau) = -1$
- χ_2 given by $\chi_2(\sigma) = -1$, $\chi_2(\tau) = 1$
- χ_3 given by $\chi_3(\sigma) = -1$, $\chi_3(\tau) = -1$

All four characters are rational, and the rational group algebra decomposes as

$$\mathbb{Q}[K_4] = \chi_0 \oplus \chi_1 \oplus \chi_2 \oplus \chi_3.$$

The Schur index of each representation is 1. Hence a group algebra decomposition coincides with the isotypical decomposition of $\widetilde{J}$.

Proposition 5.2.2 *The isotypical component of $\widetilde{J}$ corresponding to the representation*

$$\chi_0 \ is \ f^* J, \quad \chi_1 \ is \ k_\sigma^* P(h_\sigma), \quad \chi_2 \ is \ k_\tau^* P(h_\tau), \quad \chi_3 \ is \ k_{\sigma\tau}^* P(h_{\sigma\tau}).$$

In particular the isotypical as well as group algebra decompositions of $\widetilde{J}$ are given by the addition map

$$\mu : f^* J \times k_\sigma^* P(h_\sigma) \times k_\tau^* P(h_\tau) \times k_{\sigma\tau}^* P(h_{\sigma\tau}) \longrightarrow \widetilde{J}.$$

One can also prove the proposition by Corollary 3.5.10. Here we give a different, a more direct proof.

Proof As always $f^* J$ is the isotypical component corresponding to the trivial representation. The element σ acts on the whole of $k_\sigma^* J C_\sigma$, and so in particular on its subvariety $k_\sigma^* P(C_\sigma/C)$, as the identity. On the other hand, τ goes down to an involution $\overline{\tau}$ on C_σ, which gives the map $h_\sigma : C_\sigma \to C$. Hence $\overline{\tau}$ acts on $P(C_\sigma/C)$ by -1 and consequently τ acts on $k_\sigma^* P(C_\sigma/C)$ by -1. This implies that K_4 acts on $k_\sigma^* P(C_\sigma/C)$ by the representation χ_1. In the same way, one shows that K_4 acts on $k_\tau^* P(C_\tau/C)$ by the representation χ_2 and on $k_{\sigma\tau}^* P(C_{\sigma\tau}/C)$ by the representation χ_3.

Hence, in order to complete the proof of the proposition, it suffices to show that the map μ is an isogeny. For this it suffices to show that the kernel Ker μ is finite, since by Proposition 5.2.1 we know that both abelian varieties are of the same dimension. But this is clear, since the group K_4 acts on the components by different representations.   □

In the next subsection, we show a bit more than just that μ is an isogeny; namely, we compute its degree deg μ.

5.2.2   Degrees of Some Isogenies

In this subsection we compute the degree of some isogenies related to the isotypical decomposition μ. The main difficulty is the proof of Proposition 5.2.5, which we prove in the next subsection.

Proposition 5.2.3 *The map*

$$\varphi_{\sigma\tau} : P(h_\sigma) \times P(h_\tau) \longrightarrow P(k_{\sigma\tau}), \quad (x_1, x_2) \mapsto k_\sigma^*(x_1) + k_\tau^*(x_2).$$

is an isogeny of degree

$$\deg \varphi_{\sigma\tau} = \deg k_\sigma^*|_{P(h_\sigma)} \cdot \deg k_\tau^*|_{P(h_\tau)} \cdot |k_\sigma^*(P(h_\sigma)[2]) \cap k_\tau^*(P(h_\tau)[2])|.$$

An analogous result is valid for $\varphi_\sigma : P(h_\tau) \times P(h_{\sigma\tau}) \longrightarrow P(k_\sigma)$ and similarly φ_τ which are analogously defined.

Proof To show that the image of $\varphi_{\sigma\tau}$ is contained in $P(k_{\sigma\tau})$, it suffices to show that

$$(1 + \sigma\tau)\varphi_{\sigma\tau}(x_1, x_2) = 0 \quad \text{for all} \quad (x_1, x_2) \in P(h_\sigma) \times P(h_\tau),$$

since $P(h_\sigma) \times P(h_\tau)$ is connected and $P(k_{\sigma\tau}) = \text{Ker}(1 + \sigma\tau)^0$ by Eq. (3.5).

But $\sigma\tau$ induces the involution -1 on $k_\sigma^* P(h_\sigma)$ and on $k_\tau^* P(h_\tau)$ which gives

$$(1 + \sigma\tau)\varphi_{\sigma\tau}(x_1, x_2) = (1 + \sigma\tau)(k_\sigma^*(x_1) + k_\tau^*(x_2))$$
$$= (1 + (-1))(k_\sigma^*(x_1) + k_\tau^*(x_2)) = 0.$$

It remains to show that $\varphi_{\sigma\tau}$ is an isogeny of the indicated degree. From Proposition 5.2.1 we get that the source and target of $\varphi_{\sigma\tau}$ are of the same dimension. Hence it suffices to compute the kernel of $\varphi_{\sigma\tau}$.

For this we need some auxiliary maps: According to Theorem 3.2.12, the map

$$\psi : P(h_{\sigma\tau}) \times P(k_{\sigma\tau}) \longrightarrow P(f), \quad (x, y) \mapsto k_{\sigma\tau}^*(x) + y$$

is an isogeny of degree

$$\deg \psi = |P(h_{\sigma\tau})[2]| \frac{|h_{\sigma\tau}^* J \cap \text{Ker } k_{\sigma\tau}^*|}{|\text{Ker } k_{\sigma\tau}^*|}. \tag{5.5}$$

Moreover, consider the map

$$\varphi : P(h_{\sigma\tau}) \times P(h_\sigma) \times P(h_\tau) \to P(f), \tag{5.6}$$

defined by

$$(x_1, x_2, x_3) \mapsto k_{\sigma\tau}^*(x_1) + k_\sigma^*(x_2) + k_\tau^*(x_3).$$

Each summand maps into $P(f)$, hence so does their sum. Again source and target of φ are of the same dimension.

The homomorphism φ factorizes as follows:

$$\varphi = \psi \circ (\text{id}_{P(h_{\sigma\tau})} \times \varphi_{\sigma\tau})$$

It follows that

$$\deg \varphi = \deg \varphi_{\sigma\tau} \cdot \deg \psi = |\operatorname{Ker} \varphi_{\sigma\tau}| \cdot |P(h_{\sigma\tau})[2]| \frac{|h_{\sigma\tau}^* J \cap \operatorname{Ker} k_{\sigma\tau}^*|}{|\operatorname{Ker} k_{\sigma\tau}^*|}. \qquad (5.7)$$

Hence for the degree of $\varphi_{\sigma\tau}$, it suffices to compute the degree of φ. So suppose $(x_1, x_2, x_3) \in \operatorname{Ker} \varphi$. From the factorization of φ and Eq. (5.7), it follows that $x_1 \in P(h_{\sigma\tau})[2]$. By symmetry we get

$$\operatorname{Ker} \varphi \subseteq P(h_{\sigma\tau})[2] \times P(h_{\sigma})[2] \times P(h_{\tau})[2].$$

Thus φ has finite kernel and hence also $\varphi_{\sigma\tau}$ has finite kernel and thus is an isogeny. In order to compute the order of this kernel, note that we may write

$$\operatorname{Ker} \varphi_{\sigma\tau} = \{(x_1, x_2) \in P(h_\sigma)[2] \times P(h_\tau)[2] \mid k_\sigma^*(x_1) = k_\tau^*(x_2)\}$$
$$= (k_\sigma^* \times k_\tau^*)^{-1}(\{(x, x) \mid x \in k_\sigma^*(P(h_\sigma)) \cap k_\tau^*(P(h_\tau))\}).$$

This gives

$$\deg \varphi_{\sigma\tau} = |\operatorname{Ker} k_\sigma^*|_{P(h_\sigma)}| \cdot |\operatorname{Ker} k_\tau^*|_{P(h_\tau)}| \cdot |k_\sigma^* P(h_\sigma)[2] \cap k_\tau^* P(h_\tau)[2]|$$

which completes the proof of Proposition 5.2.13.   □

Proposition 5.2.4 *The isogeny* $\psi : P(h_{\sigma\tau}) \times P(k_{\sigma\tau}) \to P(f), \quad (x, y) \mapsto k_{\sigma\tau}^*(x) + y$ *is of degree*

$$\deg \psi = \begin{cases} 2^{2g-3+s+t} & \text{for } r = 0, st > 0; \\ 2^{2g-2+s+t} & \text{otherwise.} \end{cases}$$

Proof If $r = 0$ and $st > 0$, then the only non-injective homomorphism between Jacobians induced from diagram (5.3) is $k_{\sigma\tau}^*$. Let $\ker(k_{\sigma\tau}^*) = \{0, \eta\}$; since f^* is injective, $\eta \notin h_{\sigma\tau}^* J$, and hence

$$\frac{|h_{\sigma\tau}^* J \cap \operatorname{Ker} k_{\sigma\tau}^*|}{|\operatorname{Ker} k_{\sigma\tau}^*|} = \frac{1}{2} \quad \text{if} \quad r = 0, \ st > 0.$$

If $r = 0$ and $st = 0$, we have $\operatorname{Ker} k_{\sigma\tau}^* = h_{\sigma\tau}^*(\operatorname{Ker} f^*) \subset h_{\sigma\tau}^* J$ in this case. Hence

$$\frac{|h_{\sigma\tau}^* J \cap \operatorname{Ker} k_{\sigma\tau}^*|}{|\operatorname{Ker} k_{\sigma\tau}^*|} = 1 \quad \text{if} \quad r = 0, \ st = 0.$$

If $r > 0$, this is valid anyway, $k_{\sigma\tau}^*$ being injective for $r > 0$.

Hence Eq. (5.5) gives the assertion, since $\dim P(h_{\sigma\tau}) = g - 1 + \frac{s+t}{2}$.   □

Our main difficulty is the following proposition,

Proposition 5.2.5 *The isogeny* $\varphi_{\sigma\tau} : P(h_\sigma) \times P(h_\tau) \longrightarrow P(k_{\sigma\tau})$ *of Proposition 5.2.3 is of degree*

$$\deg \varphi_{\sigma\tau} = \begin{cases} 2^{2g-2} & \text{if } r = s = t = 0; \\ 2^{2g} & \text{if } r = 0, st > 0; \\ 2^{2g+r-1} & \text{otherwise.} \end{cases}$$

The proof is fairly long. Before we give it in the next section, we derive some consequences.

The first corollary follows combining Propositions 5.2.4 and 5.2.5 with Eq. (5.7).

Corollary 5.2.6 *The isogeny* $\varphi : P(h_{\sigma\tau}) \times P(h_\sigma) \times P(h_\tau) \to P(f)$ *is of degree*

$$|\deg \varphi| = \begin{cases} 2^{4g-4} & \text{if } r = s = t = 0; \\ 2^{4g-3+r+s+t} & \text{otherwise.} \end{cases}$$

Finally, we have the following corollary.

Corollary 5.2.7 *The homomorphism*

$$\phi : \begin{cases} J \times P(h_{\sigma\tau}) \times P(h_\sigma) \times P(h_\tau) \longrightarrow \tilde{J} \\ (x, x_1, x_2, x_3) \mapsto f^*(x) + k_{\sigma\tau}^*(x_1) + k_\sigma^*(x_2) + k_\tau^*(x_3) \end{cases}$$

is a K_4-equivariant isogeny of degree

$$\deg \phi = \begin{cases} 2^{8g-6} & \text{if } r = s = t = 0; \\ 2^{8g-4+r+s+t} & \text{if exactly 2 of } r, s, t \text{ are zero; } . \\ 2^{8g-3+r+s+t} & \text{if at most one of } r, s, t \text{ is zero.} \end{cases}$$

Proof According to Corollary 3.2.11, the isogeny $\beta : J \times P(f) \to \tilde{J}$ is of degree

$$\deg \beta = \begin{cases} 2^{4g-2} & \text{if } r = s = t = 0; \\ 2^{4g-1} & \text{if exactly two of } r, s, t \text{ are zero; } \\ 2^{4g} & \text{otherwise.} \end{cases}$$

The K_4-equivariance follows from Proposition 5.2.2. So the assertion on the degree of ϕ is a consequence of Corollary 5.2.6 and the following commutative diagram

$$J \times P(h_\sigma) \times P(h_\tau) \times P(h_{\sigma\tau}) \xrightarrow{1_J \times \varphi} J \times P(f)$$

$$\phi \searrow \qquad \downarrow \beta$$

$$\tilde{J}$$

$\square$

In order to deduce the degree of the isotypical decomposition, we need two lemmas. The first lemma is stated for σ and s; by symmetry the analogous results for τ and t and for $\sigma\tau$ and r are also valid.

Lemma 5.2.8 *Suppose not all numbers r, s, t are zero.*

$$|\operatorname{Ker} k_\sigma^*|_{P(h_\sigma)}| = \begin{cases} 1, & \text{if } s > 0; \\ 2, & \text{if } s = 0. \end{cases}$$

Proof The assertion is clear for $s > 0$, since k_σ^* is then injective.

So let $s = 0$. Then $\operatorname{Ker} k_\sigma^*$ consists of two elements, 0 and a 2-division point, say η.

If exactly one of t and r is positive, then $|\operatorname{Ker} f^*| = 2$, and

$$\eta \in \operatorname{Ker} k_\sigma^* = h_\sigma^*(\operatorname{Ker} f^*) \subset h_\sigma^*(J[2]) \subset P(h_\sigma)[2],$$

as follows from Proposition 3.3.1, since h_σ^* is injective.

If both r and t are positive, then $|\operatorname{Ker} f^*| = 1$ by Proposition 3.2.2, and $0 = \tau(k_\sigma^*(\eta)) = k_\sigma^*(\tilde\tau(\eta))$, where $\tilde\tau$ is the involution on C_σ induced by τ. Hence

$$f^*(\operatorname{Nm}_{h_\sigma}(\eta)) = k_\sigma^*(\eta + \tilde\tau(\eta)) = 0.$$

Since f^* is injective, it follows that $\operatorname{Nm}_{h_\sigma}(\eta) = 0$; that is, $\eta \in \ker \operatorname{Nm}_{h_\sigma}$; but since h_σ is ramified, $\ker \operatorname{Nm}_{h_\sigma}$ is connected, and therefore $\eta \in P(h_\sigma)$. This gives the assertion in the case $s = 0$. □

Consider now the case $r = s = t = 0$, in which all induced morphisms between the Jacobians are non-injective. For $\iota \in \{\sigma, \tau, \sigma\tau\}$ denote $H_\iota = \{0, \eta_\iota\}$. Since $\operatorname{Ker} f^*$ consists of four elements, we have

$$\operatorname{Ker} f^* = H_\sigma + H_\tau = \{0, \eta_\sigma, \eta_\tau, \eta_\sigma + \eta_\tau\}.$$

In particular $\eta_{\sigma\tau} = \eta_\sigma + \eta_\tau$. Hence the Klein group K_4 is naturally isomorphic to the group $\operatorname{Ker} f^*$. We call $\operatorname{Ker} f^*$ the (associated) *embedding of K_4 in $J[2]$*. We distinguish two cases, according to the values of the Weil form on $\operatorname{Ker} f^*$.

Lemma 5.2.9 *If $r = s = t = 0$, then for any $\iota \in \{\sigma, \tau, \sigma\tau\}$, we have*

$$|\operatorname{Ker} k_\iota^*|_{P(h_\iota)}| = \begin{cases} 1 & \text{if } \operatorname{Ker} f^* \text{ is not isotropic for the Weil form of } J[2]; \\ 2 & \text{if } \operatorname{Ker} f^* \text{ is isotropic for the Weil form of } J[2]. \end{cases}$$

Proof Let η_ι' be the element in J defining h_ι. Let $\iota = \sigma$. Since all squares in diagram (5.3) are cartesian, we have

$$\eta_\sigma = h_\sigma^*(\eta_\tau') = h_\sigma^*(\eta_{\sigma\tau}')$$

and similarly for τ and $\sigma\tau$. Clearly $\operatorname{Ker} k_i^* |_{P(h_i)} \subset P(h_\sigma[2])$ which is equal to $h_\sigma^*((\eta'_\sigma)^\perp)$ by Proposition 3.3.1. So either the Weil form is isotropic or non-isotropic on the whole of $\operatorname{Ker} f^* \subset J[2]$. Moreover, this implies the assertion by Proposition 2.4.1. $\square$

Theorem 5.2.10 *The isotypical decomposition*

$$\mu : f^*J \times k_\sigma^* P(h_\sigma) \times k_\tau^* P(h_\tau) \times k_{\sigma\tau}^* P(h_{\sigma\tau}) \longrightarrow \widetilde{J}$$

is of degree

$$\deg \mu = \begin{cases} 2^{8g-3+r+s+t} & \text{if } rst > 0; \\ 2^{8g-4+r+s+t} & \text{if exactly one of } s, r, t \text{ is zero}; \\ 2^{8g-7+r+s+t} & \text{if exactly two of } s, r, t \text{ are zero}; \\ 2^{8g-8} & \text{if } r = s = t = 0 \text{ and } \operatorname{Ker} f^* \text{ is not isotropic in } J[2]; \\ 2^{8g-11} & \text{if } r = s = t = 0 \text{ and } \operatorname{Ker} f^* \text{ is isotropic in } J[2]. \end{cases}$$

Proof Consider the following diagram

$$
\begin{array}{c}
f^*J \times k_\sigma^* P(h_\sigma) \times k_\tau^* P(h_\tau) \times k_{\sigma\tau}^* P(h_{\sigma\tau}) \overset{\mu}{\longrightarrow} \widetilde{J} \\
{\scriptstyle f^* \times k_\sigma^*|_{P(h_\sigma)} \times k_\tau^*|_{P(h_\tau)} \times k_{\sigma\tau}^*|_{P(h_{\sigma\tau})}} \Big\uparrow \qquad \nearrow {\scriptstyle \phi} \\
J \times P(h_\sigma) \times P(h_\tau) \times P(h_{\sigma\tau}).
\end{array}
$$

Since we know that all maps are isogenies, we get

$$\deg \mu = \frac{\deg \phi}{\deg f^* \cdot \deg(k_\sigma^*|_{P(h_\sigma)}) \cdot \deg(k_\tau^*|_{P(h_\tau)}) \cdot \deg(k_{\sigma\tau}^*|_{P(h_{\sigma\tau})})}$$

So inserting Corollary 5.2.7, Lemmas 5.2.8 and 5.2.9, as well as the degree of f^*, we obtain the assertion. $\square$

5.2.3 Proof of Proposition 5.2.5

Let the notation be as in the last section and in particular as in diagram (5.3). In this subsection we want to compute the degree of the isogeny

$$\varphi_{\sigma\tau} : P(C_\sigma/C) \times P(C_\tau/C) \longrightarrow P(\widetilde{C}/C_{\sigma\tau}), \qquad (x_1, x_2) \mapsto k_\sigma^*(x_1) + k_\tau^*(x_2).$$

We consider the following four cases separately:

(i) $r = s = t = 0$.

(ii) Exactly one of the numbers r, s, t is non-zero.

(iii) Exactly two of the numbers r, s, t are non-zero.

(iv) $rst > 0$.

Recall that diagram (5.3) with the ramification degrees instead of the names of the maps is

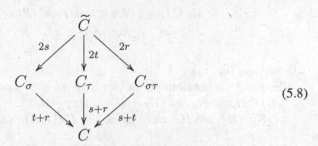

$$\text{(5.8)}$$

It follows from this diagram and Proposition 3.2.2 that all homomorphisms between Jacobians induced from diagram (5.3) are injective in case (iv), exactly one is non-injective in case (iii), exactly three are non-injective in case (ii), and all are non-injective in case (i). As for f^*, it is injective in cases (iv) and (iii), it has degree two in case (ii), and its kernel is a Klein group in case (i).

For the computation of $\deg \varphi_{\sigma\tau}$, we consider cases (i)–(iv) of above separately.

Case (i): Let $r = s = t = 0$.

Proposition 5.2.11 *If* $r = s = t = 0$, *then*

$$\deg \varphi_{\sigma\tau} = 2^{2g-2}.$$

Proof For $\iota \in \{\sigma, \tau, \sigma\tau\}$ let the double cover h_ι be given by the element $\eta_\iota \in J$. So $H_\iota := \operatorname{Ker} h_\iota = \{0, \eta_\iota\}$ and $\operatorname{Ker} f^* = H_\sigma + H_\tau = \{0, \eta_\sigma, \eta_\tau, \eta_\sigma + \eta_\tau\}$. Then we have, by Proposition 3.3.1, $P(h_\iota)[2] = h_\iota^*(\eta_\iota^\perp)$ which implies

$$|P(h_\sigma)[2]| = |P(h_\tau)[2]| = 2^{2g-2}.$$

Let $(\cdot, \cdot)$ denote the Weil form on $J[2]$. We distinguish two cases:

(a): Suppose $(\eta_\sigma, \eta_\tau) = 1$.

So $\operatorname{Ker} f^*$ is isotropic and moreover, $\operatorname{Ker} f^* \subset \eta_\sigma^\perp \cap \eta_\tau^\perp$. Since $\operatorname{Ker} k_\iota^* = h_\iota^*(\operatorname{Ker} f^*) \subset P(h_\iota)$ for $\iota = \sigma$ and τ, we get

$$|k_\sigma^*(P(h_\sigma)[2]) \cap k_\tau^*(P(h_\tau)[2])| = \left| \frac{H_\sigma^\perp}{H_\sigma + H_\tau} \cap \frac{H_\tau^\perp}{H_\sigma + H_\tau} \right|$$

$$= \left| \frac{(H_\sigma + H_\tau)^\perp}{H_\sigma + H_\tau} \right| = 2^{2g-4}.$$

Since by Lemma 5.2.9,

$$\deg k_\sigma^*|_{P(h_\sigma)} = \deg k_\tau^*|_{P(h_\tau)} = 2,$$

we get from Proposition 5.2.3,

$$\deg \varphi_{\sigma\tau} = \deg k_\sigma^*|_{P(h_\sigma)} \cdot \deg k_\tau^*|_{P(h_\tau)} \cdot |k_\sigma^*(P(h_\sigma)[2]) \cap k_\tau^*(P(h_\tau)[2])|$$
$$= 2 \cdot 2 \cdot 2^{2g-4} = 2^{2g-2}.$$

(b): Suppose $(\eta_\sigma, \eta_\tau) \neq 1$.

So Ker f^* is not isotropic and we have $\eta_\sigma \notin \eta_\tau^\perp$ and $\eta_\tau \notin \eta_\sigma^\perp$. Hence $(H_\sigma + H_\tau) \cap H_\sigma^\perp = H_\sigma$ and $(H_\sigma + H_\tau) \cap H_\tau^\perp = H_\tau$.

Now $H_\sigma^\perp$ and $H_\tau^\perp$ are of index 2 in $J[2]$ and hence

$$f^*(H_\sigma^\perp) = f^* J[2] = f^*(H_\tau^\perp),$$

and therefore by Proposition 3.3.1,

$$|k_\sigma^*(P(h_\sigma)[2]) \cap k_\tau^*(P(h_\tau)[2]))| = |f^*(H_\sigma^\perp) \cap f^*(H_\tau^\perp)| = |f^* J[2]| = 2^{2g-2}.$$

On the other hand, Lemma 5.2.9 gives

$$\deg k_\sigma^*|_{P(h_\sigma)} = \deg k_\tau^*|_{P(h_\tau)} = 1.$$

So Proposition 5.2.3 gives

$$\deg \varphi_{\sigma\tau} = \deg k_\sigma^*|_{P(h_\sigma)} \cdot \deg k_\tau^*|_{P(h_\tau)} \cdot |k_\sigma^*(P(h_\sigma)[2]) \cap k_\tau^*(P(h_\tau)[2])|$$
$$= 1 \cdot 1 \cdot 2^{2g-2} = 2^{2g-2}.$$

which completes the proof of the proposition.   □

Case (ii):   Exactly one of the numbers r, s, t is non-zero.

Proposition 5.2.12 *If exactly one of r, s, t is non-zero, then*

$$\deg \varphi_{\sigma\tau} = 2^{2g+r-1}.$$

Proof Again we consider two cases, either $r = 0$ of $r \neq 0$.

(a): Suppose $r = 0$. By symmetry of s and t, we may assume $t = 0$ and $s \neq 0$. According to Proposition 5.2.1, $s = s + r$ is even. As follows from diagram (5.8), the only injective homomorphisms between Jacobians are k_σ, h_τ and $h_{\sigma\tau}$.

By Proposition 3.3.1 we have $P(h_\sigma)[2] = h_\sigma^*(\eta_\sigma^\perp)$ where $\perp$ denotes the complement on $J[2]$. This implies $k_\sigma^* P(h_\sigma)[2] = f^*(\eta_\sigma^\perp) \subset f^*(J[2])$.

For any $\eta \in J[2]$, we have $\mathrm{Nm}_{h_\tau} h_\tau^*(\eta) = 2\eta = 0$. Hence $h_\tau^* J[2] \subset P(h_\tau)[2]$, which gives $f^* J[2] \subset k_\tau^* P(h_\tau)[2]$. Together this implies

$$|k_\sigma^* P(h_\sigma)[2] \cap k_\tau^* P(h_\tau)[2])| = |k_\sigma^* P(h_\sigma)[2]| = 2^{2g-2},$$

since h_σ is étale and k_σ^* is injective. Moreover, $\deg k_\sigma^*|_{P(h_\sigma)} = 1$ and $\deg k_\tau^*|_{P(h_\tau)} = 2$. So Proposition 5.2.3 gives, since $r = 0$,

$$\deg \varphi_{\sigma\tau} = \deg k_\sigma^*|_{P(h_\sigma)} \cdot \deg k_\tau^*|_{P(h_\tau)} \cdot |k_\sigma^*(P(h_\sigma)[2]) \cap k_\tau^*(P(h_\tau)[2])|$$

$$= 1 \cdot 2 \cdot 2^{2g-2} = 2^{2g+r-1}.$$

(b): Suppose $r > 0, s = t = 0$.

Let $q_1, \ldots, q_r \in C$ denote the branch points of f. So, f is ramified exactly over the q_i. Let $f^{-1}(q_i) = \{p_i, p_i'\}$. Then $k_\sigma(p_i) = k_\sigma(p_i')$ are exactly the ramification points of h_σ for $i = 1, \ldots, r$. Similarly $k_\tau(p_i) = k_\tau(p_i')$ are exactly the ramification points of h_τ for $i = 1, \ldots, r$. ˉ

We want to apply Proposition 3.3.3. So for $i = 2, \ldots, r$, let $u_i \in \mathrm{Pic}^0(C)$ such that

$$u_i^2 = \mathcal{O}_C(q_1 - q_i)$$

and for $\rho \in \{\sigma, \tau\}$ define

$$\mathcal{F}_i^\rho = \mathcal{O}_{C_\rho}(k_\rho(p_i) - k_\rho(p_1)) \otimes h_\rho^*(u_i) \quad \text{for} \quad i = 2, \ldots, r.$$

Then according to Proposition 3.3.3, we have

$$P(h_\rho)[2] = h_\rho^* J[2] \oplus_{i=2}^{r-1} \mathcal{F}_i^\rho \mathbb{Z}/2\mathbb{Z}.$$

Since $\ker k_\rho^* \subset h_\rho^* J[2]$, this implies

$$k_\rho^*(P(h_\rho)[2]) = f^* J[2] \oplus_{i=2}^{r-1} k_\rho^* \mathcal{F}_i^\rho \mathbb{Z}/2\mathbb{Z}.$$

Then we have for all $i = 2, \ldots, r$,

$$k_\tau^*(\mathcal{F}_i^\tau) = \mathcal{O}_{\tilde{C}}(p_i + p_i' - p_1 - p_1') \otimes f^*(u_i) = k_\sigma^*(\mathcal{F}_i^\sigma).$$

This implies

$$|k_\sigma^* P(h_\sigma) \cap k_\tau^* P(h_\tau)| = |f^* J[2] \oplus_{i=2}^{r-1} k_\sigma^* \mathcal{F}_i^\sigma \mathbb{Z}/2\mathbb{Z}| = 2^{2g-1+r-2} = 2^{2g-3+r}.$$

So we get by Proposition 5.2.3 for $r > 0, s = t = 0$,

$$\deg \varphi_{\sigma\tau} = \deg(k_\sigma^*|_{P(h_\sigma)}) \cdot \deg(k_\tau^*|_{P(h_\tau)}) \cdot |k_\sigma^*(P(h_\sigma)[2]) \cap k_\tau^*(P(h_\tau)[2])|$$

$$= 2 \cdot 2 \cdot 2^{2g-3+r} = 2^{2g+r-1}$$

This completes the proof of Proposition 5.2.12.   □

Case (iii):   Exactly two of the numbers r, s, t are non-zero.

Proposition 5.2.13 *If exactly two of r, s, t are non-zero, then*

$$\deg \varphi_{\sigma\tau} = \begin{cases} 2^{2g} & \text{if } r = 0; \\ 2^{2g+r-1} & \text{if } r > 0. \end{cases}$$

Proof By symmetry of s and t, we may assume $t > 0$.

Note first that all maps h_σ, h_τ, and $h_{\sigma\tau}$ are ramified in this case. So we can always apply Proposition 3.3.3 and Corollary 3.3.5 in order to compute $|k_\sigma^*(P(h_\sigma)[2]) \cap k_\tau^*(P(h_\tau)[2]))|$. For this it suffices to consider the two cases: $r = 0$, $st > 0$ and $s = 0$, $rt > 0$.

(a): $r = 0$, $st > 0$.

Let $p_1, \ldots, p_s, \tau p_1, \ldots, \tau p_s$ denote the ramification points of k_σ and $p_1', \ldots, p_t', \sigma p_1', \ldots, \sigma p_t'$ the ramification points of k_τ. Then $k_\sigma(p_1'), \ldots, k_\sigma(p_t') \in C_\sigma$ are the ramification points of h_σ, and $k_\tau(p_1), \ldots, k_\tau(p_s) \in C_\tau$ are the ramification points of h_τ.

For $i = 2, \ldots, t$ (respectively, $j = 2, \ldots, s$), let $u_i \in \text{Pic}^0(C)$ (respectively, $v_j \in \text{Pic}^0(C)$) such that

$$u_i^2 = \mathcal{O}_C(f(p_1') - f(p_i')) \quad \text{and} \quad v_j^2 = \mathcal{O}_C(f(p_1) - f(p_i)).$$

If we define

$$\mathcal{F}_i^\sigma = \mathcal{O}_{C_\sigma}(k_\sigma(p_i') - k_\sigma(p_1')) \otimes h_\sigma^*(u_i) \quad \text{for} \quad i = 2, \ldots, t$$

and

$$\mathcal{F}_j^\tau = \mathcal{O}_{C_\tau}(h_\tau(p_j) - h_\tau(p_1)) \otimes h_\tau^*(v_j) \quad \text{for} \quad j = 2, \ldots, s,$$

then we have, according to Proposition 3.3.3,

$$P(h_\sigma)[2] = h_\sigma^* J[2] \oplus_{i=2}^{t-1} \mathcal{F}_i^\sigma \mathbb{Z}/2\mathbb{Z} \quad \text{and} \quad P(h_\tau)[2] = h_\tau^* J[2] \oplus_{j=2}^{s-1} \mathcal{F}_j^\tau \mathbb{Z}/2\mathbb{Z}.$$

Since k_σ^* and k_τ^* are injective and the branch points of h_σ and h_τ are disjoint from each other, this implies

$$k_\sigma^* P(h_\sigma)[2] \cap k_\tau^* P(h_\tau)[2] = f^* J[2].$$

Since $\deg k_\sigma^* |_{P(h_\sigma)} = \deg k_\tau^* |_{P(h_\tau)} = 1$ by Lemma 5.2.8, Proposition 5.2.3 gives

$$\deg \varphi_{\sigma\tau} = \deg k_\sigma^* |_{P(h_\sigma)} \cdot \deg k_\tau^* |_{P(h_\tau)} \cdot |k_\sigma^* (P(h_\sigma)[2]) \cap k_\tau^* (P(h_\tau)[2])|$$
$$= 1 \cdot 1 \cdot 2^{2g},$$

since k_σ^* and k_τ^* are injective in this case.

(b): $s = 0$, $rt > 0$.

Let $p_1'', \ldots, p_r'', \tau p_1'', \ldots, \tau p_r''$ denote the ramification points of $k_{\sigma\tau}$ and $p_1', \ldots, p_t', \sigma p_1', \ldots, \sigma p_t'$ the ramification points of k_τ. Then $k_\tau(p_1''), \ldots, k_\tau (p_r'') \in C_\tau$ are the ramification points of h_τ and the ramification points of h_σ are the $r + t$ points $k_\sigma(p_1''), \ldots, k_\sigma(p_r''), k_\sigma(p_1'), \ldots, k_\sigma(p_t')$ in C_σ.

For $i = 2, \ldots, r$ (respectively, $j = 1, \ldots r + t - 1$), let $u_i \in \mathrm{Pic}^0(C)$ (respectively, $m_j \in \mathrm{Pic}^0(C)$) such that

$$u_i^2 = \mathcal{O}_C(f(p_1'') - f(p_i'')) \quad \text{and}$$

$$m_j^2 = \begin{cases} \mathcal{O}_C(f(p_i'') - f(p_{i+1}'')) & \text{for } i = 1, \ldots, r-1; \\ \mathcal{O}_C(f(p_r'') - f(p_1')) & \text{for } i = r; \\ \mathcal{O}_C(f(p_i') - f(p_{i+1}')) & \text{for } i = r+1, \ldots r+t-1. \end{cases}$$

If we define

$$\mathcal{F}_i^\tau = \mathcal{O}_{C_\tau}(k_\tau(p_i'') - k_\tau(p_1'')) \otimes h_\tau^*(u_i) \quad \text{for} \quad i = 2, \ldots, r$$

and $\mathcal{F}_i^\sigma \in P(h_\sigma)[2]$ by

$$\mathcal{F}_i^\sigma = \begin{cases} \mathcal{O}_{C_\sigma}(k_\sigma(p_{i+1}'') - k_\sigma(p_i'')) \otimes h_\sigma^*(m_i) & \text{for } i = 1, \ldots, r-1; \\ \mathcal{O}_{C_\sigma}(k_\sigma(p_1') - k_\sigma(p_r'')) \otimes h_\sigma^*(m_r) & \text{for } i = r; \\ \mathcal{O}_{C_\sigma}(k_\sigma(p_{i+1}') - k_\sigma(p_i')) \otimes h_\sigma(m_{r+i}) & \text{for for } i = r+1, \ldots r+t-1. \end{cases}$$

Then we have, according to Proposition 3.3.3, respectively, a slightly modified version of Corollary 3.3.5,

$$P(h_\tau)[2] = h_\tau^* J[2] \oplus_{i=2}^{r-1} \mathcal{F}_i^\tau \mathbb{Z}/2\mathbb{Z} \quad \text{and}$$

$$P(h_\sigma)[2] = h_\sigma^* J[2] \oplus_{i=1}^{r-1} \mathcal{F}_i^\sigma \mathbb{Z}/2\mathbb{Z} \oplus_{i=1}^{t-1} \mathcal{F}_{r+i}^\sigma \mathbb{Z}/2\mathbb{Z}.$$

This implies

$$k_\sigma^* P(h_\sigma)[2]) \cap k_\tau^* P(h_\tau)[2]) = f^* J[2] \oplus_{i=2}^{r-1} k_\sigma^* \mathcal{F}_i^\sigma \mathbb{Z}/2\mathbb{Z}.$$

Using Lemma 5.2.8, this implies

$$\deg \varphi_{\sigma\tau} = \deg k_\sigma^* |_{P(h_\sigma)} \cdot \deg k_\tau^* |_{P(h_\tau)} \cdot |k_\sigma^* (P(h_\sigma)[2]) \cap k_\tau^* (P(h_\tau)[2])|$$
$$= 2 \cdot 1 \cdot 2^{2g+r-2} = 2^{2g+r-1},$$

This completes the proof of the proposition.

□

Case (iv): $rst > 0$.

Proposition 5.2.14 *If $rst > 0$, then*

$$\deg \varphi_{\sigma\tau} = 2^{2g+r-1}.$$

Proof The automorphism τ acts on the ramification points of k_σ. Hence they are of the form $\{p_1, \ldots, p_s, \tau p_1, \ldots, \tau p_s\}$. Similarly the ramification points of k_τ respectively $k_{\sigma\tau}$ are of the form $\{p_1', \ldots, p_t', \sigma p_1', \ldots, \sigma p_t'\}$, respectively, $\{p_1'', \ldots, p_r'', \sigma p_1'', \ldots, \sigma p_r''\}$.

Then the ramification points of h_σ are the $r + t$ points $\{k_\sigma(p_1''), \ldots, k_\sigma(p_r''), k_\sigma(p_1'), \ldots, k_\sigma(p_t')\}$ and the ramification points of h_τ are the $r + s$ points $\{k_\tau(p_1''), \ldots, k_\tau(p_r''), k_\tau(p_1), \ldots, k_\tau(p_s)\}$.

For $i = 1, \ldots, r - 1$, choose $m_i \in \mathrm{Pic}^0(C)$ such that

$$m_i^2 = \mathcal{O}_C(f(p_i'') - f(p_{i+1}'')) \quad \text{for} \quad i = 1, \ldots, r - 1;$$

and for $j = 1, \ldots, t - 1$, choose $n_j \in \mathrm{Pic}^0(C)$ such that

$$n_j^2 = \mathcal{O}_C(f(p_j') - f(p_{j+1}')) \quad \text{for} \quad j = 1, \ldots t - 1.$$

Then define $\mathcal{F}_i^\sigma \in P(h_\sigma)[2]$ by

$$\mathcal{F}_i^\sigma = \mathcal{O}_{C_\sigma}(k_\sigma(p_{i+1}'') - k_\sigma(p_i'')) \otimes h_\sigma^*(m_i) \quad \text{for} \quad i = 1, \ldots, r - 1;$$

and

$$\mathcal{G}_j^\sigma = \mathcal{O}_{C_\sigma}(k_\sigma(p_{i+1}') - k_\sigma(p_i')) \otimes h_\sigma(m_{r+i}) \quad \text{for} \quad j = 1, \ldots, t - 1.$$

Then a slightly modified version of Corollary 3.3.5 implies

$$P(h_\sigma)[2] = h_\sigma^* J[2] \oplus_{i=1}^{r-1} \mathcal{F}_i^\sigma \mathbb{Z}/2\mathbb{Z} \oplus_{j=1}^{t-1} \mathcal{G}_j^\sigma \mathbb{Z}/2\mathbb{Z}. \tag{5.9}$$

Defining for $i = 1, \ldots, r - 1$, $\mathcal{F}_i^\tau := \mathcal{F}_i^\sigma$ and in an analogous way as $\mathcal{G}_j^\sigma$ also $\mathcal{G}_j^\tau$ for $j = 1, \ldots, s - 1$, we obtain that with these elements a slight modification of Corollary 3.3.5 implies

$$P(h_\tau)[2] = h_\tau^* J[2] \oplus_{i=1}^{r-1} \mathcal{F}_i^\tau \mathbb{Z}/2\mathbb{Z} \oplus_{j=1}^{s-1} \mathcal{G}_j^\tau \mathbb{Z}/2\mathbb{Z}. \tag{5.10}$$

Using Corollary 3.3.5 for the map $h_{\sigma\tau}$ which is ramified in the $r + s$ points $\{k_\tau(p_1''),$ $\ldots, k_\tau(p_r''), k_\tau(p_1), \ldots, k_\tau(p_s)\}$, we get from Lemma 3.3.4 that there is no relation between the line bundles $\mathcal{F}_i^\sigma$ and $\mathcal{G}_j^\sigma$ (respectively, $\mathcal{F}_i^\tau$ and $\mathcal{G}_j^\tau$).

Since all pullback maps of Jacobians in this case are injective and $k_\sigma^* \mathcal{F}_i^\sigma = k_\tau^* \mathcal{F}_i^\tau$ for $i = 1, \ldots, r - 1$, it follows that

$$k_\sigma^*(P(h_\sigma)[2]) \cap k_\tau^*(P(h_\tau)[2]) = f^* J[2] \oplus_{i=1}^{r-1} k_\sigma^* \mathcal{F}_i^\sigma \mathbb{Z}/2\mathbb{Z}.$$

Since all pullback maps of Jacobians in these case are injective, using Lemma 5.2.8, Proposition 5.2.3 gives

$$\begin{aligned}
\deg \varphi_{\sigma\tau} &= \deg k_\sigma^*|_{P(h_\sigma)} \cdot \deg k_\tau^*|_{P(h_\tau)} \cdot |k_\sigma^*(P(h_\sigma)[2]) \cap k_\tau^*(P(h_\tau)[2])| \\
&= 1 \cdot 1 \cdot 2^{2g+r-1} = 2^{2g+r-1},
\end{aligned}$$

which completes the proof of the proposition. $\qquad\square$

Combining Propositions 5.2.11, 5.2.12, 5.2.13, and 5.2.14 completes the proof of Proposition 5.2.5.

5.3 The Dihedral Group of Order 8

For the rest of this chapter, we consider the non-Galois covers $f' : C' \to C$ of degree 4. So its Galois closure is either the dihedral group D_4, the alternating group $\mathcal{A}_4$, or the symmetric group $\mathcal{S}_4$. This section deals with the first case.

5.3.1 Ramification and Genera

Let $f : \widetilde{C} \to C$ be a Galois cover with Galois group the dihedral group of order 8,

$$D_4 := \langle \sigma, \tau \mid \sigma^4 = \tau^2 = (\sigma\tau)^2 = 1 \rangle.$$

As always we denote for any element $\gamma \in D_4$ by C_γ the quotient curve $C_\gamma := \widetilde{C}/\langle\gamma\rangle$.

Denote the two Klein subgroups of D_4 by

$$K_\tau := \langle \sigma^2, \tau \rangle \quad \text{and} \quad K_{\sigma\tau} := \langle \sigma^2, \sigma\tau \rangle$$

and the corresponding quotients of $\widetilde{C}$ by

$$C_{K_\tau} = \widetilde{C}/K_\sigma \quad \text{and} \quad C_{K_{\sigma\tau}} := \widetilde{C}/K_{\sigma\tau}.$$

Then we have the following diagram of curves where all maps are double covers.

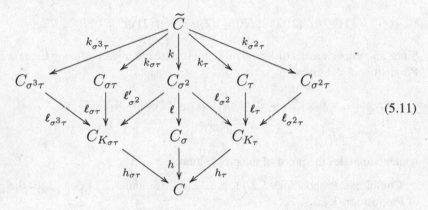

$$(5.11)$$

Note that C_τ and $C_{\sigma^2\tau}$, respectively, $C_{\sigma\tau}$ and $C_{\sigma^3\tau}$ are conjugate subcovers. All other subcovers are non-conjugate. So our starting curve C' is one of these.

Now we denote

$$2s := |\operatorname{Fix}\sigma|, \quad 2t := |\operatorname{Fix}\tau| = |\operatorname{Fix}\sigma^2\tau|, \quad 2r := |\operatorname{Fix}(\sigma\tau)| = |\operatorname{Fix}\sigma^3\tau|$$

and noting that $\operatorname{Fix}\sigma \subset \operatorname{Fix}\sigma^2$, we will see in the proof of Lemma 5.3.1 below that $|\operatorname{Fix}\sigma^2 \setminus \operatorname{Fix}\sigma|$ is divisible by 4. So let

$$4u := |\operatorname{Fix}\sigma^2 \setminus \operatorname{Fix}\sigma|$$

Lemma 5.3.1 *With these notations the ramification degrees of the corresponding double covers are as follows:*

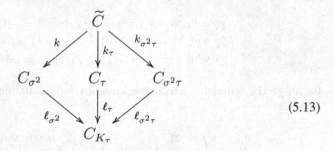

$$\tilde{C}$$

(5.12)

Moreover, the image points in C of all the pairs of fixed points corresponding to 2r, 2s, 2t, respectively, quadruples corresponding to 4u are pairwise different in C.

Proof We will define the ramification degrees differently and show that they are as in the diagram. The cover $C_{\sigma^2} \to C$ is Galois with group the Klein group K_4 and lower subsquare of (5.12) as the corresponding diagram. So as in Proposition 5.2.1, let $2t$ (respectively, $2s$, respectively, $2r$) the ramification degree of $C_{\sigma^2} \to C_{K_\tau}$ (respectively, $C_{\sigma^2} \to C_\sigma$, respectively, $C_{\sigma^2} \to C_{K_{\sigma\tau}}$). According to diagram (5.4), the ramification degree of the lower square is as indicated.

Clearly $|\operatorname{Fix}\sigma^2 \setminus \operatorname{Fix}\sigma|$ is even. So let

$$2m := |\operatorname{Fix}\sigma^2 \setminus \operatorname{Fix}\sigma|.$$

Then $k : \tilde{C} \to C_{\sigma^2}$ is of ramification degree $2m + 2s$. Since by Proposition 5.2.1 $g_{C_{\sigma^2}} = 4g - 3 + r + s + t$, this implies $g_{\tilde{C}} = 8g - 7 + 2r + 3s + 2r + m$ and thus

$$\dim P(k) = 4g - 4 + r + t + 2s + m.$$

Now note that the cover $\tilde{C} \to C_{K_\tau}$ is Galois with group the Klein group K_τ and corresponding diagram the following subdiagram of (5.11).

$$\tilde{C}$$

(5.13)

Now Proposition 5.2.3 and the conjugacy of the covers $k_{\sigma^2\tau}$ and k_τ imply

$$2 \dim P(\ell_\tau) = \dim P(k) = 4g - 4 + r + t + 2s + m. \tag{5.14}$$

Since $r + t$ is even as the ramification degree of the double cover ℓ_τ, this implies that m is even, say

$$m = 2u.$$

In particular the number $|\operatorname{Fix}\sigma^2 \setminus \operatorname{Fix}\sigma|$ is divisible by 4.

Using the Galois cover $\widetilde{C} \to C_{K_{\sigma\tau}}$ in the same way, one gets the remaining ramification degrees.

The last assertion follows by working out the ramification in all cases and using the analogous statement for the two involved subdiagrams with K_4-Galois cover. This completes the proof of the lemma.   □

Using the Hurwitz formula, the diagram implies

Proposition 5.3.2 *Let the notation be as above. If g denotes the genus of C, the signature of the Galois cover is* $(g; 4, \ldots, 4, 2, \ldots, 2)$, *with s numbers* 4 *and* $r + t + u$ *numbers* 2. *We have for the genera of the curves in diagram* (5.12)

$$g_{\widetilde{C}} = 8g - 7 + 3s + 2r + 2t + 2u,$$

$$g_{C_{\sigma^2}} = 4g - 3 + r + s + t, \qquad g_{C_\sigma} = 2g - 1 + \tfrac{r+t}{2},$$

$$g_{C_\tau} = 4g - 3 + r + u + \tfrac{3s+t}{2}, \qquad g_{C_{\sigma\tau}} = 4g - 3 + t + u + \tfrac{3s+r}{2},$$

$$g_{C_{K_\tau}} = 2g - 1 + \tfrac{r+s}{2}, \qquad g_{C_{K_{\sigma\tau}}} = 2g - 1 + \tfrac{s+t}{2}.$$

5.3.2   Decompositions of $\widetilde{J}$

Let the notations be as in diagrams (5.11) and (5.12). According to [34] the irreducible complex representations of D_4 are 4 of degree 1, namely, the trivial representation χ_0 and the representations

$$\begin{aligned}
&\chi_1 \text{ with } \quad \chi_1(\sigma^j) = 1, \ \chi_1(\tau\sigma^j) = -1 \\
&\chi_2 \text{ with } \quad \chi_2(\sigma^j) = (-1)^j, \ \chi_2(\tau\sigma^j) = (-1)^j \\
&\chi_3 \text{ with } \chi_3(\sigma^j) = (-1)^j, \ \chi_3(\tau\sigma^j) = (-1)^{j+1}
\end{aligned} \tag{5.15}$$

for all j. The remaining irreducible complex representation is V of degree 2 and given by

$$V(\sigma^j) = \begin{pmatrix} i^j & 0 \\ 0 & i^{-j} \end{pmatrix} \quad \text{and} \quad V(\tau^j) = \begin{pmatrix} 0 & i^j \\ i^{-j} & 0 \end{pmatrix} \quad \text{for all } j$$

where i denotes the complex number $\sqrt{-1}$. Its field of definition coincides with its character field, which is the field $\mathbb{Q}$ of rational numbers. Hence the representation V is defined over $\mathbb{Q}$,

$$V = W \otimes \mathbb{C}$$

with an irreducible rational representation W.

Theorem 5.3.3 *Let $\widetilde{C}$ be a curve with an action of the dihedral group D_4 of order 8, and let the notation be as in diagram (5.11). Then the isotypical component of $\widetilde{J}$ with respect to this action corresponding to the representation*

(i) χ_0 *is* $f^* J$, χ_1 *is* $k^* \ell^* P(h)$, χ_2 *is* $k_\tau^* \ell_\tau^* P(h_\tau)$, χ_3 *is* $k_{\sigma\tau}^* \ell_{\sigma\tau}^* P(h_{\sigma\tau})$,
(ii) W *is* $P(k)$.

The isotypical decomposition of $\widetilde{J}$ is given by the addition map

$$\mu : f^* J \times k^* \ell^* P(h) \times k_\tau^* \ell_\tau^* P(h_\tau) \times k_{\sigma\tau}^* \ell_{\sigma\tau}^* P(h_{\sigma\tau}) \times P(k) \longrightarrow \widetilde{J}.$$

Proof The assertion on χ_0 is clear. Next we show that D_4 acts on $k^* \ell^* P(h)$ by the representation χ_1. Let τ be any non-central involution of D_4. Since $\langle \sigma \rangle$ is a normal subgroup of D_4, τ induces an involution τ' on C_σ such that

$$\tau^* \circ k_\sigma^* = k_\sigma^* \circ \tau'^*,$$

where k_σ denotes the composition $k_\sigma = \ell \circ k : \widetilde{C} \to C_\sigma$.

By Corollary 3.5.2(b) we have $P(h_\sigma) = \mathrm{Ker}(1+\tau')^0$. It follows that $\tau'(y) = -y$ for any $y \in P(h_\sigma)$. Hence

$$\tau^*(k_\sigma^*(y)) = -k_\sigma^*(y)$$

for any $y \in P(h_\sigma)$ and any non-central involution $\tau \in D_4$. We also have $k_\sigma^* P(h_\sigma) \subset k_\sigma^* J C_\sigma \subset \widetilde{J}^{\langle \sigma \rangle}$; that is, σ acts trivially on $P(h_\sigma)$. So D_4 acts on $k_\sigma^* P(h_\sigma)$ by the representation χ_1.

The proof of the assertions on χ_2 and χ_3 is very similar to the previous proof, where we now we use that K_τ (respectively, $K_{\sigma\tau}$) is normal of index 2 in D_4. This time σ induces -1 and τ (respectively, $\sigma\tau$) acts trivially on $P(h_\tau)$ (respectively, $P(h_{\sigma\tau})$). This completes the proof of (i).

For the proof of (ii), we first show that the addition map

$$\mu : f^* J \times k^* \ell^* P(h) \times k_\tau^* \ell_\tau^* P(h_\tau) \times k_{\sigma\tau}^* \ell_{\sigma\tau}^* P(h_{\sigma\tau}) \times P(k) \to \widetilde{J}$$

is an isogeny.

To see this we apply successively Corollary 3.2.10 to the covers k, ℓ, and h, to get the isogenies

$$\widetilde{J} \sim P(k) \times k^* J C_{\sigma^2}$$
$$\sim P(k) \times k^* P(\ell) \times k^* \ell^* J_{C_\sigma}$$
$$\sim P(k) \times k^* P(\ell) \times k^* \ell^* P(h) \times f^* J.$$

Corollary 3.2.10 also implies that an inverse isogeny is given by the addition map. Since the subgroups generated by σ and σ^2 are both normal in D_4, the addition map is clearly equivariant.

Now according to (i), the first four factors of the left-hand side of the addition map are the isotypical components of the representations χ_0, χ_1, χ_4, and χ_3, respectively. This implies in particular that there is not other positive dimensional abelian subvariety on which D_4 acts by a character. Since there is only one other irreducible representation, namely W, and since the addition map is equivariant, it follows that D_4 can act on $P(k)$ only by W. Since we know already that μ is an isogeny, this implies assertion (ii).   □

In order to determine a group algebra decomposition of $\widetilde{J}$, note that if W is the rational irreducible representation of D_4 corresponding to a complex irreducible representation V with $\frac{\dim V}{s} > 1$ where s denotes the Schur index of V, then the corresponding isotypical component A_W decomposes further, namely, there is an abelian subvariety B_W, not uniquely determined, such that

$$A_W \sim B_W^{\frac{\dim V}{s}}.$$

Hence the isotypical component corresponding to W; that is, $P(k)$, is the only component which decomposes, namely $P(k) \sim B_W^2$. The following proposition gives a particular such B_W.

Proposition 5.3.4 *Let the notations be as in diagram* (5.11) *and g the genus of C. Then the homomorphism*

$$\psi : P(\ell_{\sigma^2 \tau}) \times P(\ell_\tau) \to P(k), \quad (x, y) \mapsto k_{\sigma^2 \tau}^*(x) + k_\tau^*(y)$$

is an isogeny of degree

$$\deg \psi = \begin{cases} 2^{4g-4+r} & \text{if } s = t = u = 0; \\ 2^{4g-3+r+2s+2u} & \text{otherwise.} \end{cases}$$

Proof Considering $\widetilde{C}$ with the action of the subgroup $K_\tau = \langle \sigma^2, \tau \rangle \subset D_4$, we have the following subdiagram of (5.11) with corresponding diagram of ramification degrees,

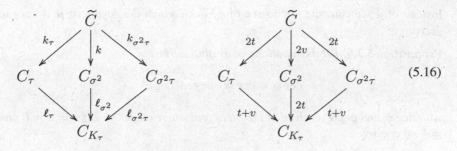

$$(5.16)$$

where

$$2v := 4u + 2s$$

is the number of fixed points of σ^2. Note that the first diagram coincides with (5.13) written in an equivalent way. It follows from the second diagram that the Prym varieties $P(\ell_\tau)$ and $P(\ell_{\sigma^2\tau})$ are of the same dimension. This is clear anyway, since the involutions τ and $\sigma^2\tau$ are conjugate in D_4.

We apply Proposition 5.2.5 to diagram (5.13) to get

$$\deg \psi = \begin{cases} 2^{2g(C_{K_\tau})-2} & if\ t = v = 0 \\ 2^{2g(C_{K_\tau})+v-1} & otherwise. \end{cases}$$

Note that the middle case of Proposition 5.2.5 does not occur.

Inserting $g(C_{K\tau}) = 2g - 1 + \frac{r+s}{2}$ from Proposition 5.3.2, gives the assertion. $\square$

Corollary 5.3.5 *Let the notations be as above. A group algebra decomposition is given by an isogeny*

$$J \times P(h) \times P(h_\tau) \times P(h_{\sigma\tau}) \times P(\ell_\tau)^2 \longrightarrow \widetilde{J}.$$

Proof According to Theorem 5.3.3, we only have to show that there is an isogeny $k^* P(\ell_\tau)^2 \to P(k)$. But the Prym varieties $P(\ell_{\sigma^2\tau})$ and $P(\ell_\tau)$ are isomorphic, since the covers $C_{\sigma^2\tau} \to C_{K_\tau}$ and $C_\tau \to C_{K\tau}$ are conjugate. Combining an isomorphism with the isogeny of Proposition 5.3.4 gives the assertion. $\square$

Finally in this section, we want to compute the degree of the isotypical decomposition μ. For this we consider the commutative diagram

$$\begin{array}{c} J \times P(h) \times P(h_\tau) \times P(h_{\sigma\tau}) \times P(k) \\ {\scriptstyle f^* \times k^* \ell^* \times k_\tau^* \ell_\tau^* \times k_{\sigma\tau}^* \ell_{\sigma\tau}^* \times 1_{P(k)}} \Big\downarrow \qquad \qquad \searrow \phi \\ f^* J \times k^* \ell^* P(h) \times k_\tau^* \ell_\tau^* P(h_\tau) \times k_{\sigma\tau}^* \ell_{\sigma\tau}^* P(h_{\sigma\tau}) \times P(k) \xrightarrow[\mu]{} \widetilde{J}. \end{array}$$

Instead of μ we compute the degree of ϕ, from which the degree of μ is easy to derive.

Proposition 5.3.6 *Let the notations be as above. Then*

$$\deg \phi = \deg \psi \cdot \deg \alpha \cdot \deg k^*$$

with $\deg \psi$ *and* $\deg \alpha$ *as given in the proof and where* $\deg k^* = 2$ *if* $s = u = 0$ *and* $= 1$ *otherwise.*

Proof Applying Corollary 5.2.7 to diagram (5.17) below, we get that the isogeny

$$\psi : J \times P(h) \times P(h_\tau) \times P(h_{\sigma\tau}) \to JC_{\sigma^2}$$

given by $(x, x_1, x_2, x_3) \mapsto \ell^* h^*(x) + \ell^*(x_1) + \ell^*_{\sigma^2}(x_2) + {\ell'_{\sigma^2}}^*(x_3)$ is of degree

$$\deg \psi = \begin{cases} 2^{8g-6} & \text{if } r = s = t = 0; \\ 2^{8g-4+r+s+t} & \text{if exactly two of } r, s, t \text{ are zero}; \\ 2^{8g-3+r+s+t} & \text{if at most one of } r, s, t \text{ is zero.} \end{cases}$$

On the other hand, according to Corollary 3.2.10, the addition map $\alpha : k^* JC_{\sigma^2} \times P(k) \to \tilde{J}$ is an isogeny of degree $\frac{|JC_{\sigma^2}[2]|}{|\operatorname{Ker} k^*|^2}$; that is,

$$\deg \alpha = \begin{cases} 2^{8g-8+2r+2t} & \text{if } s + u = 0; \\ 2^{8g-6+2s+2r+2t} & \text{if } s + u > 0. \end{cases}$$

Hence the assertion follows from the commutative diagram

$$J \times P(h) \times P(h_\tau) \times P(h_{\sigma\tau}) \times P(k) \xrightarrow{\psi \times 1_{P(k)}} JC_{\sigma^2} \times P(k)$$

$$\downarrow{k^* \times 1_{P(k)}}$$

$$k^* JC_{\sigma^2} \times P(k)$$

$$\phi \qquad \qquad \downarrow{\alpha}$$

$$\tilde{J}$$

using that $\operatorname{Ker} k^*$ is of order 2 if $s + u = 0$ and 1 otherwise.   □

5.3.3 An Isogeny Coming from an Action of a Quotient Group

Consider the quotient group $D_4/\langle \sigma^2 \rangle$. It is isomorphic to the Klein group K of order 4, $\langle \sigma^2 \rangle$ being normal in D_4. Consider the following subdiagram of diagram (5.11), which corresponds to the action of K on the Jacobian JC_{σ^2}.

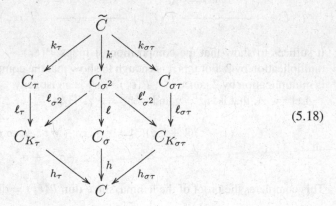

Then Proposition 5.2.3 gives an isogeny $\ell_{\sigma^2}^* P(h_\tau) \times \ell_{\sigma^2}'^* P(h_{\sigma\tau}) \sim P(\ell)$ and hence, using the commutativity of diagram (5.11), an isogeny

$$k_\tau^* \ell_\tau^* P(h_\tau) \times k_{\sigma\tau}^* \ell_{\sigma\tau}^* P(h_{\sigma\tau}) \sim k^* P(\ell).$$

5.3.4 A Generalization of the Bigonal Construction

Given again a curve $\widetilde{C}$ with an action of the group D_4, we let the notations be as in diagrams (5.11) and (5.12). We consider the following subdiagram of diagram (5.11) with ramification degrees as in diagram (5.12).

$$(5.18)$$

The aim of this section is to show that the map

$$\varphi := \operatorname{Nm} k_{\sigma\tau} \circ k_\tau^* |_{P(\ell_\tau)} : P(\ell_\tau) \to P(\ell_{\sigma\tau}) \qquad (5.19)$$

is an isogeny (see Proposition 5.3.7) and compute its degree (see Theorem 5.3.12). In the next section, we will see that a special case of this is the well-known bigonal construction. For this see Remark 5.4.6.

Proposition 5.3.7 *The map φ is an isogeny with $\operatorname{Ker} \varphi \subset P(\ell_\tau)[2]$.*

Proof Let $x \in P(\ell_\tau)$. It follows from Proposition 5.2.3 that $k_\tau^*(x) \in P(k)$. Hence $\mathrm{Nm}\, k(k_\tau^*(x)) = 0$ and therefore

$$\mathrm{Nm}\, \ell_{\sigma\tau}(\varphi(x)) = \mathrm{Nm}\, \ell_{\sigma\tau} \circ \mathrm{Nm}\, k_{\sigma\tau}(k_\tau^*(x)) = \mathrm{Nm}\, \ell'_{\sigma\tau} \circ \mathrm{Nm}\, k(k_\tau^*(x)) = 0,$$

which gives $\varphi(P(\ell_\tau)) \subset P(\ell_{\sigma\tau})$.

In order to see that φ is an isogeny, denote

$$A := k_\tau^*(P(\ell_\tau)) \quad \text{and} \quad B := k_{\sigma\tau}^*(P(\ell_{\sigma\tau}))$$

Now $\sigma^2\tau$ (respectively, σ^2) induces an involution on JC_τ (respectively, $JC_{\sigma\tau}$) which we denote by the same letter. Thus the Prym variety $P(\ell_\tau)$ is $\mathrm{Ker}(1 + \sigma^2\tau)^0$ and $P(\ell_{\sigma\tau})$ is $\mathrm{Ker}(1 + \sigma^2)^0$. Hence we have by Corollary 3.5.2(b):

$$A = \{z \in \tilde{J}^\tau \mid z + \sigma^2\tau z = 0\}^0 \quad \text{and} \quad B = \{w \in \tilde{J}^{\sigma\tau} \mid w + \sigma^2 w = 0\}^0.$$

Moreover, the following diagram commutes

(5.20)

It suffices to show that the composition of maps $P(\ell_\tau) \to P(\ell_{\sigma\tau}) \to P(\ell_\tau)$ is multiplication by 2. For this it is enough to show that the composition $A \to B \to A$ is multiplication by 2, since $k_\tau^* : P(\ell_\tau) \to A$ is an isogeny.

Let $z \in A$, that is, $\tau z = z$ and $\sigma^2 \tau z = -z$. Then

$$(1+\tau)(1+\sigma\tau)(z) = z + \tau z + \sigma\tau z + \tau\sigma\tau z$$
$$= z + z - \sigma^3 z + \sigma^3 z = 2z.$$

This completes the proof of the lemma, since $\dim P(\ell_\tau) = \dim P(\ell_{\sigma\tau})$. □

We will now compute the degree of φ. We have just proved that $\mathrm{Ker}\,\varphi \subset P(\ell_\tau)[2]$. According to diagram (5.12), the ramification degree of the double cover ℓ_τ is

$$v := s + t + 2u.$$

Hence Proposition 3.3.3, applied to the double cover ℓ_τ gives

$$P(\ell_\tau)[2] = \ell_\tau^*(JC_{K_\tau}[2]) \oplus \bigoplus_{i=2}^{v-1} \mathcal{F}_i \mathbb{Z}/2\mathbb{Z}, \tag{5.21}$$

with the line bundles $\mathcal{F}_i \in P(\ell_\tau)[2]$ as defined in Proposition 3.3.3 for the double cover ℓ_τ. In particular, if $v = 0$ or if $v = 2$; that is, if ℓ_τ is étale or has two branch points, there are no $\mathcal{F}_i$.

We will see that $\operatorname{Ker} \varphi$ is closely related to the $(\mathbb{Z}/2\mathbb{Z})$-vector space of points x in $P(\ell_\tau)[2]$ such that $k_\tau^*(x)$ is D_4-invariant; it is therefore convenient to define the corresponding subspaces of the decomposition (5.21)

$$\Gamma := \{x \in \ell_\tau^*(JC_{K_\tau}[2]) : k_\tau^*(x) \text{ is } D_4 - \text{invariant}\} \tag{5.22}$$

and

$$\Delta := \{y \in \bigoplus_{i=2}^{v-1} \mathcal{F}_i \mathbb{Z}/2\mathbb{Z} : k_\tau^*(y) \text{ is } D_4 - \text{invariant}\}. \tag{5.23}$$

Moreover, the involution $\sigma\tau$ on $\widetilde{C}$ induces an involution on C_{K_τ} which we denote by the same letters, i.e., $\sigma\tau : C_{K_\tau} \to C_{K_\tau}$.

Then we have

Proposition 5.3.8

$$\deg \varphi = \frac{|\Gamma| + |\Delta|}{\deg(k_{\sigma\tau}^*|_{P(\ell_{\sigma\tau})})} \qquad \text{with}$$

$\deg(k_{\sigma\tau}^*|_{P(\ell_{\sigma\tau})})$

$$= \begin{cases} 1, & \text{if } r > 0; \\ 2, & \text{if } r = 0 \text{ and } s + u > 0; \\ 1, & \text{if } r = s = u = 0 \text{ and } \operatorname{Ker}((\ell_{\sigma\tau} k_{\sigma\tau})^*) \text{ is not isotropic in } JC_{K_{\sigma\tau}}[2]; \\ 2, & \text{if } r = s = u = 0 \text{ and } \operatorname{Ker}((\ell_{\sigma\tau} k_{\sigma\tau})^*) \text{ is isotropic in } JC_{K_{\sigma\tau}}[2]. \end{cases}$$

Furthermore,

$$\Gamma = \ell_\tau^*((JC_{K_\tau})^{\sigma\tau}) \cap P(\ell_\tau)[2].$$

Proof According to diagram (5.20), we have

$$\deg \varphi = \frac{\deg((1 + \sigma\tau) \circ k_\tau^*|_{P(\ell_\tau)})}{\deg(k_{\sigma\tau}^*|_{P(\ell_{\sigma\tau})})}.$$

From the definition of A, one checks that

$$\operatorname{Ker}((1 + \sigma\tau)|_A) = \{z \in k_\tau^*(P(\ell_\tau))[2] \mid z \text{ is } D_4 - \text{invariant}\}.$$

This implies, together with Proposition 5.3.7, that

$$\mathrm{Ker}((1 + \sigma\tau) \circ k_\tau^*|_{P(\ell_\tau)}) = \{x \in P(\ell_\tau)[2] \mid k_\tau^*(x) \text{ is } D_4 - \text{invariant}\} = |\Gamma| + |\Delta|,$$

and the first assertion is proved.

The assertion on $\deg(k_{\sigma\tau}^*|_{P(\ell_{\sigma\tau})})$ follows directly from Lemmas 5.2.8 and 5.2.9, applied to the left-hand subdiagram of (5.11) corresponding to the Galois cover $\tilde{C} \to C_{K_{\sigma\tau}}$ with Galois group the Klein group of order 4 generated by $\sigma\tau$ and σ^2.

For the last assertion, it remains to show that if $x \in \ell_\tau^*(JC_{K_\tau}[2])$, then

$$k_\tau^*(x) \text{ is } D_4\text{-invariant} \Leftrightarrow x \in \ell_\tau^*((JC_{K_\tau})^{\sigma\tau}) \cap P(\ell_\tau)[2].$$

This is a straightforward computation using the definition of A.   □

Lemma 5.3.9 *With the notations of above, we have*

(i) *If $r = s = 0$, that is, if h_τ is étale, then*

$$(JC_{K_\tau})^{\sigma\tau}[2] = (h_\tau^* J)[2] \quad and \quad |(JC_{K_\tau})^{\sigma\tau}[2]| = 2^{2g}$$

(ii) *If $r + s > 0$, that is, if h_τ is ramified, then*

$$(JC_{K_\tau})^{\sigma\tau}[2] = P(h_\tau)[2] \quad and \quad |(JC_{K_\tau})^{\sigma\tau}[2]| = 2^{2g-2+r+s}$$

(iii) *In any case,*

$$(JC_{K_\tau})^{\sigma\tau} \cap \mathrm{Ker}\, \ell_\tau^* = 0$$

In particular $P(h_\tau) \cap \mathrm{Ker}\, \ell_\tau^* = 0$, and $\ell_\tau^*|_{(JC_{K_\tau})^{\rho\tau}} : (JC_{K_\tau})^{\sigma\tau} \to \ell_\tau^*((JC_{K_\tau})^{\sigma\tau})$ is an isomorphism.

Proof Note that $\sigma\tau$ induces an involution on K_τ, denoted by the same letter. Hence Corollary 3.5.4 applied to the cover h_τ gives $(JC_{K_\tau})^{\sigma\tau} = h_\tau^* J + P(h_\tau)[2]$ and therefore

$$(JC_{K_\tau})^{\sigma\tau}[2] = (h_\tau^* J)[2] + P(h_\tau)[2]. \tag{5.24}$$

But by Proposition 3.3.1 we have

$$P(h_\tau)[2] \subsetneq h_\tau^*(J[2]) \subset (h_\tau^* J)[2] \quad \text{if} \quad h_\tau \text{ is étale}, \tag{5.25}$$

and by Proposition 3.3.3 we have

$$(h_\tau^* J)[2] = h_\tau^*(J[2]) \subset P(h_\tau)[2] \quad \text{if} \quad h_\tau \text{ is ramified}.$$

Together with (5.24) this gives (i), since the only element of $J[2]$ going to 0 under h_τ^* is the line bundle defining the double cover h_τ, and gives (ii) by using Proposition 5.3.2.

(iii): If ℓ_τ is ramified, then ℓ_τ^* is injective, so there is nothing to show.

If ℓ_τ is étale, that is, if $s = t = u = 0$, then all maps in diagram (5.16) are étale. The following diagram is commutative

$$
\begin{array}{ccc}
\widetilde{C} & \xrightarrow{\;\sigma\tau\;} & \widetilde{C} \\
\downarrow{\scriptstyle k_{\sigma^2\tau}} & & \downarrow{\scriptstyle k_\tau} \\
C_{\sigma^2\tau} & & C_\tau \\
\downarrow{\scriptstyle \ell_{\sigma^2\tau}} & & \downarrow{\scriptstyle \ell_\tau} \\
C_{K_\tau} & \xrightarrow{\;\sigma\tau\;} & C_{K_\tau}
\end{array}
$$

If $\eta_{\ell_\tau} \in JC_{K_\tau}$ is the element defining the cover ℓ_τ, then $(\sigma\tau)^*(\eta_{\ell_\tau})$ defines the cover $\ell_{\sigma^2\tau}$, which implies $(\sigma\tau)^*(\eta_{\ell_\tau}) \neq \eta_{\ell_\tau}$, i.e., $\eta_{\ell_\tau} \notin (JC_{K_\tau})^{\sigma\tau}$. This implies the first assertion in the étale case. The last assertion follows from Eq. (5.24).   □

Proposition 5.3.10 *Suppose ℓ_τ is étale, that is, if $s = t = u = 0$. Then*

$$
|\Gamma| + |\Delta| = |\Gamma| = \left\{ \begin{array}{ll} 2^{2g-3+r} & \text{if } r > 0; \\ 2^{2g-1} & \text{if } r = 0. \end{array} \right\}.
$$

Proof The first equality is clear from the definition of Γ and Δ.

Suppose ℓ_τ is given by the line bundle $\eta_{\ell_\tau} \in JC_{K_\tau}[2]$. Then we have, according to Proposition 3.3.1,

$$
P(\ell_\tau)[2] = \ell_\tau^*(\eta_{\ell_\tau}^\perp).
$$

Hence, according to Theorem 3.2.12, the kernel of the map

$$
\psi : P(\ell_\tau) \times P(h_\tau) \to P(h_\tau \circ \ell_\tau), \qquad (x, y) \mapsto x + \ell_\tau^*(y)
$$

is of order $\frac{1}{2}|P(h_\tau)[2]|$, since $\mathrm{Im}\, h_\tau^* \cap \mathrm{Ker}\, k_\tau^* = 0$. Clearly this kernel is isomorphic to

$$
K := \{ y \in P(h_\tau)[2] \mid \ell_\tau^*(y) \in P(\ell_\tau)[2] \}
$$

where the isomorphism $K \to \mathrm{Ker}(\psi)$ is given by $y \mapsto (-\ell_\tau(y), y)$. We distinguish two cases:

(i): Suppose h_τ is ramified, i.e., $r > 0$. Then from Lemma 5.3.9 (ii), we know that $(JC_{K_\tau})^{\sigma\tau}[2] = P(h_\tau)[2]$, and Lemma 5.3.9(iii) gives

$$K \simeq \ell_\tau^*(K) = \ell_\tau^*(JC_\tau)^{\sigma\tau}[2] \cap P(\ell_\tau)[2] = \Gamma.$$

Hence $|\Gamma| = 2^{2\dim P(h_\tau)-1} = 2^{2g-3+r}$, if $r > 0$ and $s = t = u = 0$.

(ii): Suppose h_τ is étale, i.e., $r = 0$. Then from Lemma 5.3.9(i) we know that $(JC_\tau)^{\sigma\tau}[2] = (h_\tau^*J)[2]$. Since $\ell_\tau^*(\eta_\tau^\perp) = P(\ell_\tau)[2]$, we have

$$K = \eta_\tau^\perp \cap P(h_\tau)[2]$$

is of index 2 in $P(h_\tau)[2]$. So $P(h_\tau)[2]$ is not contained in $\eta_{K_\tau}^\perp$. Since by (5.25) we have $P(\ell_\tau)[2] \subset (h_\tau^*J)[2]$, we get

$$(h_\tau^*J)[2] \not\subset \eta_{K_\tau}^\perp.$$

Moreover, $\eta_\tau^\perp$ is of index 2 in $JC_{K_\tau}[2]$, and hence

$$(h_\tau^*J)[2] \cap \eta_\tau^\perp \text{ is of index 2 in } (h_\tau^\vee J)[2].$$

Now we claim

$$\ell_\tau^*((h_\tau^*J)[2] \cap \eta_\tau^\perp) = \ell_\tau^*((h_\tau^*J)[2]) \cap \ell_\tau^*(\eta_\tau^\perp) = \Gamma.$$

For the proof note that if z is contained in the right-hand side of the first equality, then $z = \ell_\tau^*(x) = \ell_\tau^*(y)$ with $x \in (h_\tau^*J)[2]$ and $y \in (\mathrm{Ker}\,\ell_\tau^*)^\perp$. This implies $x = y + v$ with $v \in \mathrm{Ker}\,\ell_\tau^*$, which shows that $x \in (\mathrm{Ker}\,\ell_\tau^*)^\perp$, and therefore z is contained in the left-hand side of the first equality.

Putting everything together, we get $|\Gamma| = \frac{1}{2}|(h_\tau^*J)[2]| = \frac{1}{2}|J[2]| = 2^{2g-1}$, if $r = s = t = u = 0$. $\square$

Proposition 5.3.11 *Suppose ℓ_τ is ramified, that is, if $s + t + u > 0$. Then*

$$|\Gamma| + |\Delta| = \begin{cases} 2^{2g-1} & \text{if } r = s = 0, \ t > 0; \\ 2^{2g-2+r} & \text{if } s = 0, \ rt > 0; \\ 2^{2g-3+r+2s} & \text{if } su > 0; \\ 2^{2g-4+r+2s} & \text{if } s > 0, \ u = 0. \end{cases}$$

Proof

(a): We claim

$$|\Gamma| = |\ell_\tau^*((JC_{K_\tau})^{\sigma\tau}) \cap P(\ell_\tau)[2]| = \begin{cases} 2^{2g-1} & \text{if } r = s = 0 \\ 2^{2g-2+r+s} & \text{if } r + s > 0. \end{cases} \tag{5.26}$$

Here we have $t + u > 0$ if $r = s = 0$, and $s + t + u > 0$ in any case, since ℓ_τ is supposed to be ramified.

For the proof note that by Proposition 3.3.1 we know that ℓ_τ^* injects $JC_{K_\tau}[2]$ into $P(\ell_\tau)[2]$. From Lemma 5.3.9(iii), we know that

$$\ell_\tau^*((JC_{K_\tau})^{\sigma\tau}[2]) = \ell_\tau^*((JC_{K_\tau})^{\sigma\tau})[2] \subset P(\ell_\tau)[2].$$

Hence

$$\ell_\tau^*((JC_{K_\tau})^{\sigma\tau}) \cap P(\ell_\tau)[2] = \ell_\tau^*((JC_{K_\tau})^{\sigma\tau}[2]) \simeq (JC_{K_\tau})^{\sigma\tau}[2]$$

and the assertion follows from Lemma 5.3.9 and Proposition 5.3.2.

(b): Here we compute the cardinality $|\Delta|$ of Δ as defined just before Proposition 5.3.8.

We claim

$$|\Delta| = \begin{cases} 2^{s-1} & \text{if } su > 0; \\ 2^{s-2} & \text{if } s > 0,\ u = t = 0; \\ 1 & \text{if } u = s = 0,\ t > 0. \end{cases} \tag{5.27}$$

Here we use Proposition 3.3.3 and its corollaries.

Recall diagram (5.16) and the notation $v = s + t + 2u$: the double cover k, respectively, $k_{\sigma^2\tau}$ is given by the involution σ^2, respectively, $\sigma^2\tau$ whose fixed points can be written as $p_1, \ldots p_v, \tau p_1, \ldots \tau p_v$, respectively, $p'_1, \ldots p'_t, \tau p'_1, \ldots, p'_t$. Hence the ramification poins of ℓ_τ are the $v + t$ points:

- $q_1, \ldots, q_v$ with $q_i = k_\tau(p_i)$ for all i
- $q'_1, \ldots, q'_t$ with $q'_i = k_\tau(p'_i)$ for all i

Suppose first $vt > 0$.

Then we define the line bundles $\mathcal{G}_i$ for the double cover ℓ_τ as in Corollary 3.3.6. To be more precise, for $i = 1, \ldots, v + t - 1 = s + t + 2u$, choose $n_i \in \text{Pic}^0(C_{K_\tau})$ such that

$$n_i^2 = \begin{cases} \mathcal{O}_{K_\tau}(\ell_\tau(q_i) - \ell_\tau(q_{i+1})) & \text{for } i = 1, \ldots, v - 1; \\ \mathcal{O}(\ell_\tau(q'_i) - \ell_\tau(q'_{i+1})) & \text{for } i = v + 1, \ldots, v + t - 1; \end{cases}$$

and define, for $i = 1, \ldots, v - t - 1$, $i \neq v$,

$$\mathcal{G}_i := \begin{cases} \mathcal{O}_{C_\tau}(q_{i+1} - q_i) \otimes \ell_\tau^*(n_i) & \text{for } i = 1, \ldots, v - 1; \\ \mathcal{O}_{C_\tau}(q'_{i+1} - q'_i) \otimes \ell_\tau^*(n_i) & \text{for } i = v + 1, \ldots, v + t - 1; \end{cases}$$

Then Corollary 3.3.6 gives

$$P(\ell_\tau)[2] = \ell_\tau^* J_{K_\tau}[2] \oplus_{i=1}^{v-1} \mathcal{G}_i \oplus_{i=v+1}^{v+t-1} \mathcal{G}_i.$$

Moreover, note that $k_\tau^*(\mathcal{G}_i)$ is D_4-invariant precisely if $\sigma(\tau^*\mathcal{G}_i) = \tau^*\mathcal{G}_i$, since it is τ-invariant anyway and this is the case if and only if p_i and p_{i+1} are fixed points of σ. So label $p_1, \ldots, p_{v-1}$ in such a way that $p_1, \ldots p_s$ are the fixed points of σ and the rest of the p_i are fixed points of σ^2 only. Then exactly $k_\tau^*\mathcal{G}_1, \ldots, k_\tau^*\mathcal{G}_{s-1}$ are the D_4-invariant line bundles among the $k_\tau^*\mathcal{G}_i$. So Eq. (5.23) implies

$$|\Delta| = 2^{s-1} \quad \text{if} \quad su > 0.$$

Now suppose $v > 0$ and $t = 0$.

In this case the ramification points of ℓ_τ are $q_1, \ldots, q_v$. Then let $\mathcal{F}_2, \cdots, \mathcal{F}_v$ be the line bundles of Proposition 3.3.3 for these ramification points. Then this proposition gives $P(\ell_\tau[2]) = \ell_\tau^* JC_{K_\tau}[2] \oplus_{i=2}^{v-1} \mathcal{F}_i \mathbb{Z}/2\mathbb{Z}$. Labelling the p_i as above, i.e., such that $p_1, \ldots p_s$ are the fixed points of σ and noting that the line bundle defining the étale double cover k_τ cannot be among the $\mathcal{F}_i$, we conclude with Eq. (5.23),

$$|\Delta| = \begin{cases} 2^{s-1} & \text{if } su > 0, \ t = 0. \\ 2^{s-2} & \text{if } s > 0, \ t = u = 0. \end{cases}$$

Note that if $s > 0$, it is automatically ≥ 2, since it is an even number.

Finally we are left with the case $s = 0$ and $t > 0$.

In this case there are no D_4-invariant line bundles $\mathcal{F}_i$. Hence Eq. (5.23) gives $|\Delta| = 1$.

Adding the equations for $|\Gamma|$ and $|\Delta|$ completes the proof of the proposition.   □

Combining everything, we obtain the following theorem.

Theorem 5.3.12 *Let $\widetilde{C}$ be a curve with action of the group $D_4 = \langle \sigma, \tau \rangle$ with $2s = |\operatorname{Fix} \sigma|$, $2t = |\operatorname{Fix} \tau|$, $2r = |\operatorname{Fix} \sigma\tau|$, and $4u = |\operatorname{Fix} \sigma^2 \setminus \operatorname{Fix} \sigma|$. Then the map*

$$\varphi := \operatorname{Nm} k_{\sigma\tau} \circ k_\tau^* |_{P(\ell_\tau)} : P(\ell_\tau) \to P(\ell_{\sigma\tau})$$

of (5.19) is an isogeny with kernel in the 2-division points and degree given by Propositions 5.3.8, 5.3.10, and 5.3.11.

5.4   The Bigonal Construction

As mentioned in the last section, Theorem 5.3.12 can be considered as a generalization of the bigonal construction. In this section we want to explain why this is the case. For this we first have to explain the bigonal construction itself. It was introduced by Pantazis in [28].

5.4.1 Definition and First Properties

Let

$$Y \xrightarrow{\ell} X \xrightarrow{h} C$$

be a double cover of a double cover of a smooth curve C. We denote by f the composition $f = h\ell$. The bigonal construction associates to it another such pair of double covers

$$Y' \xrightarrow{\ell'} X' \xrightarrow{h'} C.$$

The name comes from the special case $C = \mathbb{P}^1$, when such covers are called *bigonal covers*. Following Donagi [11] we use this name also for the more general case, where C is an arbitrary curve.

Given $Y \xrightarrow{\ell} X \xrightarrow{h} C$ as above, Y' is defined as follows: Recall from [15, p. 282] that a *correspondence* of degree 2 between two curves X and C is given by a holomorphic map $C \to X^{(2)}$ of C into the second symmetric product of X. Let σ denote the involution defining the double cover h. It induces a correspondene of degree 2, namely,

$$C \to X^{(2)} \qquad t \mapsto x + \sigma(x), \text{ if } h^{-1}(t) = \{x, \sigma(x)\}.$$

Note that the map $C \to X^{(2)}$ is an embedding. We denote its image by

$$\gamma_h := \{x + \sigma(x) \in X^{(2)}\}$$

and identify $C = \gamma_h$ under this map. Sometimes it is also called an algebraic system, since it is a generalization of a linear system $\mathbb{P}^1 = g_2^1 \to X^{(2)}$.

Then Y' is defined by the cartesian square

$$
\begin{array}{ccc}
Y' = \ell^{(2)-1}(\gamma_h) & \hookrightarrow & Y^{(2)} \\
\downarrow f' & & \downarrow \ell^{(2)} \\
C = \gamma_h & \hookrightarrow & X^{(2)}
\end{array}
$$

The fibre of f' over a point $p \in C$ consists of the four sections of ℓ over p:

$$s : h^{-1}(p) \to \ell^{-1}h^{-1}(p) \quad \text{with} \quad \ell \circ s = id.$$

Locally over a point p, we describe the elements of Y' as follows:

For $p \in C$ let $h^{-1}(p) = \{x, y = \sigma(x)\}$ and, if ι denotes the involution of Y defining ℓ, $f^{-1}(p)$ is of the form

$$f^{-1}(p) = \ell^{-1}(\{x, y\}) = \{x_1, x_2 = \iota(x_1), y_1, y_2 = \iota(y_1)\}. \tag{5.28}$$

So the elements of the fibre of $f' : Y' \to C$ over $p \in C$ can be written as

$$f'^{-1}(p) = \{x_1 + y_1, x_1 + y_2, x_2 + y_1, x_2 + y_2\}.$$

In particular the induced map $f' : Y' \to C$ is of degree 4 and étale over an open dense set of C. Moreover, ι induces an involution on Y', denoted by the same letter:

$$\iota : Y' \to Y'. \qquad x_i + y_j \mapsto \iota(x_i) + \iota(y_j).$$

Let

$$X' := Y'/\langle \iota \rangle$$

the corresponding quotient curve. Note that for $\deg h = 2$, the equivalence relation on the fibres of f' defined in [11] coincides with the equivalence relation defined by the involution ι.

Then the map $f' : Y' \to C$ factorizes as a product of two double covers

$$Y' \xrightarrow{\ell'} X' \xrightarrow{h'} C.$$

Given the local behavior of $f : Y \to C$ over a point $p \in C$, it is straightforward to compute the local behavior of $f' : Y' \to C$ over the same point. Using the notation of (5.28), we get the following possibilities:

(1) h and ℓ unramified $\Rightarrow h'$ and ℓ' are also unramified.
(2) h ramified, i.e., $x = y$, ℓ unramified $\Rightarrow h'$ unramified and ℓ' ramified at $x_1 + x_2$ and unramified at the other two points.
(3) h unramified, ℓ ramified at exactly one point, say $x_1 = x_2 \Rightarrow h'$ ramified, and ℓ' unramified.
(4) h unramified, ℓ ramified at both points, i.e., $x_1 = x_2$ and $y_1 = y_2 \Rightarrow h'$ ramified and ℓ' ramified, i.e., $x_1 + y_1$ is the only point of Y' over t.
(5) h and ℓ ramified, i.e., $f^{-1}(t)$ consists of one point $\Rightarrow h'$ and ℓ' ramified, i.e., $f'^{-1}(t)$ consists of one point.

We call the corresponding fibre of f a *fibre of type* (j) for $j = 1, \ldots, 5$. Moreover, we call a corresponding branch point $p \in C$ a *branch point of type* (j). Note that we call a double cover ramified over a point, if the fibre over this point consists only of one point (which might be a singular point).

The following lemma is obvious.

Lemma 5.4.1

(a) *A fibre of $f : Y \to C$ is of type (1) $\Leftrightarrow$ the corresponding fibre of $f' : Y' \to C$ is of type (1).*

(b) *A fibre of $f : Y \to C$ is of type (2) $\Leftrightarrow$ the corresponding fibre of $f' : Y' \to C$ is of type (3).*

(c) *A fibre of f is of type (4) or (5) $\Rightarrow$ the corresponding fibre of $f' : Y' \to C$ is of type (5).*

Here are pictures of the four local ramification types (2), ..., (5) of $Y \xrightarrow{\ell} X \xrightarrow{h} C$:

Case (2) (3) (4) (5)

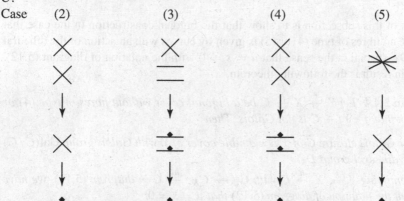

Lemma 5.4.1 immediately implies the following two corollaries.

Corollary 5.4.2 *The involution $\iota : Y' \to Y'$ admits a fixed point in a fibre over C if and only if the corresponding fibre of $f : Y \to C$ is of type (2), (4), or (5).*

For any cover of curves $\alpha : D \to C$, let $B(\alpha) \subset C$ denote the branch locus.

Corollary 5.4.3 *Suppose f does not admit fibres of type (4). Then the bigonal construction exchanges the branch loci in the following sense:*

$$B(h') = h_* B(\ell) \qquad B(h) = h'_* B(\ell').$$

Given any bigonal cover $Y \to X \to C$, we associated above another bigonal cover $Y' \to X' \to C$. In general Y' is not always smooth and irreducible. However, in any case we call the following map the *bigonal construction*:

$$Y \xrightarrow{\ell} X \xrightarrow{h} C \ \mapsto \ Y' \xrightarrow{\ell'} X' \xrightarrow{h'} C.$$

Note that, according to Lemma 5.4.1, if f admits a fibre of type (4), then the bigonal construction cannot be symmetric.

The bigonal cover $Y' \xrightarrow{\ell'} X' \xrightarrow{h'} C$ is sometimes called the *bigonal construct* of $Y \xrightarrow{\ell} X \xrightarrow{h} C$.

We always assume that the branch locus of ℓ is disjoint from the ramification locus of h. In other words, there are no fibres of type (5).

5.4.2 Determination of the Bigonal Construction in the Non-Galois Case

The aim of this subsection is to show that the bigonal construction in the case that there are no fibres of type (4) or (5) is given by curves with an action of the dihedral group D_4, special in the sense that $u = s = 0$ with the notation of diagram (5.12). The main result is the following theorem.

Theorem 5.4.4 *Let $Y \xrightarrow{\ell} X \xrightarrow{h} C$ be a bigonal cover without fibres of type (4) or (5), such that $f : Y \to C$ is not Galois. Then*

(i) *The Galois closure $\widetilde{C}$ of f is a double cover of Y with Galois group $\mathrm{Gal}(\widetilde{C}/C)$ the dihedral group D_4.*

(ii) *Identifying $Y \xrightarrow{\ell} X \xrightarrow{h} C$ with $C_\tau \xrightarrow{\ell_\tau} C_{K_\tau} \xrightarrow{h_\tau} C$ in diagram (5.11), we have with the notation of diagram (5.12) that $u = s = 0$.*

(iii) *The bigonal cover $Y' \xrightarrow{\ell'} X' \xrightarrow{h'} C$ given by the bigonal construction coincides with the bigonal cover $C_{\sigma\tau} \xrightarrow{\ell_{\sigma\tau}} C_{K_{\sigma\tau}} \xrightarrow{h_{\sigma\tau}} C$ of diagram (5.11) up to an automorphism.*

Proof

(i): Since $f : Y \to C$ is of degree 4, the Galois group $\mathrm{Gal}(\widetilde{C}/C)$ is a subgroup of the symmetric group S_4 of degree 4. Now S_4 itself and the alternating group A_4 of degree 4 do not admit a tower of subgroups $H_2 \subset H_1 \subset G$ with $G = S_4$ or A_4 such that $H_1 \subset G$ and $H_2 \subset H_1$ are both of index 2 (see diagrams (5.36) and (5.44) below). Hence $G = D_4$, $G =$ the Klein group of order 4 or G cyclic of order 4. But G cannot be one of the last two cases, since f is non-Galois.

(ii): We may identify $Y \xrightarrow{\ell} X \xrightarrow{h} C$ with $C_\tau \xrightarrow{\ell_\tau} C_{K_\tau} \xrightarrow{h_\tau} C$ in diagram (5.11). Then according to diagram (5.12), $s \neq 0$ means that the bigonal cover $Y \xrightarrow{\ell} X \xrightarrow{h} C$ admits fibres of type (5) and $u \neq 0$ means that it admits fibres of type (4).

(iii): With the notation of diagram (5.11), the composition $\mathrm{Nm}\, k_{\sigma\tau} \circ k_\tau^*$ is in terms of the curves is given by a correspondence of degree 2. Since the double cover $\widetilde{C} \to Y$ is Galois over C, the double cover of Y, given by the correspondence defining the bigonal construction, coincides with $\widetilde{C}$ up to an automorphism. Hence up to an automorphism, the correspondence $\mathrm{Nm}\, k_{\sigma\tau} \circ k_\tau^*$ coincides with the correspondence giving the bigonal construction. Hence the bigonal cover

$C_{\sigma\tau} \overset{\ell_{\sigma\tau}}{\to} C_{K_{\sigma\tau}} \overset{h_{\sigma\tau}}{\to} C$ is given by the bigonal construction and thus coincides with $Y' \to X' \to C$ up to an automorphism. □

Identifying $Y \overset{\ell}{\to} X \overset{h}{\to} C$ with $C_\tau \overset{\ell_\tau}{\to} C_{K_\tau} \overset{h_\tau}{\to} C$ as in Theorem 5.4.4 (ii), the diagram (5.12) simplifies to the following diagram.

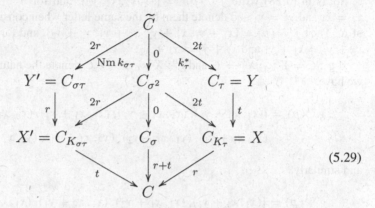

$$(5.29)$$

Note that if $C = \mathbb{P}^1$ as in the following section, we have $rt > 0$. In general this need not be the case.

Let us give a second, more direct proof of parts (i) and (iii) of Theorem 5.4.4 in the case $C = \mathbb{P}^1$, which we think clarifies the situation.

Proof According to Propositions 5.4.7 and 5.4.8 below, Y and X are smooth and irreducible. According to Corollary 5.4.3, the branch loci of h' and h are disjoint. This implies that the fibre product

$$D := X \times_C X'$$

is a smooth and irreducible curve. Moreover, it is easy to see, using the definition of X' (but we do not need this fact) that the Klein group K_4 acts on D with X and X' as quotients of degree 2.

Now define

$$\widetilde{C} := Y \times_X D.$$

Looking at the ramification picture over a branch point q of h', we see that $X \times_C X'$ is ramified over both points of $h^{-1}(q)$ whereas ℓ is ramified over exactly one of them. This implies that the ramification cancels when taking the fibre product. Hence $\widetilde{C}$ is a smooth and irreducible double cover of Y. Now we have

$$\widetilde{C} = Y \times_X D = Y \times_X X \times_{\mathbb{P}^1} X' = Y \times_C X'.$$

Using the symmetry of the situation, we can also define a smooth and irreducible $\widetilde{C}'$ by the fibre product $\widetilde{C}' = D \times_{X'} Y'$. Then we have

$$\widetilde{C}' = D \times_{X'} Y' = X \times_{\mathbb{P}^1} X' \times_{X'} Y' = X \times_C Y'.$$

We claim that there is a canonical isomorphism $\widetilde{C} \simeq \widetilde{C}'$.

We show that there is a canonical isomorphism between the fibres $f^{-1}(p)$ and $f'^{-1}(p)$ for every $p \in C$ which extends to a holomorphic map.

So as in (5.28) write $f^{-1}(p) = \{x_1, x_2, y_1, y_2\}$, and for $h^{-1}(p)$ we identify $x_2 = x_1$ and $y_2 = y_1$ and denote them by the same letter when considered as points of X. Then $f'^{-1}(p) = \{x_1 + y_1, x_1 + y_2, x_2 + y_1, x_2 + y_2\}$, and for X' we identify $x_2 + y_2 = x_1 + y_1$ and $x_2 + y_1 = x_1 + y_2$.

If $g : \widetilde{C} = Y \times_C X' \to C$ and $g' : X \times_C Y' \to C$ denote the natural projections, we have

$$g^{-1}(p) = \{(x_1, x_1 + y_1), (x_1, x_1 + y_2), (x_2, x_1 + y_1), (x_2, x_1 + y_2),$$

$$(y_1, x_1 + y_1), (y_1, x_1 + y_2), (y_2, x_1 + y_1), (y_2, x_1 + y_2)\}$$

and similarly

$$g'^{-1}(p) = \{(x_1, x_1 + y_1), (x_1, x_1 + y_2), (x_1, x_2 + y_1), (x_1, x_2 + y_2),$$

$$(y_1, x_1 + y_1), (y_1, x_1 + y_2), (y_1, x_2 + y_1), (y_1, x_2 + y_2)\}$$

If one uses the above identifications for X and X', one immediatly checks that both sets coincide. For example, the third element of $g^{-1}(p)$ equals the fourth element of $g'^{-1}(p)$: $(x_2, x_1 + y_1) = (x_1, x_2 + y_2)$. Varying $p \in C$, this gives clearly a holomorphic map. So $\widetilde{C}'$ and $\widetilde{C}$ are canonically isomorphic. We identify them and write $\widetilde{C}$ for both of them.

Since $\widetilde{C}$ is the Galois closure of $f : Y \to C$, the Galois group $\mathrm{Gal}(\widetilde{C}/C)$ is necessarily D_4, since it has to be a subgroup of S_4 and is the only subgroup admitting a non-Galois bigonal cover $Y \to X \to C$. Clearly we can choose the generators σ and τ of D_4 in such a way that $Y \to X \to C$ coincides with $C_\tau \to C_{K_\tau} \to C$ and $Y' \to X' \to C$ coincides with $C_{\sigma\tau} \to C_{K_{\sigma\tau}} \to C$. $\qquad\square$

Corollary 5.4.5 *Let* $Y \xrightarrow{\ell} X \xrightarrow{h} C$ *be a bigonal cover without fibres of type* (4) *and* (5), *such that* $f : Y \to C$ *is not Galois and let* $Y' \xrightarrow{\ell'} X' \xrightarrow{h'} C$ *be the bigonal construct. Then*

(i) *The curves* Y' *and* X' *are smooth and irreducible.*

(ii) *The bigonal construction is symmetric; that is,* $Y \to X \to C$ *is the bigonal construct of* $Y' \to X' \to C$.

(iii) *If in addition* $C = \mathbb{P}^1$, *then the map* $\varphi := \mathrm{Nm}\, k_{\sigma\tau} \circ k_\tau^* : P(\ell) \to P(\ell')$ *is an isogeny of degree*

$$\deg \varphi = 2^{r-2}.$$

Note that r is even and thus ≥ 2, being the ramification degree of a double cover of $\mathbb{P}^1$.

Proof (i) and (ii) follow directly from Theorem 5.4.4. (iii) is a special case of Theorem 5.3.12 using the identification of Theorem 5.4.4 using the fact that $rt > 0$ in the case $C = \mathbb{P}^1$. To be more precise, with the notation of Sect. 5.3.3, we have, by Propositions 5.3.8 and 5.3.11,

$$\deg \varphi = |\Gamma| + |\Delta|$$
$$= 2^{2g-2+r} = 2^{r-2},$$

since $\deg(k_{\sigma\tau}|_{P(\ell_{\sigma\tau}^*)}) = 1$ and $s = u = 0, r \geq 2$. $\qquad\square$

Note that Theorem 5.3.12 also gives the degree of φ for an arbitrary base curve C. We omit it, since it is more complicated to state and we do not need it.

Remark 5.4.6 We mentioned in Sect. 5.3.3 that Theorem 5.3.12 may be considered as a generalization of the bigonal construction. It is in the sense of part (iii) of the Corollary 5.4.5 that this was meant. In fact, Theorem 5.3.12 is valid also for $s > 0$ or $u > 0$.

5.4.3 The Bigonal Construction over $C = \mathbb{P}^1$

In the case $C = \mathbb{P}^1$, we can say a bit more, due to the following two propositions. So let $Y \xrightarrow{\ell} X \xrightarrow{h} \mathbb{P}^1$ be a bigonal cover with bigonal construct $Y' \xrightarrow{\ell'} X' \xrightarrow{h'} \mathbb{P}^1$. We mentioned already that Y' is not always smooth and irreducible. The next two propositions give the cases when it is not. Again we always assume that the branch locus of ℓ is disjoint from the ramification locus of h. In other words, there are no fibres of type (5).

Proposition 5.4.7 *Under these assumptions we have*

(1) *if ℓ is étale, then Y' is not connected*
(2) *if ℓ is ramified, then Y' is connected*

Proof It is well known (proved in [36], see also [3]) that Y' consists of two (or four) connected components if ℓ is étale. The reason is that the monodromy respects the equivalence relation in the sense of Donagi [11], which is given by the involution $\iota : Y' \to Y'$ here. In the bigonal case, the monodromy group of f' consists of even permutations. The existence of a ramification point of ℓ adds also odd permutations to the monodromy group. This implies that Y' is connected. $\qquad\square$

The proposition generalizes to more general bigonal covers (see [11]), but this is slightly more complicated. We omit this, because we are mainly interested in a proof of Pantazis' Theorem in the next section, which assumes $C = \mathbb{P}^1$.

Proposition 5.4.8 *If the bigonal cover* $Y \to X \to \mathbb{P}^1$ *contains fibers of type* (4), *then* Y' *contains a singularity over each branch point of this type.*

The proof is very analogous to the proof of [23, Proposition 2.4], where an analogous statement is proved in the case of a double cover of a trigonal cover.
Proof Let $t \in \mathbb{P}^1$ be a branch point of $f = h\ell : Y \to \mathbb{P}^1$ of type (4), and let

$$h^{-1}(t) = \{x, y\}.$$

So $x + y$ is the corresponding point of $X^{(2)}$, and

$$\tilde{x} + \tilde{y} = x_i + y_j$$

is the only point of $Y' \subset Y^{(2)}$ over t for all $i, j = 1, 2$.
The Zariski tangent spaces yield the cartesian diagram

$$
\begin{array}{ccc}
T_{Y'}(\tilde{x} + \tilde{y}) & \hookrightarrow & T_{Y^{(2)}}(\tilde{x} + \tilde{y}) \\
{\scriptstyle df'}\downarrow & & \downarrow{\scriptstyle d\ell^{(2)}} \\
T_{\mathbb{P}_1}(x + y) & \underset{dj}{\hookrightarrow} & T_{X^{(2)}}(x + y)
\end{array}
\tag{5.30}
$$

where $j : \mathbb{P}_1 = g_2^1 \hookrightarrow X^{(2)}$ denotes the inclusion map. The curve Y' will be smooth at $\tilde{x} + \tilde{y}$ if and only if $\dim T_{Y'}(\tilde{x} + \tilde{y}) = 1$. It is easy to see [36, page 108] that this is the case if and only if

$$T_{X^{(2)}}(x + y) = dj T_{\mathbb{P}_1}(x + y) + d\ell^{(2)} T_{Y^{(2)}}(\tilde{x} + \tilde{y}). \tag{5.31}$$

According to deformation theory, the lower-right triangle of diagram (5.30) is given by

$$
\begin{array}{c}
H^0(\mathcal{O}_{\tilde{x}+\tilde{y}}(\tilde{x} + \tilde{y})) \\
\downarrow{\scriptstyle \beta} \\
H^0(\mathcal{O}_X(x + y))/H^0(\mathcal{O}_X) \overset{\alpha}{\hookrightarrow} H^0(\mathcal{O}_{x+y}(x + y))
\end{array}
\tag{5.32}
$$

where α is the canonical map. In order to define the map β precisely, consider $\mathcal{O}_X(x + y)$ (respectively, $\mathcal{O}_Y(\tilde{x} + \tilde{y})$) as a subsheaf of the function field $\mathcal{R}_X$ of X (respectively, $\mathcal{R}_Y$ of Y) by putting for each point $x \in X$ and effective divisor D on X and similarly for each point $\tilde{x} \in Y$ and effective divisor $\tilde{D}$ on Y,

$$\mathcal{O}_X(D)_x := \mathbf{m}_{X,x}^{-\nu_x(D)} \quad \text{and} \quad \mathcal{O}_Y(\tilde{D})_{\tilde{x}} := \mathbf{m}_{Y,\tilde{x}}^{-\nu_{\tilde{x}}(\tilde{D})}$$

where $\mathbf{m}_{X,x}$, respectively, $\mathbf{m}_{Y,\tilde{x}}$, is the maximal ideal in $\mathcal{O}_{X,x}$, respectively, $\mathcal{O}_{Y,\tilde{x}}$. Translating diagram (5.32) to $\mathcal{O}_X$ and $\mathcal{O}_Y$ gives the diagram

$$\mathbf{m}_{Y,\tilde{x}}^{-1}/\mathcal{O}_{Y,\tilde{x}} \oplus \mathbf{m}_{Y,\tilde{y}}^{-1}/\mathcal{O}_{Y,\tilde{y}}$$

$$\downarrow \beta$$

$$H^0(\mathcal{O}_X(x+y))/H^0(\mathcal{O}_X) \overset{\alpha}{\hookrightarrow} \mathbf{m}_{X,x}^{-1}/\mathcal{O}_{X,x} \oplus \mathbf{m}_{X,y}^{-1}/\mathcal{O}_{X,y}$$

where β is given by the transposition of the natural map

$$\beta^* : \omega_{X,x}/\mathbf{m}_{X,x}^1 \oplus \omega_{X,y}/\mathbf{m}_{X,y}^1 \to \omega_{Y,\tilde{x}}/\mathbf{m}_{Y,\tilde{x}}^1 \oplus \omega_{Y,\tilde{y}}/\mathbf{m}_{Y,\tilde{y}}^1$$

induced by the cover ℓ.

Let t_x and t_y be local parameters around x and y, respectively. Then $t_{\tilde{x}} = t_x^2$ and $t_{\tilde{y}} = t_y^2$ since ℓ has ramification index 2 at $\tilde{x}$ and $\tilde{y}$. Then, since dt_x and dt_y are generators of the one-dimensional spaces $\omega_{Y,\tilde{x}}/\mathbf{m}_{Y,\tilde{x}}^1$ and $\omega_{Y,\tilde{y}}/\mathbf{m}_{Y,\tilde{y}}^1$, we have

$$\beta^*(dt_x) = dt_{\tilde{x}} = d(t_x^2) = 2t_x dt_x \quad \text{and} \quad \beta^*(dt_y) = dt_{\tilde{y}} = d(t_y^2) = 2t_y dt_y.$$

Hence β^* vanishes at dt_x and dt_y which implies that the map β is the zero map. Since α is a map of a one-dimensional space into a two-dimensional space, Eq. (5.31) cannot be satisfied. This completes the proof of the proposition. $\qquad\square$

Now let $Y \overset{\ell}{\to} X \overset{h}{\to} \mathbb{P}^1$ be a bigonal cover such that Y is Galois over $\mathbb{P}^1$. Certainly the bigonal construct $Y' \to X' \to \mathbb{P}^1$ exists. However in this case we have

Proposition 5.4.9 *Let $Y \overset{\ell}{\to} X \overset{h}{\to} \mathbb{P}^1$ be a bigonal cover with $Y/\mathbb{P}^1$ Galois. Then the bigonal cover $Y' \to X' \to \mathbb{P}^1$ is not smooth and irreducible.*

Proof By assumption and Proposition 5.4.8, we may assume that $Y \overset{\ell}{\to} X \overset{h}{\to} C$ does not admit branch points of type (4) or (5). Hence, since $Y/\mathbb{P}^1$ is Galois, any branch point of ℓ would give a branch point of type (2). So ℓ has to be étale. But then according to Proposition 5.4.7, Y' is not irreducible. $\qquad\square$

Combining Theorem 5.4.4 and Proposition 5.4.9, we obtain the following theorem.

Theorem 5.4.10 *Let $Y \overset{\ell}{\to} X \overset{h}{\to} \mathbb{P}^1$ be a bigonal cover such that the cover $Y' \to X' \to \mathbb{P}^1$ given by the bigonal construction is smooth and irreducible. Then*

(i) *The Galois cover $\tilde{C}$ of $Y \to \mathbb{P}^1$ has Galois group $\mathrm{Aut}(\tilde{C}/\mathbb{P}^1)$ the dihedral group D_4 of order 8, and we may choose the generators σ and τ of D_4 in such a way that with the notation of diagram (5.11), $Y \to X \to C$ coincides with $C_\tau \to C_{K_\tau} \to \mathbb{P}^1$ and $Y' \to X' \to \mathbb{P}^1$ coincides with $C_{\sigma\tau} \to C_{K_{\sigma\tau}} \to \mathbb{P}^1$.*

(ii) *The diagram (5.12) simplifies to diagram (5.29) with $rt > 0$.*

Proof (i) is a combination of Theorem 5.4.4 and Proposition 5.4.9. For (ii) it remains only to show that $rt > 0$, but this follows from diagram (5.29) with $C = \mathbb{P}^1$, since any irreducible cover of $\mathbb{P}^1$ is ramified. $\qquad\square$

5.4.4 Pantazis' Theorem

Consider the special case $C = \mathbb{P}^1$, and let $Y \overset{\ell}{\to} X \overset{h}{\to} \mathbb{P}^1$ be a non-Galois bigonal cover of $\mathbb{P}^1$ with bigonal cover $Y' \overset{\ell'}{\to} X' \overset{h'}{\to} \mathbb{P}^1$, given by the bigonal construction. Pantazis showed in [28] the following theorem:

Theorem 5.4.11 *Let* $Y \overset{\ell}{\to} X \overset{h}{\to} \mathbb{P}^1$ *be a general bigonal cover with bigonal construct* $Y' \overset{\ell'}{\to} X' \overset{h'}{\to} \mathbb{P}^1$. *Then*

(i) *The curves* Y *and* Y' *admit a joint double cover* $\widetilde{C}$ *which is Galois over* $\mathbb{P}^1$ *with Galois group* D_4.
(ii) *The Prym varieties* $P(\ell)$ *and* $P(\ell')$ *are dual abelian varieties.*

The proof uses the tetragonal construction for the allowable double cover (in the sense of Beauville) applied to the tetragonal cover given by doubling the bigonal construction. In this subsection we give a direct proof of the theorem with precise assumptions applying the results of above.

In order to state our version of Pantazis' Theorem, we use the following assumptions and notations. Let $Y \overset{\ell}{\to} X \overset{h}{\to} \mathbb{P}^1$ be a bigonal cover without fibres of type (4) and (5). According to Theorem 5.4.4, the bigonal construct $Y' \overset{\ell'}{\to} X' \overset{'h}{\to} \mathbb{P}^1$ is smooth and irreducible. According to Theorem 5.4.10 the Galois cover $\widetilde{C}$ of the cover $Y \to \mathbb{P}^1$ has Galois group D_4, and we may choose the generators σ and τ of it in such a way that with the notation of diagram (5.11) we may identify $Y \to X \to \mathbb{P}^1$ with $C_\tau \overset{\ell_\tau}{\to} C_{K_\tau} \overset{h_\tau}{\to} \mathbb{P}^1$ and $Y' \to X' \to \mathbb{P}^1$ with $C_{\sigma\tau} \overset{\ell_{\sigma\tau}}{\to} C_{K_{\sigma\tau}} \overset{h_{\sigma\tau}}{\to} \mathbb{P}^1$. Moreover with the notation of diagram (5.12), we have $rt > 0$ and $u = s = 0$.

With these notations Pantazis' Theorem is:

Theorem 5.4.12 *Suppose* $\widetilde{C}$ *is a curve with an action of the dihedral group* D_4 *such that* $\widetilde{C}/D_4 = \mathbb{P}^1$ *and* $s = u = 0$ *and* $rt > 0$. *Then with the notation of diagram* (5.11) *we have*

(i) *The Prym varieties* $P(\ell_\tau)$ *and* $P(\ell_{\sigma\tau})$ *are dual abelian varieties.*
(ii) *Let* $\lambda_{P(\ell_\tau)}$ *respectively* $\lambda_{P(\ell_{\sigma\tau})}$ *be the polarization induced by the canonical principal polarization of* JC_τ *respectively* $JC_{\sigma\tau}$. *Then* $\lambda_{P(\ell_\tau)}$ *and* $\lambda_{P(\ell_{\sigma\tau})}$ *are dual polarizations.*

Proof Note first that since r and t are positive, the maps $k_\tau^* : JC_\tau \to \tilde{J}, \ell_\tau^* : JC_{K_\tau} \to JC_\tau$ and $k_{\sigma\tau}^* : JC_{\sigma\tau} \to \tilde{J}$ are injective. According to Corollary 5.4.5, the isogeny

$$\varphi = \text{Nm} \, k_{\sigma\tau} \circ k_\tau^* |_{P(\ell_\tau)} : P(\ell_\tau) \to P(\ell_{\sigma\tau})$$

has degree $2^{r-2} = 2^{2g(C_{K_\tau})}$ with kernel $\ell_\tau^* P(h_\tau)[2] = \ell_\tau^* JC_{K_\tau}[2]$. Moreover, defining analogously the map

$$\tilde{\varphi} : k_{\sigma\tau}^* \circ \text{Nm} \, k_\tau |_{P(\ell_{\sigma\tau})} : P(\ell_{\sigma\tau}) \to P(\ell_\tau),$$

we saw in the proof of Proposition 5.3.7 that the composed map $\tilde{\varphi} \circ \varphi : P(\ell_\tau) \to P(\ell_\tau)$ is multiplication by 2. So $\text{Ker}(\tilde{\varphi} \circ \varphi) = P(\ell_\tau)[2]$.

On the other hand, according to Proposition 3.2.9, the polarization

$$\lambda_{P(\ell_\tau)} : P(\ell_\tau) \to \widehat{P(\ell_\tau)},$$

induced by the principal polarization of JC_τ has kernel

$$\text{Ker} \, \lambda_{P(\ell_\tau)} = \ell_\tau^* JC_{K_\tau} \cap P(\ell_\tau) = \ell_\tau^* JC_{K_\tau}[2].$$

This implies

$$\text{Ker} \, \lambda_{P(\ell_\tau)} = \text{Ker} \, \varphi. \tag{5.33}$$

This gives (i).

(ii): Analogously to Eq. (5.33), we get

$$\text{Ker} \, \lambda_{P(\ell_{\sigma\tau})} = \text{Ker} \, \tilde{\varphi} = \ell_{\sigma\tau}^* P(h_{\sigma\tau})[2] = \ell_{\sigma\tau}^* JC_{K_{\sigma\tau}}[2]. \tag{5.34}$$

Equations (5.33) and (5.34) mean that there are isomorphisms $\gamma : \widehat{P(\ell_\tau)} \to P(\ell_{\sigma\tau})$ and $\delta : \widehat{P(\ell_{\sigma\tau})} \to P(\ell_\tau)$ such that the following diagram commutes

$$
\begin{array}{ccc}
P(\ell_\tau) & \xrightarrow{\varphi} P(\ell_{\sigma\tau}) & \xrightarrow{\tilde{\varphi}} P(\ell_\tau) \, . \\
{\scriptstyle \lambda_{P(\ell_\tau)}} \downarrow & \nearrow {\scriptstyle \gamma} \quad \downarrow {\scriptstyle \lambda_{P(\ell_{\sigma\tau})}} & \nearrow {\scriptstyle \delta} \\
\widehat{P(\ell_\tau)} & \widehat{P(\ell_{\sigma\tau})} &
\end{array}
$$

If we denote by $\lambda : \widehat{P(\ell_\tau)} \to P(\ell_\tau)$ the polarization induced by $\lambda_{P(\ell_{\sigma\tau})}$ via γ, we can complete the above diagram to get the following commutative diagram

$$
\begin{array}{ccc}
P(\ell_\tau) \xrightarrow{\ \varphi\ } P(\ell_{\sigma\tau}) \xrightarrow{\ \widetilde{\varphi}\ } P(\ell_\tau) \\
\end{array}
$$

$$
\begin{array}{ccc}
\lambda_{P(\ell_\tau)} \downarrow & \nearrow^{\gamma} & \downarrow \lambda_{P(\ell_{\sigma\tau})} \nearrow_{\delta} \\
\widehat{P(\ell_\tau)} & \widehat{P(\ell_{\sigma\tau})} \\
\lambda \downarrow & \nearrow^{\gamma'} \\
P(\ell_\tau)
\end{array}
$$

with an isomorphism γ'. But this gives

$$
\mathrm{Ker}(\lambda \circ \lambda_{P(\ell_\tau)}) = \lambda_{P(\ell_\tau)}^{-1}(\mathrm{Ker}\,\lambda) = \varphi^{-1}(\mathrm{Ker}\,\lambda_{P(\ell_{\sigma\tau})}) \tag{5.35}
$$

$$
= \varphi^{-1}(\mathrm{Ker}\,\widetilde{\varphi}) = \mathrm{Ker}(\widetilde{\varphi} \circ \varphi) = P(\ell_\tau)[2].
$$

Now for $t = 2$ we know by Theorem 3.2.6 that $\lambda_{P(\ell_\tau)}$ is twice a principal polarization, which immediately implies the assertion. So suppose $t \geq 4$. Then $\dim P(\ell_\tau) > g(C_\tau)$, and hence the polarization $\lambda_{P(\ell_\tau)}$ is of type $(1, \ldots, 1, 2, \ldots, 2)$ with $g(C_\tau)$ numbers 2. Then according to Proposition 2.2.2, Eq. (5.35) implies that λ coincides with the dual polarization $\lambda_{\widehat{P(\ell_\tau)}}$ on $\widehat{P(\ell_\tau)}$ and hence that, identifying $\widehat{P(\ell_\tau)} = P(\ell_{\sigma\tau})$, we have

$$
\lambda_{\widehat{P(\ell_\tau)}} = \lambda_{P(\ell_{\sigma\tau})}
$$

which was to be shown.   □

5.5   The Alternating Group of Degree 4

As we mentioned at the beginning of Sect. 5.3, the Galois closure of a non-Galois cover $f' : C' \to C$ of degree 4 can be one of the groups D_4, $\mathcal{A}_4$, or $\mathcal{S}_4$. In this section we study the decomposition of the corresponding Jacobian in the $\mathcal{A}_4$ case. Moreover, we compute the degrees of some isogenies between Prym varieties of some intermediate covers, among them a generalization of the trigonal construction.

5.5.1   Ramification and Genera

Let $f : \widetilde{C} \to C$ be a Galois cover with Galois group the alternating group $\mathcal{A}_4$ of order 12,

$$
\mathcal{A}_4 = \langle (12)(34), (123) \rangle = \langle \tau, \sigma \rangle
$$

where τ can also denote any element of order 2 and σ any element of order 3 of $\mathcal{A}_4$. Let K denote the Klein subgroup of order 4 of $\mathcal{A}_4$. Then we have the following diagram of curves

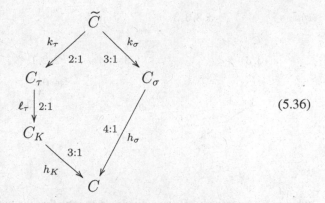

(5.36)

Note that up to conjugacy these are all the subcovers, since all subgroups of order two, as well as all subgroups of order three, are conjugate among themselves. h_K is cyclic, since K is normal in $\mathcal{A}_4$, whereas h_σ is non-Galois, since $\mathcal{A}_4$ does not admit a subgroup of order 6.

Let

$$s := |\operatorname{Fix}\sigma| \qquad \text{and} \qquad 2t := |\operatorname{Fix}\tau|.$$

Note that the number $|\operatorname{Fix}\tau|$ is even, since k_τ is a double cover.

If $x \in \operatorname{Fix}\sigma$ and $p = f(x)$, then each point in the fibre $f^{-1}(p)$ has stabilizer conjugate to the subgroup generated by σ, and conversely, any subgroup conjugate to the subgroup generated by σ will have a fixed point in the fibre $f^{-1}(p)$. Also, since the order of the subgroup generated by σ is three, there will be $12/3 = 4$ points in this fiber, hence one fixed point in the fiber for each subgroup conjugate to the subgroup generated by σ (there are exactly four subgroups satisfying this condition). Moreover, this implies that the map k_σ has exactly one ramified point and is unramified at the three other points in the fibre. The other possibility would be that k_σ is ramified at all four points. But then all four maps conjugate to k_σ would be ramified at these points, so the stabilizer at these points would be of degree > 3, which is impossible. So the ramification diagram for the point p is the first diagram below. We call it and also the branch point p *of type s*.

Similarly, if $y \in \operatorname{Fix}\tau$ and $q = f(y)$, then each point in the fibre $f^{-1}(q)$ has stabilizer conjugate to the subgroup generated by τ, and conversely, any subgroup conjugate to the subgroup generated by τ will have a fixed point in the fibre $f^{-1}(q)$. Also, since the order of the subgroup generated by τ is two, there will be $12/2 = 6$ points in this fiber, hence two fixed points in the fiber for each subgroup conjugate to the subgroup generated by τ (there are exactly three subgroups satisfying this condition). So the ramification diagram for the point q is the second diagram below. We call it and also the branch point q *of type t*.

Clearly the branch points of type s are disjoint from the branch points of type t.

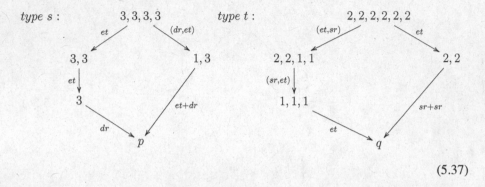

$$(5.37)$$

where et, respectively sr, respectively dr, means that the corresponding map is étale, respectively simply ramified, respectively doubly ramified at the corresponding point. The corresponding points are denoted by 1, respectively 2, respectively 3.

Lemma 5.5.1 *Let $\widetilde{C}$ be a curve with an action of the alternating group $\mathcal{A}_4$ with s and t as above. Then the ramification degrees of the covers of diagram* (5.36) *are as follows:*

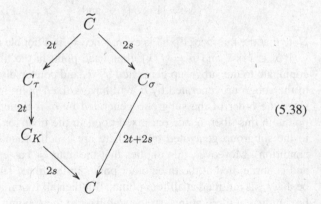

$$(5.38)$$

Proof The lemma follows by combining the above two diagrams.   □

Since the branch points of type s are disjoint from the branch points of type t and there are s branch points of type s and t branch points of type t, we have for the genera of the curves in diagram (5.36) using the Hurwitz formula:

Corollary 5.5.2 *Let the notation be as above. If g denotes the genus of C, the signature of the Galois cover is $(g; 3, \ldots, 3, 2, \ldots, 2)$ with s numbers 3 and t numbers 2. We have for the genera of the curves in diagram* (5.36)

$$g(\widetilde{C}) = 12g - 11 + 4s + 3t, \quad g(C_\sigma) = 4g - 3 + s + t,$$

$$g(C_\tau) = 6g - 5 + 2s + t, \quad g(C_K) = 3g - 2 + s.$$

The conditions on the data in this case are the following: If $C = \mathbb{P}^1$, then $s \geq 2$ and $s + t \geq 3$, since C_K and C_σ must be connected covers.

5.5.2 Decompositions of $\widetilde{J}$

Let the notations be as in diagrams (5.37) and (5.38). According to [34] the irreducible complex representations of $\mathcal{A}_4$ are the trivial representation χ_0, the characters χ_1 and χ_2 given by

$$\chi_i(\tau) = 1 \quad \text{and} \quad \chi_i(\sigma) = \xi_3^i$$

for $i = 1$ and 2 and the natural three-dimensional representation V. The direct sum $\chi_1 \oplus \chi_2$ and V are rational say

$$\chi_1 \oplus \chi_2 = W_1 \otimes_{\mathbb{Q}} \mathbb{C} \quad \text{and} \quad V = W_2 \otimes_{\mathbb{Q}} \mathbb{C}.$$

Hence the irreducible rational representations of $\mathcal{A}_4$ are χ_0, W_1 of degree 2, and W_2 of degree 3. If A_{χ_0}, A_{W_1}, and A_{W_2} denote the corresponding isotypical components of the Jacobian $\widetilde{J}$, then the isotypical decomposition is given by the addition map

$$\mu : A_{\chi_0} \times A_{W_1} \times A_{W_2} \to \widetilde{J}.$$

The following proposition works out the isotypical components and the group algebra components up to isogeny. For this denote by ϵ the composition $\epsilon = \ell_\tau k_\tau : \widetilde{C} \to C_K$.

Proposition 5.5.3

$$A_{\chi_0} = f^* J, \quad A_{W_1} = \epsilon^* P(h_K) \quad \text{and} \quad A_{W_2} = P(\epsilon) \sim k_\tau^* P(\ell_\tau)^3.$$

Proof For any subcover $C_U = \widetilde{C}/U$ of $\widetilde{C}$, let ρ_U denote the representation induced by the trivial representation of the subgroup U. Then we have

$$\rho_K = \chi_0 \oplus W_1, \quad \rho_{\langle \tau \rangle} = \rho_K \oplus W_2, \quad \rho_{\langle \sigma \rangle} = \chi_0 \oplus W_2 \quad \text{and} \quad \rho_{\mathcal{A}_4} = \chi_0. \quad (5.39)$$

Clearly we have $A_{\chi_0} = f^* J$. Since χ_1 is a character, we know by Theorem 2.9.2 and Eq. (2.30) that $A_{W_1} = B_{W_1}$, and from Corollary 3.5.10 applied to $\rho_K - \rho_{\mathcal{A}_4} = W_1$, we conclude that $B_{W_1} \sim P(h_K)$. Hence $A_{W_1} = \epsilon^* P(h_K)$.

It follows immediately from the natural decomposition

$$\widetilde{J} \sim J \times P(h_K) \times P(\epsilon)$$

that $A_{W_2} = P(\epsilon)$.

Since dim $V = 3$ with Schur index 1, we know by Theorem 2.9.2 and Eq. (2.30) that $A_{W_2} \sim B_{W_2}^3$ with an abelian subvariety B_{W_2} of A_{W_2}. From Corollary 3.5.10 applied to $\rho_{\langle\tau\rangle} - \rho_K = W_2$, we obtain that there is an isogeny $B_{W_2} \sim P(\ell_\tau)$. This gives the last assertion $A_{W_2} \sim k_\tau^* P(\ell_\tau)^3$.   □

Corollary 5.5.4 *There is an isogeny*

$$P(\ell_\tau) \sim P(h_\sigma).$$

Proof From Eq. (5.39) we conclude also that $\rho_{\langle\sigma\rangle} - \rho_{\mathcal{A}_4} = W_2$. As above we get an isogeny $B_{W_2} \sim P(h_\sigma)$. Combined with the isogeny $B_{W_2} \sim P(\ell_\tau)$, we obtain the isogeny of the assertion.   □

The following theorem gives an explicit version of a group algebra decomposition and works out the degree of the corresponding isogeny.

Theorem 5.5.5 *The map*

$$\varphi : J \times P(h_K) \times 3P(\ell_\tau) \to \widetilde{J}, \quad (c, x, y_0, y_1, y_2) \mapsto f^*(c) + \epsilon^*(x) + \sum_{i=0}^{2} \sigma^i k_\tau^*(y_i)$$

is an isogeny of degree

$$\deg(\varphi) = \begin{cases} 2^{24g-22} \cdot 3^{2g-1} & \text{if } s = t = 0; \\ 2^{24g-22+8s} \cdot 3^{2g} & \text{if } t = 0, s > 0; \\ 2^{24g-19+3t} \cdot 3^{2g-1} & \text{if } s = 0, t > 0; \\ 2^{24g-19+8s+3t} \cdot 3^{2g} & \text{if } st > 0. \end{cases}$$

In particular the isogeny $\varphi : J \times P(h_\tau) \times 3P(\ell_\tau) \to \widetilde{J}$ is a group algebra decomposition of $\widetilde{J}$ with respect to the action of $\mathcal{A}_4$.

Proof Recall that $\tau = (12)(34)$ and $\sigma = (123)$. Hence $\tau_1 := \sigma\tau\sigma^{-1}$ and $\tau_2 := \sigma^2\tau\sigma^{-2}$ are the other two involutions of $\mathcal{A}_4$. Considering $\widetilde{C}$ with the action of the subgroup K of $\mathcal{A}_4$, we have the following diagram where $C_{\tau_i} = \widetilde{C}/\langle\tau_i\rangle$

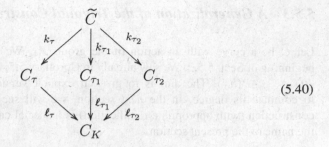

$$(5.40)$$

According to Proposition 5.2.2, Corollary 5.2.7, and Corollary 5.5.2, the map

$$\varphi_K : JC_K \times P(\ell_\tau) \times P(\ell_{\tau_1}) \times P(\ell_{\tau_2}) \to \widetilde{J},$$

$$(x, y_0, y_1, y_2) \mapsto \epsilon^*(x) + k_\tau^*(y_0) + k_{\tau_1}^*(y_1) + k_{\tau_2}^*(y_2)$$

is an isogeny of degree

$$\deg \varphi_K = \begin{cases} 2^{24g-22+8s} & \text{if } t = 0; \\ 2^{24g-19+8s+3t} & \text{if } t > 0. \end{cases} \qquad (5.41)$$

Moreover, considering the isomorphism

$$\nu : 3P(\ell_\tau) \xrightarrow{\simeq} P(\ell_\tau) \times P(\ell_{\tau_1}) \times P(\ell_{\tau_2}), \quad (y_0, y_1, y_2) \mapsto (y_0, \overline{\sigma}(y_1), \overline{\sigma}^2(y_2)),$$

where $\overline{\sigma}$, respectively $\overline{\sigma}^2$, is the isomorphism $JC_\tau \to JC_{\tau_1}$, respectively $JC_\tau \to JC_{\tau_2}$, induced by σ, and using the isogeny

$$\phi_K : J \times P(h_K) \to JC_K, \quad (c, x) \mapsto h_K^*(c) + x$$

we can write

$$\varphi(c, x, y_0, y_1, y_2) = \varphi_K(\phi_K(c, x), \nu(y_0, y_1, y_2)).$$

It follows that φ is an isogeny of degree

$$\deg \varphi = \deg \varphi_K \cdot \deg \phi_K.$$

According to Corollary 3.2.11, we have

$$\deg \phi_K = \begin{cases} 3^{2g-1} & \text{if } s = 0; \\ 3^{2g} & \text{if } s > 0; \end{cases}$$

Combining this with (5.41) gives the assertion on $\deg \varphi$. $\qquad \square$

5.5.3   A Generalization of the Trigonal Construction

Let $\widetilde{C}$ be a curve with an action of the group $\mathcal{A}_4$. We use the notation of the beginning of Sect. 5.5.2. We saw already in Corollary 5.5.4 that there is an isogeny $P(\ell_\tau) \sim P(h_\sigma)$. The aim is to give an explicit version of this isogeny and to compute its degree. In the next section, we will see that the usual trigonal construction (with appropriate ramification) is a special case of it, which explains the name of the present section.

Theorem 5.5.6 *Let the notations be as in diagrams* (5.36) *and* (5.38). *The natural map*

$$\psi := \operatorname{Nm} k_\sigma \circ k_\tau^* |_{P(\ell_\tau)} : P(\ell_\tau) \to P(h_\sigma)$$

is an isogeny of degree

$$\deg \psi = \begin{cases} 2^{4g-6+2s} & \text{if } t = 0 \text{ and } P(h_K)[2] \not\subset \operatorname{Ker}(\ell_\tau^*)^\perp; \\ 2^{4g-5+2s+t} & \text{otherwise.} \end{cases}$$

and kernel contained in $P(\ell_\tau)[2]$.

We prove the theorem in a series of lemmas.

Lemma 5.5.7 *The map* ψ *is an isogeny with kernel contained in the 2-division points* $P(\ell_\tau)[2]$.

Proof First we show

$$\operatorname{Im} \psi \subset P(h_\sigma).$$

Let $x \in P(\ell_\tau)$. Then

$$\operatorname{Nm} h_\sigma(\psi(x)) = \operatorname{Nm} h_K(\operatorname{Nm} \ell_\tau(\operatorname{Nm} k_\tau(k_\tau^*(x))))$$
$$= 2\operatorname{Nm} h_K(\operatorname{Nm} \ell_\tau(x)) = 2\operatorname{Nm} h_K(0) = 0,$$

which implies the assertion.

For the proof of the lemma, denote

$$A := k_\tau^*(P(\ell_\tau)) \quad \text{and} \quad B := k_\sigma^*(P(h_\sigma)).$$

By Proposition 3.5.5 we know

$$A = \{z \in \widetilde{J} \mid \tau(z) = z, \ \tau_1^{-1}(z) = -z\}^0$$

where τ_1 denotes any element of order 2 in K different from τ and

$$B = \{w \in \tilde{J} \mid \sigma(w) = w, \sum_{\kappa \in K} \kappa(w) = 0\}^0.$$

Note that, since we assumed $\tau = (12)(34)$, we have $K = \{1, \tau, \sigma\tau\sigma^{-1}, \sigma^{-1}\tau\sigma\}$ with $\sigma\tau\sigma^{-1} = (13)(24)$ and $\sigma^{-1}\tau\sigma = (14)(23)$.

Now, since $k_\sigma^* \circ \mathrm{Nm}\,k_\sigma : A \to B$ equals $1 + \sigma + \sigma^2$ and similarly for τ, we have the following commutative diagram:

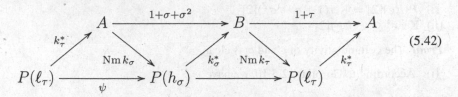

$$(5.42)$$

Since $k_\tau^* : P(\ell_\tau) \to A$ is an isogeny, it suffices to show

$$(1 + \tau) \circ (1 + \sigma + \sigma^2) = 2_A. \qquad (5.43)$$

So let $z \in A$, i.e., $\tau(z) = z$ and $\tau_1(z) = -z$. Then

$$(1 + \tau) \circ (1 + \sigma + \sigma^2)(z) = z + \sigma(z) + \sigma^2(z) + \tau(z) + \tau\sigma(z) + \tau\sigma^2(z)$$
$$= 2z + \sigma(z) + \sigma^2(z) + \tau\sigma(z) + \tau\sigma^2(z).$$

So we have to show

$$\sigma(z) + \sigma^2(z) + \tau\sigma(z) + \tau\sigma^2(z) = 0.$$

But $\tau\sigma = \sigma\tau_1$ and $\tau\sigma^2 = \sigma^2\tau_2$, where τ, τ_1, and τ_2 are the three elements of order 2 in G, which implies

$$\sigma(z) + \tau\sigma(z) + \sigma^2(z) + \tau\sigma^2(z) = \sigma(z) - \sigma(z) + \sigma^2(z) - \sigma^2(z) = 0.$$

This completes the proof of the lemma.   □

From diagram (5.42) we can conclude the following lemma.

Lemma 5.5.8

$$\mathrm{Ker}\,\psi = \{z \in P(\ell_\tau)[2] \mid (1 + \sigma + \sigma^2)(k_\tau^*(z)) = 0\}.$$

Proof It follows from $(1 + \tau) \circ (1 + \sigma + \sigma^2) = 2_A$ that $\mathrm{Ker}(1 + \sigma + \sigma^2)|_A \subseteq A[2]$. Since $\deg k_\sigma = 3$, it follows that any $z \in P(h_\sigma) \cap \mathrm{Ker}\,k_\sigma^*$ satisfies $2z = 0$ and $3z = 0$ and hence $z = 0$.

Hence $\mathrm{Ker}\,k_\sigma^*|_{P(h_\sigma)} = 0$ and the commutativity of diagram (5.42) gives the assertion.   □

Since K is a normal subgroup of $\mathcal{A}_4$, the automorphism σ descends to an automorphism on C_K which we denote by the same letter. Moreover, we denote by ϵ the composed cover

$$\epsilon := k_\tau \circ \ell_\tau : \widetilde{C} \to C_K.$$

Lemma 5.5.9 ϵ *commutes with* σ *and we have*

(i) $P(h_K)[2] = \mathrm{Ker}(1 + \sigma + \sigma^2)[2]$
(ii) $\mathrm{Ker}\,\epsilon^* \subseteq P(h_K)[2]$

Proof The commutativity of ϵ and σ is clear.

(i): According to Corollary 2.5.10, we have

$$\mathrm{Ker}(1 + \sigma + \sigma^2) = P(h_K) + (h_K^* J)[3].$$

Therefore, if $y \in (\mathrm{Ker}(1 + \sigma + \sigma^2)[2]$, then $y = x + z$ with $x \in P(h_K)$ and $z \in (h_K^* J)[3]$. But then $0 = 2y = 2x + 2z = 2x - z$. Hence $z \in P(h_K)$. This proves (i).

(ii): Note that the ramification degrees of all maps in diagram (5.40) are $2t$. This implies

$$\mathrm{Ker}\,\epsilon^* = \begin{cases} 0 & \text{if } t \neq 0; \\ \{0, \eta_{\ell_\tau}, \eta_{\ell_{\tau_1}}, \eta_{\ell_{\tau_2}} = \eta_{\ell_\tau} + \eta_{\ell_{\tau_1}}\} & \text{if } t = 0. \end{cases}$$

where $\eta_\ell, \eta_{\ell_{\tau_1}}$, and $\eta_{\ell_{\tau_2}}$ are the 2-division points defining the corresponding covers. Moreover, in the second case, σ permutes the three non-trivial 2-division points. So (i) implies assertion (ii).   □

Lemma 5.5.10 *If* $t = 0$, *then*

$$\deg \psi = \begin{cases} 2^{4g+2s-6} \text{ if } g \geq 0, t = 0 \text{ and } P(h_K)[2] \nsubseteq \eta_{\ell_\tau}^\perp = \mathrm{Ker}(\ell_\tau)^\perp; \\ 2^{4g+2s-5} \text{ if } g > 0, t = 0 \text{ and } P(h_K)[2] \subseteq \eta_{\ell_\tau}^\perp = \mathrm{Ker}(\ell_\tau)^\perp. \end{cases}$$

Proof We keep the notation $\mathrm{Ker}\,\epsilon^* = \{0, \eta_{\ell_\tau}, \eta_{\ell_{\tau_1}}, \eta_{\ell_{\tau_2}}\} \subseteq P(h_K)[2]$ of the proof of the preceding lemma. We know from Proposition 3.3.1 that $P(\ell_\tau)[2] = \ell_\tau^*(\eta_{\ell_\tau}^\perp)$.

Suppose first $g = 0$. Then $P(h_K) = JC_K$, and therefore $(1 + \sigma + \sigma^2)(x) = 0$ for all $x \in JC_K[2]$ according to Lemma 5.5.9. This implies that for any $z \in P(\ell_\tau)[2]$ we have

$$\psi(z) = (1 + \sigma + \sigma^2)(k_\tau^*(z)) = 0,$$

since $P(\ell_\tau)[2] = \ell_\tau^*(\eta_{k_\tau}^\perp)$ and $\epsilon^*\sigma = \sigma\epsilon^*$. Hence in this case we have $\operatorname{Ker}\psi = P(\ell_\tau)[2]$ which implies the assertion by Corollary 5.5.2. Note that if $g = 0$, then the condition $P(h_K)[2] \not\subseteq \eta_{\ell_\tau}^\perp$ is automatically satisfied.

Now suppose $g > 0$ and let $z \in \operatorname{Ker}\psi$, i.e., $z \in P(\ell_\tau)[2]$ with $(1 + \sigma + \sigma^2)(k_\tau^*(z)) = 0$ by Lemma 5.5.8. Choose any $y \in \eta_{\ell_\tau}^\perp$ with $\ell_\tau^*(y) = z$. Then

$$y + \sigma y + \sigma^2 y \in \ker\epsilon^* \subseteq P(h_K)[2].$$

But we have also $y + \sigma y + \sigma^2 y \in JC_K^\sigma \cap \eta_{\ell_\tau}^\perp \subseteq P(\ell_\tau)[3]$ by Corollary 2.5.10(i). This implies $y + \sigma y + \sigma^2 y = 0$; that is, $y \in P(h_K)[2] \cap \eta_{\ell_\tau}^\perp$. Since $\eta_{\ell_\tau}^\perp$ is a subgroup of index 2 in $JC_K[2]$, this gives by Corollary 5.5.2

$$|P(h_K)[2] \cap \eta_{\ell_\tau}^\perp| = \begin{cases} 2^{2\dim P(h_K)-1} = 2^{4g-5+2s} & \text{if } P(h_K)[2] \not\subseteq \eta_{\ell_\tau}^\perp; \\ 2^{2\dim P(h_K)} = 2^{4g-4+2s} & \text{if } P(h_K)[2] \subseteq \eta_{\ell_\tau}^\perp. \end{cases}$$

Since $\operatorname{Ker}\ell_\tau^* = \langle\eta_\tau\rangle \subset P(\ell_\tau)$, this gives the assertion. $\qquad\square$

Lemma 5.5.11 *If $t > 0$, then*

$$\deg\psi = 2^{4g-5+2s+t}.$$

Proof In this case all pullback maps of Jacobians induced by the covers of diagram (5.40) are injective.

We first describe $P(\ell_\tau)[2]$. Let $\{p_1, \ldots, p_t, \tau p_1, \ldots, \tau p_t\}$ in $\tilde{C}$ be the ramification points of the cover $k_{\tau_1} : \tilde{C} \to C_{k_{\tau_1}}$ of diagram (5.40), where τ is the involution giving the cover $k_\tau : \tilde{C} \to C_\tau$. Then the ramification points of $\ell_\tau : C_\tau \to C_K$ are $\{q_1 = k_\tau(p_1), q_2 = k_\tau(\sigma(q_1)), \ldots, q_{2t-1} = k_\tau(p_t), q_{2t} = k_\tau(\sigma(p_t))\}$.

We now apply Corollary 3.3.5 to the cover $\ell_\tau : C_\tau \to C_K$. For this note first that

$$\mathcal{O}_{C_K}(\epsilon p_i - \epsilon\sigma p_i) \in \operatorname{Ker}(1 + \sigma + \sigma^2) \subseteq JC_K.$$

Hence for $i = 1, \ldots t$ we can choose

$$n_i \in \operatorname{Ker}(1 + \sigma + \sigma^2) \quad \text{such that} \quad n_i^2 = \mathcal{O}_{C_K}(\epsilon p_i - \epsilon\sigma p_i).$$

Defining $\mathcal{G}_i = \mathcal{O}_{C_\tau}(q_{2i} - q_{2i-1}) \otimes \ell_\tau^*(n_i) \in P(\ell_\tau)[2]$, we conclude that each $\mathcal{G}_i$ satisfies $(1+\sigma+\sigma^2)k_\tau^*\mathcal{G}_i = 0$, and since $\mathcal{G}_1 \otimes \cdots \otimes \mathcal{G}_t \in \ell_\tau^* JC_K[2] \cap \operatorname{Ker}(1+\sigma+\sigma^2)$, we obtain

$$\mathcal{G}_1 \otimes \cdots \otimes \mathcal{G}_t \in \ell_\tau^* P(h_K)[2].$$

If we define $\mathcal{H}_1, \ldots, \mathcal{H}_{t-1}$ as in Corollary 3.3.5, we obtain

$$P(\ell_\tau')[2] = \ell_\tau^* J C_K[2] \oplus_{i=1}^{t-1} \mathcal{G}_i \mathbb{Z}/2\mathbb{Z} \oplus_{i=1}^{t-1} \mathcal{H}_i \mathbb{Z}/2\mathbb{Z}.$$

Now observe that ℓ_τ^* gives an isomorphism

$$P(h_K)[2] = J C_K[2] \cap \mathrm{Ker}(1 + \sigma + \sigma^2) \xrightarrow{\ell_\tau^*} \{z \in \ell_\tau^* J C_K[2] \mid (1 + \sigma + \sigma^2)(k_\tau^* z) = 0\}$$
$$= \ell_\tau^* J C_K[2] \cap \mathrm{Ker}\,\psi.$$

Since no non-trivial combination of the $\mathcal{H}_i$ is in $\mathrm{Ker}\,\psi$, we can conclude if $t > 0$, then

$$\mathrm{Ker}\,\psi = \ell_\tau^* P(h_K)[2] \oplus_{i=1}^{t-1} \mathcal{G}_i \mathbb{Z}/2\mathbb{Z}.$$

So Corollary 5.5.2 implies the assertion.   □

Combining Lemmas 5.5.7, 5.5.10, and 5.5.11 completes the proof of Theorem 5.5.6.   □

5.6 The Trigonal Construction for Covers with Group $\mathcal{A}_4$

The trigonal construction of Recillas [30] says that the Prym variety of an étale double cover of a trigonal curve is canonically isomorphic to the Jacobian of a tetragonal curve. An analogous theorem for double covers of a trigonal which are ramified in exactly 2 points in the $\mathcal{A}_4$ and the $\mathcal{S}_4$-case was given in [23]. Another proof of this fact was given by O. Debarre (private communication). In this section we will give a different proof of both these results in the special case assuming that the trigonal cover is cyclic or to be more precise that the corresponding Galois group is $\mathcal{A}_4$.

So let $\widetilde{C}$ be a curve with an action of $\mathcal{A}_4$ with the notation of the previous section and the additional hypotheses $C = \widetilde{C}/\mathcal{A}_4 = \mathbb{P}^1$ and $t = 0$ or 1. Note that in order to get connected curves, this implies $s + t \geq 3$. We could also start equivalently as follows:

Lemma 5.6.1 *Given a tetragonal cover $h : Y \to \mathbb{P}^1$ with s fibres of ramification type $(3, 1)$ and $t = 0$ or 1 fibres of ramification type $(2, 2)$ with $s + t \geq 3$ and no other ramification types.*

Then the Galois closure $\widetilde{C}$ of $Y/\mathbb{P}^1$ has Galois group $\mathcal{A}_4 = \langle \sigma, \tau \rangle$, and we may assume the diagram (5.36) with the identifications $C_\sigma = Y$ and $h_\sigma = h$.

Proof Since all ramifications come from even permutations , the monodromy group and thus the Galois group are contained in $\mathcal{A}_4$. It cannot be D_4, since $s > 0$. This implies the assertion.   □

Note that the triple cover $h_K : C_K \to \mathbb{P}^1$ is then cyclic, and if $t = 1$, the two ramification points of $\ell_\tau : C_\tau \to C_K$ are in the same fibre of h_K. In the notation of [23], this means that ℓ_τ is a special ramified double cover.

Theorem 5.6.2 *Let $\widetilde{C}$ be a curve with an action of $\mathcal{A}_4$ such that $C = \widetilde{C}/\mathcal{A}_4 = \mathbb{P}^1$, and suppose that $t = 0$ or 1. Then the isogeny $\psi : P(\ell_\tau) \to JC_\sigma$ of Theorem 5.5.6 induces an isomorphism of principally polarized abelian varieties $\widehat{P(\ell_\tau)} \to JC_\sigma$.*

Proof The double cover ℓ_τ is étale if $t = 0$ and ramified in two points if $t = 1$. Hence according to Theorem 3.2.6, the Prym variety $P(\ell_\tau)$ is principally polarized. On the other hand, according to Theorem 5.5.6, the isogeny $\psi : P(\ell_\tau) \to JC_\sigma$ is of degree 2^{2s-6} for $t = 0$ and 2^{2s-4} for $t = 1$ with kernel consisting of 2-division points. Hence by Corollary 5.5.2 it consists of all 2-division points. Since in both cases the polarization λ_P induced by the principal polarization of JC_τ is twice a principal polarization, this implies that ψ factorizes as follows with $\overline{\psi}$ an isomorphism,

$$
\begin{array}{ccc}
P(\ell_\tau) & \xrightarrow{\ \psi\ } & JC_\sigma \\
{\scriptstyle \lambda_P}\downarrow & {\scriptstyle \simeq}\nearrow & \\
\widehat{P(\ell_\tau)} & \overline{\psi} &
\end{array}
$$

It remains to show that $\overline{\psi}$ respects the principal polarization. If we denote by λ_1 the polarization of $\widehat{P(\ell_\tau)}$ induced by the canonical polarization λ_{JC_σ} via $\overline{\psi}$, we may complete the diagram to the following one:

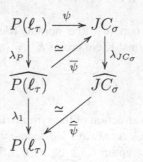

It follows from the commutativity of this diagram that λ_1 is principal and that $\mathrm{Ker}(\lambda_1 \circ \lambda_P) = P(\ell_\tau)[2]$. Hence λ_1 is the canonical principal polarization as claimed. $\qquad\square$

5.7   The Symmetric Group $\mathcal{S}_4$

The last case of the Galois closure of a non-Galois cover $f' : C' \to C$ of degree 4 to be considered is the symmetric group $\mathcal{S}_4$. In this section we study the decomposition of the corresponding Jacobian in this case. Moreover, we compute the degrees of some isogenies between Prym varieties of some covers arising from actions of sub- or quotient groups of $\mathcal{S}_4$.

5.7.1   Ramification and Genera

Let $\mathcal{S}_4$ denote the symmetric group of degree 4; that is, the group of permutations of the elements 1,2,3,4. It consists of five classes of conjugacy:

- The unit element 1
- Six transpositions: (ij)
- Three elements of order 2: $(ij)(kl)$
- Eight elements of order 3: (jkl)
- Six elements of order 4: $(ijkl)$

The group $\mathcal{S}_4$ is generated by the following three elements:

$$\sigma := (1234), \quad \tau := (12)(34) \quad \text{and} \quad \kappa := (123).$$

Up to conjugacy $\mathcal{S}_4$ admits the following non-trivial subgroups:

- Cyclic of order 2: $\langle \tau \rangle$ and $\langle \sigma\tau = (13) \rangle$
- Cyclic of order 3: $\langle \kappa \rangle$
- Of order 4: the cyclic group $\langle \sigma \rangle$, the normal Klein group $K := \langle \tau, \sigma^2 = (13)(24) \rangle$, and the non-normal Klein group $K' := \langle \sigma\tau, \sigma^2 \rangle$
- Of order 6: $\mathcal{S}_3 := \langle \kappa, \sigma\tau \rangle$
- Of order 8: $D_4 := \langle \sigma, \tau \rangle$
- Of order 12: $\mathcal{A}_4 = \langle \kappa, \tau \rangle$

Now let $\widetilde{C}$ be a curve with an action of the symmetric group $\mathcal{S}_4$ with corresponding quotient map $f : \widetilde{C} \to C$. As always we denote the quotient of $\widetilde{C}$ by a cyclic group $\langle \alpha \rangle$ by C_α and by a subgroup $U \subset \mathcal{S}_4$ by C_U. Then we have the following diagram of subquotients of $\widetilde{C}$:

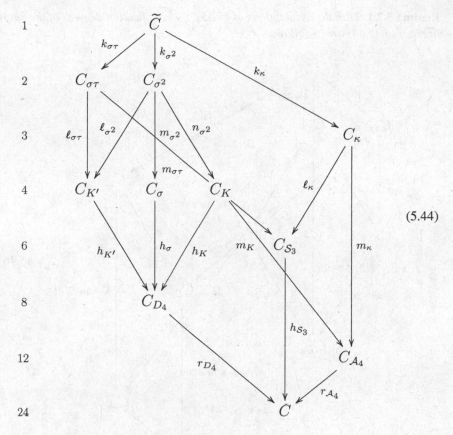

$$(5.44)$$

Here the back-side pentagon with the vertices $\widetilde{C}, C_{\sigma^2}, C_K, C_{\mathcal{A}_4}$, and C_κ coincides with the diagram (5.36), due to the fact that the corresponding subgroups of $\mathcal{A}_4$ are non-conjugate in $\mathcal{A}_4$ if and only if they are non-conjugate in $\mathcal{S}_4$.

The square with the vertices $\widetilde{C}, C_{\sigma\tau}, C_{\mathcal{S}_3}$, and C_κ coincides with the diagram (4.1), due to the action of the subgroup $\mathcal{S}_3$.

Similarly, the subdiagram consisting of all covers of $C_{\mathcal{D}_4}$ coincides with the inner subdiagram of (5.11) with the difference that the outer vertex C_τ of the inner subdiagram and the corresponding maps in diagram (5.11) are missing, since here we are considering only subcovers which are non-conjugate in $\mathcal{S}_4$. In fact, the elements τ and σ^2 are non-conjugate in D_4, but conjugate in $\mathcal{S}_4$.

According to Hurwitz the number of fixed points of any transposition is even; according to Lemma 5.3.1 for any element σ of order 4, the number $|\operatorname{Fix}(\sigma)|$ is even, whereas the number $|\operatorname{Fix}(\sigma) \setminus \operatorname{Fix}(\sigma^2)|$ is divisible by 4. Finally, according to Lemma 4.2.5, the number of fixed points of any element κ of order 3 is even, since $\kappa \in \mathcal{S}_3$. Hence we may denote

$$2r := |\operatorname{Fix}(\sigma\tau)|, \quad 2s := |\operatorname{Fix}(\kappa)|, \quad 4u := |\operatorname{Fix}(\sigma^2) \setminus \operatorname{Fix}(\sigma)|, \quad 2v := |\operatorname{Fix}(\sigma)|.$$
$$(5.45)$$

Lemma 5.7.1 *With the assumptions of (5.45), the ramification degrees of the maps in diagram (5.44) are as follows:*

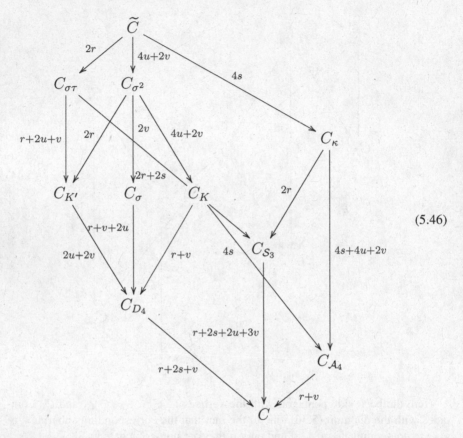

(5.46)

Proof Most of the ramification degrees follow from Lemma 4.2.4 with the action of the subgroup $\mathcal{S}_3$, Lemma 5.3.1 with the action of the subgroup D_4, and Lemma 5.5.1 with the action of the subgroup $\mathcal{A}_4$. It remains to determine the ramification degrees of the covers r_{D_4}, $h_{\mathcal{S}_3}$, and $r_{\mathcal{A}_4}$. These follow by using the square with the vertices $C\kappa$, $C_{\mathcal{S}_3}$, C, and $C_{\mathcal{A}_4}$ and then the square with the vertices C_K, C_{D_4}, C, and $C_{\mathcal{A}_4}$.
□

Since the images of the branch points of the different types in C are pairwise different, we have for the genera of the curves in diagram (5.44) using the Hurwitz formula:

Corollary 5.7.2 *Let the notation be as above. If g denotes the genus of C, the signature type of the Galois cover is $(g; 2, \ldots, 2, 3, \ldots, 3, 4, \ldots, 4)$ with $r + v$ numbers 2, s numbers 3, and v numbers 4. The genera of the curves in diagram (5.44) are*

$$g(C_{D_4}) = 3g - 2 + s + \frac{r+v}{2}, \quad g(C_{\mathcal{A}_4}) = 2g - 1 + \frac{r+v}{2}, \quad g(C_K) = 6g - 5 + 2s + \frac{3}{2}(r+v),$$

$$g(C_{\mathcal{S}_3}) = 4g - 3 + s + u + \frac{r+3v}{2}, \quad g(C_\kappa) = 8g - 7 + 2r + 2s + 2u + 3v,$$

$$g(C_K) = 6g - 5 + 2s + \frac{3}{2}(r+v), \quad g(C_{K'}) = 6g - 5 + r + 2s + u + 2v,$$

$$g(C_\sigma) = 6g - 5 + 2s + u + \frac{3}{2}(r+v), \quad g(C_{\sigma\tau}) = 12g - 11 + \frac{1}{2}(5r + 9v) + 4s + 3u$$

$$g(C_{\sigma^2}) = 12g - 11 + 3r + 4s + 2u + 4v, \quad g(\widetilde{C}) = 24g - 23 + 6r + 8s + 6u + 9v.$$

5.7.2 Decomposition of $\widetilde{J}$

The group $\mathcal{S}_4$ is the semidirect product of subgroup $\mathcal{S}_3$ by the Klein subgroup K:

$$\mathcal{S}_4 = K \rtimes \mathcal{S}_3.$$

From this fact one gets the irreducible representations. Note that, as for all symmetric groups, all the irreducible representations are defined over the rationals. We get from [34] the following lemma.

Lemma 5.7.3 *The irreducible rational representations of $\mathcal{S}_4$ are:*

- *The trivial character χ_0.*
- *The alternating character χ_1.*
- *The representation W_1 of degree 2 defined by*

$$\rho_{W_1}(k \cdot x) = \rho(x) \quad \text{for all } k \in K \text{ and } x \in \mathcal{S}_3,$$

  where ρ denotes the irreducible representation of degree 2 of $\mathcal{S}_3$.
- *The natural representation of degree 3 of $\mathcal{A}_4$ extends to an irreducible representation W_2 of degree 3 of $\mathcal{S}_4$.*
- *The last irreducible representation of $\mathcal{S}_4$ is the product $W_3 := W_2 \otimes \chi_1$.*

Now let $\widetilde{C}$ be a curve with an action of the group $\mathcal{S}_4$. If we denote as usual by A_W the isotypical component of the Jacobian $\widetilde{J}$ corresponding to the irreducible representation W, the isotypical decomposition of $\widetilde{J}$ is given by the addition map

$$\mu : A_{\chi_0} \times A_{\chi_1} \times A_{W_1} \times A_{W_2} \times A_{W_3} \longrightarrow \widetilde{J}.$$

The following proposition determines the isotypical components in terms of Prym varieties. If $h : C_1 \to C_2$ is a cover of curves, we write instead of $P(h)$ for the sake of clarity: $P(C_1/C_2)$, since diagram (5.44) is fairly complicated. For a subcurve C_U in diagram (5.44), we write $f_{C_U} : \tilde{C} \to C_U$ for the corresponding composition of maps.

Proposition 5.7.4 *The following equalities and isogenies imply the isotypical and a group algebra decomposition of $\tilde{J}$:*

$$A_{\chi_0} = f^* J, \quad A_{\chi_1} = f^*_{C_{\mathcal{A}_4}} P(C_{\mathcal{A}_4}/C), \quad A_{W_1} = f^*_K P(C_K/C_{\mathcal{A}_4}) \sim P(C_{D_4}/C)^2,$$

$$A_{W_2} \sim P(C_{\mathcal{S}_3}/C)^3 \quad and \quad A_{W_3} \sim P(C_\sigma/C_{D_4})^3.$$

Proof As always, we denote for a subgroup $U \subset \mathcal{S}_4$ by ρ_U the induced representation of the trivial representation of U. Then one computes:

- $\rho_{\mathcal{S}_4} = \chi_0.$
- $\rho_{\mathcal{A}_4} = \chi_0 \oplus \chi_1.$
- $\rho_{D_4} = \chi_0 \oplus W_1.$
- $\rho_{\mathcal{S}_3} = \chi_0 \oplus W_2.$
- $\rho_\sigma = \rho_{D_4} \oplus W_3.$

Clearly $A_{\chi_0} = f^* J$. Since $\rho_{\mathcal{A}_4} = \chi_0 \oplus \chi_1$ and $\rho_{\{1\}} = \chi_0$, Corollary 3.5.10 implies $A_{\chi_1} = f^*_{C_{\mathcal{A}_4}} P(C_{\mathcal{A}_4}/C)$.

By Theorem 2.9.2 and Eq. (2.30), there is an abelian subvariety B_{W_1} of A_{W_1} such that $A_{W_1} \sim B_{W_1}^2$. Furthermore, $\rho_K = \rho_{\mathcal{A}_4} \oplus 2W_1$ gives $A_{W_1} = f^*_K P(C_K, C_{\mathcal{A}_4})$. Since $\rho_{D_4} = \chi_0 \oplus W_1$, we conclude from Corollary 3.5.10 that $B_{W_1} \sim P(C_{D_4}/C)$. Together this implies that $A_{W_1} \sim P(C_{D_4}/C)^2$.

Also by Theorem 2.9.2 and Eq. (2.30), for $i = 2$ and 3, there are abelian subvarieties B_{W_i} of A_{W_i} such that $A_{W_i} \sim B_{W_i}^3$. Since $\rho_{\mathcal{S}_3} = \chi_0 \oplus W_2$, by Corollary 3.5.10 this gives $B_{W_2} \sim P(C_{\mathcal{S}_3}/C)$ and hence $A_{W_2} \sim P(C_{\mathcal{S}_3}/C)^3$.

Similarly we get for $i = 3$ that $A_{W_3} \sim P(C_\sigma/C_{D_4})^3$, since $\rho_\sigma = \rho_{D_4} \oplus W_3$. $\square$

Corollary 5.7.5 *There is an isogeny $P(C_{\mathcal{S}_3}/C) \sim P(C_{K'}/C_{D_4})$.*

Proof One computes also $\rho_{K'} = \rho_{D_4} \oplus W_2$. In the same way as in the previous proof, this implies $A_{W_2} \sim P(C_{K'}/C_{D_4})^3$. So together with $A_{W_2} \sim P(C_{\mathcal{S}_3}/C)^3$, this implies the assertion. $\square$

The following theorem gives an explicit version of a group algebra decomposition and computes the degree of the corresponding isogeny. Recall that for any subgroup U of C_{S_4}, we denote by $f_{C_U} : \tilde{C} \to C_U$ the corresponding quotient map.

Theorem 5.7.6 *Let the notation be as above. There is an isogeny*

$$\varphi : J \times P(C_{\mathcal{A}_4}/C) \times P(C_{D_4}/C)^2 \times P(C_{K'}/C_{D_4})^3 \times P(C_\sigma/C_{D_4})^3 \to \tilde{J}$$

given by

$$(u, v, x_1, x_2, y_1, y_2, y_3, z_1, z_2, z_3) \mapsto f^*(u) + f^*_{C_{\mathcal{A}_4}}(v) + f^*_{C_{D_4}}(x_1) + \kappa f^*_{C_{D_4}}(x_2)$$

$$+ f^*_{C_{K'}}(y_1) + \kappa f^*_{C_{K'}}(y_2) + \kappa^2 f^*_{C_{K'}}(y_3)$$

$$+ f^*_{C_\sigma}(z_1) + \kappa f^*_{C_\sigma}(z_2) + \kappa^2 f^*_{C_\sigma}(z_3)$$

The isogeny φ is of degree

$$\deg \varphi = \deg \phi \cdot \deg \psi \cdot \deg \gamma \cdot (\deg \delta)^3$$

with $\deg \phi$, $\deg \psi$, $\deg \gamma$, and $\deg \delta$ given in Eqs. (5.47), (5.48), (5.49), and (5.51).

Proof From Corollary 3.2.11 we know the map

$$\phi : JC_K \times P(\widetilde{C}/C_K) \to \widetilde{J}, \quad (u, w) \mapsto f^*_{C_K}(u) + w$$

is an isogeny of degree

$$\deg \phi = \begin{cases} 2^{24g-22+8s}, & \text{if } r = v = 0; \\ 2^{24g-20+6r+8s+6v} & \text{otherwise.} \end{cases} \tag{5.47}$$

We now decompose each of the two factors. Since K is normal in S_4 and the action of K on $JC_K = J(\widetilde{C}/K)$ is trivial, there is a natural S_3 action on JC_K with $S_3 = \langle \sigma, \kappa \rangle$. Here we denote the induced maps on C_K by the same letter. Note that σ acts on C_K as an involution. It follows from Theorem 4.2.16 (with $\alpha = s$ and $\beta = r + v$) that the map

$$\psi : J \times P(C_{\mathcal{A}_4}/C) \times P(C_{D_4}/C)^2 \to JC_K,$$

defined by

$$(u, v, x_1, x_2) \mapsto (m_K \circ r_{\mathcal{A}_4})^*(u) + m_K^*(v) + h_K^*(x_1) + \kappa h_K^*(x_2)$$

is an isogeny of degree

$$\deg \psi = \begin{cases} 2^{2g-1} 3^{6g-3+s} & \text{if } r + v = 0; \\ 2^{2g} 3^{6g-3+r+s+v} & \text{if } r + v > 0; \end{cases} \tag{5.48}$$

Now consider the factor $P(\widetilde{C}/C_K)$. First consider the map

$$\gamma := f^*_{C_{\sigma^2}} + \kappa f^*_{C_{\sigma^2}} + \kappa^2 f^*_{C_{\sigma^2}} : P(C_{\sigma^2}/C_K)^3 \to P(\widetilde{C}/C_K)$$

It is a composition of the isomorphism $P(C_{\sigma^2}/C_K)^3 \overset{\cong}{\to} P(C_{\sigma^2}/C_K) \times P(C_{\tau\sigma^2}/C_K) \times P(C_\tau/C_K)$ with the isogeny

$$\gamma' : P(C_{\sigma^2}/C_K) \times P(C_{\tau\sigma^2}/C_K) \times P(C_\tau/C_K) \to P(\widetilde{C}/C_K).$$

Hence we may apply Corollary 5.2.6 to the diagram

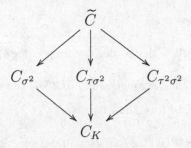

where all maps are of ramification degree $4u + 2v$. Hence γ is an isogeny of degree

$$\deg \gamma = \deg \gamma' = \begin{cases} 2^{24g-24+6r+8s}, & if\ u = v = 0; \\ 2^{24g-23+6r+8s+6u+9v} & otherwise. \end{cases} \tag{5.49}$$

In order to obtain the decomposition corresponding to irreducible representations, we further decompose $P(C_{\sigma^2}/C_K)$ as follows: Since $\langle \sigma^2 \rangle$ is normal in D_4, the following subdiagram of diagram (5.44) corresponds to the action of a Klein group on C_{σ^2}

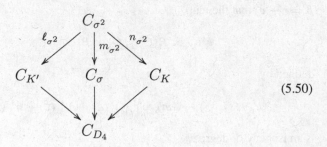

$$\tag{5.50}$$

According to Propositions 5.2.3 and 5.2.5, the map

$$\delta := \ell_{\sigma^2}^* + m_{\sigma^2}^* : P(C_{K'}/C_{D_4}) \times P(C_\sigma/C_{D_4}) \to P(C_{\sigma^2}/C_K)$$

is an isogeny of degree

$$\deg \delta = \begin{cases} 2^{6g-6+2s}, & if\ r = u = v = 0; \\ 2^{6g-3+2r+2s+v} & otherwise. \end{cases} \tag{5.51}$$

For this note that the middle case in the formula of Proposition 5.2.5 does not occur, since $2u + v = 0$ implies that $rv = 0$.

Putting everything together, note that we have

$$\varphi = \phi \circ (\beta \times (\gamma \circ \delta^3))$$

which implies

$$\deg \varphi = \deg \phi \cdot \deg \psi \cdot \deg \gamma \cdot (\deg \delta)^3.$$

which was to be shown. □

5.7.3 Isogenies Arising from Actions of Subgroups of S_4

First consider $\widetilde{C}$ with the action of the subgroup $\mathcal{A}_4$ and quotient $\widetilde{C}/\mathcal{A}_4 = C_{\mathcal{A}_4}$ and the corresponding subdiagram of (5.44). Applying Theorem 5.5.6, which is the generalization of the trigonal construction for $\mathcal{A}_4$ covers, we get with the ramification of (5.46) and Corollary 5.7.2

Proposition 5.7.7 *The map*

$$\alpha := \operatorname{Nm} k_\kappa \circ k_{\sigma^2}^* |_{P(C_{\sigma^2}/C_K)} : P(C_{\sigma^2}/C_K) \to P(C_K/C_{\mathcal{A}_4})$$

is an isogeny with kernel in the 2-division points and degree

$$\deg \alpha = \begin{cases} 2^{8g-10+2r+4s} & \text{if } u = v = 0 \text{ and } P(C_K/C_{\mathcal{A}_4})[2] \not\subset (\operatorname{Ker} n_{\sigma^2}^*)^\perp, \\ 2^{8g-9+2r+4s+2u+3v} & \text{otherwise.} \end{cases}$$

Next we consider $\widetilde{C}$ with the action of the subgroup D_4 and quotient $\widetilde{C}/D_4 = C_{D_4}$ and the corresponding subdiagram of (5.44). Applying Theorem 5.3.12 we get with the ramification of (5.46) and using Corollary 5.7.2

Proposition 5.7.8 *The map*

$$\beta := \operatorname{Nm} k_{\sigma^2} \circ k_{\sigma\tau}^* |_{P(C_{\sigma\tau}/C_{K'})} : P(C_{\sigma\tau}/C_{K'}) \to P(C_{\sigma^2}/C_K)$$

is an isogeny with kernel in the 2-division points of degree as given in Theorem 5.3.12 by replacing s by v and t by $2u + v$ and leaving r and u as they were.

Proof Just note that in the case of an S_4 action, the covers $k_\tau : \widetilde{C} \to C_\tau$ and $k_{\sigma^2} : \widetilde{C} \to C_{\sigma^2}$ are equivalent. So the cover k_τ does not occur in diagram (5.44), and we may replace k_τ by k_{σ^2}. □

5.7.4   An Isogeny Arising from the Action of a Quotient of S_4

The only non-trivial normal subgroup of S_4, apart from A_4, is the Klein group K. Its quotient is the group S_3 which acts on the curve C_K with the following diagram of subquotients

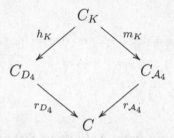

Lemmas 4.2.11 and 4.2.15 with $\alpha = s$, $\beta = r + v$, give the following isogeny, where we denote the automorphism of C_K induced by κ by the same letter.

Proposition 5.7.9 *The map*

$$\varphi : P(r_{D_4}) \times P(r_{D_4}) \to P(m_K) \quad (z_1, z_2) \mapsto h_K^*(z_1) + \kappa h_K^*(z_2)$$

is an isogeny with kernel $\{(z, -z) \in P(r_{D_4})[3] \times P(r_{D_4})[3] \mid h_K^*(z) \text{ is } S_3 - invariant\}$ *and degree*

$$\deg \varphi = \begin{cases} 3^{2g} & if \ s = 0; \\ 3^{2g+s-1} & if \ s > 0. \end{cases}$$

5.8   Another Generalization of the Trigonal Construction

5.8.1   Statement and Preparations

In Theorem 5.5.6 we had a generalization of the trigonal construction, if we consider $\tilde{C}$ as a curve with A_4 action.

There is however another generalization, which does not come from an action of A_4. For this consider the the following diagram of subquotients of (5.44) with corresponding diagram of ramification degrees.

$$
\begin{array}{ccc}
C_{\sigma\tau} & & \\
\ell_{\sigma\tau}\Big\downarrow \quad\searrow^{m_{\sigma\tau}} & & \\
C_{K'} & & C_{\mathcal{S}_3} \\
h_{K'}\Big\downarrow & & \\
C_{D_4} & \quad\searrow^{h_{\mathcal{S}_3}} & \\
\quad\searrow^{r_{D_4}} & & \Big\downarrow \\
& C &
\end{array}
\qquad\qquad
\begin{array}{ccc}
C_{\sigma\tau} & & \\
r+2u+v\Big\downarrow \quad\searrow^{2r+2s} & & \\
C_{K'} & & C_{\mathcal{S}_3} \\
2u+2v\Big\downarrow & & \\
C_{D_4} & & \Big\downarrow^{r+2s+2u+3v} \\
\quad\searrow^{r+2s+v} & & \\
& C &
\end{array}
$$

$$(5.52)$$

Although the diagram looks exactly as the corresponding diagram (5.36), there is no action of a group $\mathcal{A}_4$ on $C_{\sigma\tau}$ such that this is the corresponding diagram. In fact, even the ramification degrees are different. However, we have the following theorem, which contains the trigonal construction as a special case.

Theorem 5.8.1 *The map*

$$
\gamma := \mathrm{Nm}\, m_{\sigma\tau} \circ \ell_{\sigma\tau}^{*}|_{P(h_{K'})} : P(h_{K'}) \to P(h_{\mathcal{S}_3})
$$

is an isogeny of degree

$$
\deg \gamma = \begin{cases}
2^{2\dim P(h_{K'})} & \text{if } g = 0 \text{ and } u+v = 0 \text{ or } 1; \\
2^{4g-5+r+2s-(\delta+\epsilon)} & \text{if } g > 0 \text{ and } u+v = 0; \\
2^{4g-4+r+2s+v-\epsilon} & \text{if } g > 0 \text{ and } u+v = 1; \\
2^{4g-5+r+2s+u+2v-\epsilon} & \text{if } u+v > 1,
\end{cases}
$$

where

$$
\delta := \begin{cases} 0 \text{ if } P(r_{D_4})[2] \subset \eta_{h_{K'}}^{\perp}; \\ 1 \text{ if } P(r_{D_4})[2] \not\subset \eta_{h_{K'}}^{\perp}. \end{cases}
\quad \text{and} \quad
\epsilon = \begin{cases} 0 \text{ if } r > 0; \\ 1 \text{ if } r = 0. \end{cases}
$$

Here and in the sequel, for any étale double cover h of curves η_h denotes the line bundle defining it.

The proof of the theorem is fairly long, in fact contains the whole Sect. 5.8. First we show that γ is an isogeny and prove a series of lemmas. In Sect. 5.8.2 we prove the trigonal construction in the case $C = \mathbb{P}^1$ and $P(h_{K'})$ principally polarized, and in Sect. 5.8.3 we compute $\deg \gamma$ in the general principally polarized case and finally in Sect. 5.8.4 the remaining cases.

Lemma 5.8.2 $\mathrm{Im}\,\gamma \subset P(h_{\mathcal{S}_3})$.

Proof Let $x \in P(h_{K'})$. The commutativity of diagram (5.52) implies

$$\operatorname{Nm} h_{\mathcal{S}_3}(\gamma(x)) = \operatorname{Nm} r_{D_4}(\operatorname{Nm} h_{K'}(\operatorname{Nm} \ell_{\sigma\tau}(\ell^*_{\sigma\tau}(x))))$$

$$= \operatorname{Nm} r_{D_4}(\operatorname{Nm} h_{K'}(2x) = 2\operatorname{Nm} r_{D_4}(0) = 0.$$

This implies the assertion, since $P(h_{K'})$ is connected.   □

Lemma 5.8.3 $\gamma : P(h_{K'}) \rightarrow P(h_{\mathcal{S}_3})$ *is an isogeny with kernel in the 2-division points.*

Proof Recall that $C_{K'} = \widetilde{C}/\langle \sigma\tau, \sigma^2 \rangle = \widetilde{C}/\langle (13), (24) \rangle$ and $C_{\mathcal{S}_3} = \widetilde{C}/\langle \sigma\tau, \kappa \rangle = \widetilde{C}/\langle (13), (123) \rangle$. Moreover, $\tau = (12)(34)$ induces the involution on K' giving the double cover $h_{K'} : C_{K'} \rightarrow C_{D_4}$. We denote it by the same letter.

Now let $A := k^*_{\sigma\tau}(\ell^*_{\sigma\tau}(P(h_{K'})))$ and $B := k^*_{\sigma\tau}(m^*_{\sigma\tau}(P(h_{\mathcal{S}_3})))$. According to Proposition 3.5.6, applied to the cover $\ell_{\sigma\tau}$ and then pulled back via $k_{\sigma\tau}$, we have

$$A = \{z \in \widetilde{J}^{K'} \mid z + \tau z = 0\}^0$$

and similarly, since $K = \{1, (12)(34), (13)(24), (14)(23)\}$ is a complete set of representatives of $\mathcal{S}_4/\mathcal{S}_3$:

$$B = \{w \in \widetilde{J}^{\mathcal{S}_3} \mid \sum_{k \in K} kw = 0\}^0.$$

We claim that the following diagram commutes

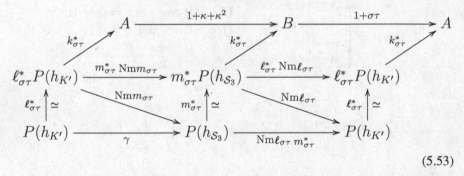

$$(5.53)$$

To see this note first that $m^*_{\sigma\tau} \operatorname{Nm} m_{\sigma\tau} = (1 + \kappa + \kappa^2)$ on $\ell^*_{\sigma\tau} P(h_{K'})$ implies the commutativity of the upper left square. Similarly we get the commutativity of the upper right square. The remaining triangles are obviously commutative.

It remains to show that the three vertical maps are isomorphisms. It follows from Lemma 5.3.9(iii) that the left- and right-hand vertical maps are injective and hence isomorphisms. The middle vertical map is an isomorphism, since the degree-3 cover $m_{\sigma\tau}$ is not cyclic by Proposition 3.2.2.

We now claim that the composition of the two maps in the top horizontal line is multiplication by 2. Suppose this is proven. Since $k^*_{\sigma\tau} \circ \ell^*_{\sigma\tau} : P(h_{K'}) \rightarrow A$ is an isogeny, this implies that also the composition of the two lower horizontal maps is multiplication by 2. Since $\dim P(h_{K'}) = \dim P(h_{S_3})$ according to Corollary 5.7.2, this implies that γ is an isogeny with kernel in the 2-division points.

It remains to show that $(1 + \tau\sigma)(1 + \kappa + \kappa^2)|_A = 2_A$.

So let $z \in A$, i.e., $\sigma\tau(z) = \tau\sigma(z) = z$ and $\tau(z) = -z$. Then

$$(1 + \tau\sigma)(1 + \kappa + \kappa^2)(z) = z + \kappa(z) + \kappa^2(z) + \tau\sigma(z) + \tau\sigma\kappa(z) + \tau\sigma\kappa^2(z).$$

But

$$\tau\sigma\kappa = (24)(123) = (132)(24)(12)(34) = \kappa^2(\tau\sigma)\tau \quad \text{and hence} \quad \tau\sigma\kappa(z) = -\kappa^2(z),$$

$$\tau\sigma\kappa^2 = (24)(132) = (123)(13)(12)(34) = \kappa(\sigma\tau)\tau \quad \text{and hence} \quad \tau\sigma\kappa^2(z) = -\kappa(z).$$

This gives

$$(1 + \tau\sigma)(1 + \kappa + \kappa^2)(z) = z + \kappa(z) + \kappa^2(z) + z - \kappa^2(z) - \kappa(z) = 2z,$$

which was to be shown. □

It follows from diagram (5.53) that the kernel of γ is contained in the following set

$$F := \{x \in P(h_{K'})[2] \mid (1 + \kappa + \kappa^2)k_{\sigma\tau}^* \ell_{\sigma\tau}^*(x) = 0\}.$$

For $r = 0$ we have moreover

Lemma 5.8.4 *If $r = 0$, there is a $y \in F$ such that $m_{\sigma\tau}^*(\gamma(y)) = \eta_{k_{\sigma\tau}}$. In particular the element $\eta_{k_{\sigma\tau}}$ defining the étale double cover $k_{\sigma\tau}$ is contained in $m_{\sigma\tau}^*(P(h_{S_3}))$.*

Proof Note first that by (5.46) $r = 0$ implies that $k_{\sigma\tau}$ and ℓ_κ are étale double covers with $m_{\sigma\tau}^*(\eta_{\ell_\kappa}) = \eta_{k_{\sigma\tau}}$. Moreover, $m_{\sigma\tau}^*$ is always injective, since $m_{\sigma\tau}$ is not cyclic.

We claim that it is enough to show that there is a $y \in P(h_{K'})[2]$ such that $\gamma(y) = \eta_{\ell_\kappa}$.

This follows from the fact that if this holds, then

$$0 = k_{\sigma\tau}^*(m_{\sigma\tau}^*(\eta_{\ell_\kappa})) = k_{\sigma\tau}^*(m_{\sigma\tau}^*(\gamma(y))) = (1 + \kappa + \kappa^2)(k_{\sigma\tau}^*(\ell_{\sigma\tau}^*(y)))$$

and hence $y \in F$.

In order to show that there is a $y \in P(h_{K'})[2]$ with $\gamma(y) = \eta_{\ell_\kappa}$, first note that

$$\mathrm{Nm}\, m_{\sigma\tau}(\eta_{k_{\sigma\tau}}) = \mathrm{Nm}\, m_{\sigma\tau}(m_{\sigma\tau}^*(\eta_{\ell_\kappa})) = 3\eta_{\ell_\kappa} = \eta_{\ell_\kappa}$$

and therefore

$$m_{\sigma\tau}^*(\mathrm{Nm}\, m_{\sigma\tau}(\eta_{k_{\sigma\tau}}) = \eta_{k_{\sigma\tau}}.$$

Also observe that since $h^*_{S_3} : J \to J C_{S_3}$ is injective (because h_{S_3} does not factor), we have

$$h^*_{S_3}(J[4]) \subset P(h_{S_3})[4],$$

since $\operatorname{Nm} h_{S_3} \circ h^*_{S_3} = 4_J$.

We distinguish two cases. In both cases we consider the following subdiagram of diagram (5.44), where $\tilde{C}$ is considered with the action of the subgroup K'.

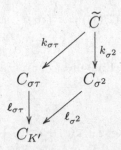

Here $k_{\sigma\tau}$ and ℓ_{σ^2} are étale double covers.

Case (1): $v > 0$.

We first claim that $\eta_{\ell_{\sigma^2}} \in P(h_{K'})$.

To see this, we use the action of the Klein group on C_{σ^2} and the corresponding subdiagram (5.50). With this, the assertion follows from Lemma 5.2.8.

Moreover, we have $\ell^*_{\sigma\tau}(\eta_{\ell_{\sigma^2}}) = \eta_{k_{\sigma\tau}}$ by the above subdiagram and the injectivity of $\ell^*_{\sigma\tau}$. But then

$$\eta_{\ell_\kappa} = \operatorname{Nm} m_{\sigma\tau}(\eta_{k_{\sigma\tau}}) = \operatorname{Nm} m_{\sigma\tau}(\ell^*_{\sigma\tau}(\eta_{\ell_{\sigma^2}})) = \gamma(\eta_{\ell_{\sigma\tau}})$$

which proves the assertion and thus the lemma in this case.

Case (2): $v = 0$.

In this case $\eta_{r_{A_4}}$ is a 2-division point in J and therefore $h^*_{S_3}(\eta_{r_{A_4}}) = \eta_{\ell_\kappa} \in P(h_{S_3})$. We also have

$$k^*_{\sigma\tau}(\ell^*_{\sigma\tau} \operatorname{Nm} \ell_{\sigma\tau}(\eta_{k_{\sigma\tau}})) = (1 + \tau\sigma)(k^*_{\sigma\tau}(\eta_{k_{\sigma\tau}})) = 0$$

where the first equality follows from diagram (5.53) and hence

$$\ell^*_{\sigma\tau} \operatorname{Nm} \ell_{\sigma\tau}(\eta_{k_{\sigma\tau}}) \in \operatorname{Ker} k^*_{\sigma\tau} = \{0, \eta_{k_{\sigma\tau}}\}.$$

We claim that $\ell^*_{\sigma\tau} \operatorname{Nm} \ell_{\sigma\tau}(\eta_{k_{\sigma\tau}}) = 0$.

Otherwise $\ell^*_{\sigma\tau} \operatorname{Nm} \ell_{\sigma\tau}(\eta_{k_{\sigma\tau}}) = \eta_{k_{\sigma\tau}}$ and it follows that

$$m^*_{\sigma\tau} \operatorname{Nm} m_{\sigma\tau}(\ell^*_{\sigma\tau} \operatorname{Nm} \ell_{\sigma\tau}(\eta_{k_{\sigma\tau}})) = m^*_{\sigma\tau} \operatorname{Nm} m_{\sigma\tau}(\eta_{k_{\sigma\tau}}) = \eta_{k_{\sigma\tau}}.$$

But we have also that $m_{\sigma\tau}^* \operatorname{Nm} m_{\sigma\tau}(\ell_{\sigma\tau}^* \operatorname{Nm} \ell_{\sigma\tau})$ is multiplication by 2 on $\ell_{\sigma\tau}^*(P(h_{K'})$ by the proof of Lemma 5.8.3, which is a contradiction and thus completes the proof of the claim.

From this we get

$$(m_{\sigma\tau}^* \operatorname{Nm} m_{\sigma\tau})^{-1}(\eta_{k_{\sigma\tau}}) \subseteq \operatorname{Ker}((\ell_{\sigma\tau}^* \operatorname{Nm} \ell_{\sigma\tau})(m_{\sigma\tau}^* \operatorname{Nm} m_{\sigma\tau})) = (\ell_{\sigma\tau}^* P(h_{K'}))[2]$$

where the last equality follows from Lemma 5.8.3. This implies that there is an element $z \in (\ell_{\sigma\tau}^* P(h_{K'}))[2]$ such that $(m_{\sigma\tau}^* \operatorname{Nm} m_{\sigma\tau})(z) = \eta_{k_{\sigma\tau}}$. Since $\ell_{\sigma\tau}^*$ is an isomorphism on $P(h_{K'})$, there exists an element $y \in P(h_{K'})[2]$ such that $(m_{\sigma\tau}^* \operatorname{Nm} m_{\sigma\tau})(\ell_{\sigma\tau}^*(y)) = \eta_{k_{\sigma\tau}}$. Then the diagram (5.53) gives

$$m_{\sigma\tau}^*(\eta_{\ell_\kappa}) = \eta_{k_{\sigma\tau}} = m_{\sigma\tau}^* \gamma(y).$$

This implies the assertion, since $m_{\sigma\tau}^*|_{P(h_{S_3})}$ is an isomorphism. □

We know from diagram (5.53) that $\operatorname{Ker}\gamma \subset F$. To be more precise, we have

Lemma 5.8.5

(i) If $r \neq 0$, then $\operatorname{Ker}\gamma = F$;
(ii) If $r = 0$, then $[F : \operatorname{Ker}\gamma] = 2$.

Proof Assume that there is a $y \in F$ with $y \notin \operatorname{Ker}\gamma$. Then $m_{\sigma\tau}^*(\gamma(y))$ is the non-zero element of $\operatorname{Ker} k_{\sigma\tau}^*$ in $m_{\sigma\tau}^*(P(h_{S_3}))$, since $m_{\sigma\tau}^*$ is always injective and by the commutativity of diagram (5.53). This implies $r = 0$.

Furthermore, it follows that $\gamma(y) \in P(h_{S_3})[2]$ is the non-zero element of $\operatorname{Ker}\ell_\kappa$. Hence the difference of two such y is an element of $\operatorname{Ker}\gamma$. Both parts of the Lemma now follow from Lemma 5.8.4. □

Lemma 5.8.5 implies that in order to compute the degree of γ, it suffices to compute the cardinality of F. For this we need the following lemma. Recall that the automorphism κ induces an automorphism on C_K which we denote by the same letter.

Lemma 5.8.6

$$h_K^{*-1}(P(m_K)[2]) = P(r_{D_4})[2] = \{x \in J_{C_{D_4}}[2] \mid (1 + \kappa + \kappa^2)k_{\sigma\tau}^* \ell_{\sigma\tau}^* h_{K'}^*(x) = 0\}.$$

Proof Since the group S_3 acts on C_K, we can apply Corollary 4.2.7 to give that $h_K^* : P(r_{D_4}) \to P(m_K)$ is injective. This implies the first equation of the lemma.

Moreover we conclude from Lemma 4.2.11 that

$$\operatorname{Ker}\left(P(r_{D_4}) \times P(r_{D_4}) \xrightarrow{h_K^* + \kappa h_K^*} P(m_K)\right)$$

$$= \{(x, -x) \mid x \in P(r_{D_4})[3], \ h_K^*(x) \ S_3\text{-invariant}\}$$

$$= \{(x, y) \mid x \in P(r_{D_4})[3], \ h_K^*(x) = \kappa h_K^*(y)\}.$$

Hence $h_K^* P(r_{D_4})[2] \cap \kappa h_K^* P(r_{D_4})[2] = 0$, and therefore by counting cardinalities using Corollary 5.7.2,

$$h_K^* P(r_{D_4})[2] + \kappa h_K^* P(r_{D_4})[2] = P(m_K)[2].$$

Noting that with $\epsilon := n_{\sigma^2} k_{\sigma^2}$ we have

$$(1 + \kappa + \kappa^2) k_{\sigma\tau}^* \ell_{\sigma\tau}^* h_{K'}^* = (1 + \kappa + \kappa^2) \epsilon^* h_K^* = \epsilon^* (1 + \kappa + \kappa^2) h_K^*.$$

Using the first equality, we obtain that

$$P(r_{D_4})[2] \subset \{x \in JC_{D_4}[2] \mid (1 + \kappa + \kappa^2) k_{\sigma\tau}^* \ell_{\sigma\tau}^* h_{K'}^*(x) = 0\}.$$

Conversely, if $x \in JC_{D_4}[2]$ with $(1 + \kappa + \kappa^2) k_{\sigma\tau}^* \ell_{\sigma\tau}^* h_{K'}^*(x) = 0$, then $(1 + \kappa + \kappa^2) h_K^*(x) \in \operatorname{Ker} \epsilon^*$. Since $\operatorname{Ker} \epsilon^* \subset \operatorname{Ker}(1 + \kappa + \kappa^2)$ by Lemma 5.5.9(ii), it follows

$$
\begin{aligned}
0 &= (1 + \kappa + \kappa^2)(1 + \kappa + \kappa^2) h_K^*(x) \\
  &= 3(1 + \kappa + \kappa^2) h_K^*(x) \\
  &= (1 + \kappa + \kappa^2) h_K^*(x) \quad \text{(since } 3x = x\text{)}.
\end{aligned}
$$

So $x \in P(r_{D_4})[2]$. Using the first equation again completes the proof of the lemma. $\qquad \square$

Corollary 5.8.7 *If the double cover $h_{K'} : C_{K'} \to C_{D_4}$ is étale, then $\eta_{h_{K'}} \in P(r_{D_4})$.*

Proof $h_{K'}$ étale means that $u = v = 0$. But then also the double cover n_{σ^2} is étale by diagram (5.46). From the commutativity of the diagram (5.44), we conclude that $\eta_{n_{\sigma^2}} = h_K^*(\eta_{h_{K'}})$. But then $h_K^*(\eta_{h_{K'}})$ belongs to $\operatorname{Ker} \epsilon^* \subset P(m_K)[2]$, and the assertion follows from Lemma 5.8.6. $\qquad \square$

Recall that we want to compute the cardinality of the set F. For this we need the following proposition.

Proposition 5.8.8 *Suppose $2u + 2v = 0$ or 2, i.e., the double cover $h_{K'}$ is étale or has two ramification points. Then*

$$F = h_{K'}^*(P(r_{D_4})[2]) \cap P(h_{K'})[2].$$

Proof First note that

$$h_{K'}^*(P(r_{D_4})[2]) \cap P(h_{K'})[2] \subseteq F.$$

To see this, let $x \in h_{K'}^*(P(r_{D_4})[2]) \cap P(h_{K'})[2]$. Then $x \in P(h_{K'})[2]$ and $x = h_{K'}^*(y)$ for some $y \in P(r_{D_4})[2]$. So by Lemma 5.8.6,

$$(1 + \kappa + \kappa^2) k_{\sigma\tau}^* \ell_{\sigma\tau}^*(x) = (1 + \kappa + \kappa^2) k_{\sigma\tau}^* \ell_{\sigma\tau}^* h_{K'}^*(y) = 0$$

that is, $x \in F$.

Conversely, let $x \in F$, i.e., $x \in P(h_{K'})[2]$ with $(1 + \kappa + \kappa^2) k_{\sigma\tau}^* \ell_{\sigma\tau}^*(x) = 0$.

If $2u + 2v = 0$, we know by Proposition 3.3.1 that $P(h_{K'}) = h_{K'}^*(\eta_{h_{K'}}^\perp)$, and if $2u + 2v = 2$, then $P(h_{K'})[2] = h_{K'}^*(JC_{K'[2]})$. In both cases there exists an element $y \in JC_{K'}[2]$ such that $h_{K'}^*(y) = x$. It follows from Lemma 5.8.6 that $y \in P(r_{D_4})$. □

Of course this does not yet prove Theorem 5.8.1. We will apply it in the next three sections in the different cases to complete the proof.

5.8.2 The Trigonal Construction

Consider the special case, $\widetilde{C}$ a curve with an action of $\mathcal{S}_4$ and quotient $C = \widetilde{C}/\mathcal{S}_4 = \mathbb{P}^1$. Otherwise let the notation be as in the previous subsection. We want to compute the degree of the isogeny

$$\gamma := \operatorname{Nm} m_{\sigma\tau} \circ \ell_{\sigma\tau}^* |_{P(h_{K'})} : P(h_{K'}) \to JC_{\mathcal{S}_3}$$

in a special case and deduce the trigonal construction.

Proposition 5.8.9 *If $C = \mathbb{P}^1$ and either $2u + 2v = 0$ or 2, then the kernel of the isogeny $\gamma : P(h_{K'}) \to JC_{\mathcal{S}_3}$ is $P(h_{K'})[2]$.*

Proof Note first that the assumption implies that $r \neq 0$, since $r_{\mathcal{A}_4}$ is ramified. It follows from Lemma 5.8.6 that

$$JC_{D_4}[2] = P(r_{D_4})[2] = \{x \in JC_{D_4}[2] \mid (1 + \kappa + \kappa^2) k_{\sigma\tau}^* \ell_{\sigma\tau}^* h_{K'}^*(x) = 0\}.$$

This implies $h_{K'}^*(JC_{D_4}[2]) \subset F$. Applying Proposition 5.8.8 we obtain

$$F = P(h_{K'}[2]).$$

Since $r \neq 0$, Lemma 5.8.5 implies $\operatorname{Ker} \gamma = P(h_{K'})[2]$ which was to be shown. □

The following theorem is the trigonal construction in the $\mathcal{S}_4$-case.

Theorem 5.8.10 *Let $\widetilde{C}$ be a curve admitting a $\mathcal{S}_4$ action with quotient $C = \mathbb{P}^1$ such that either $2u + 2v = 0$ or 2. Then the isogeny*

$$\gamma : P(h_{K'}) \to JC_{\mathcal{S}_3}$$

induces an isomorphism of principally polarized abelian varieties $\widehat{P(h_{K'})} \to JC_{\mathcal{S}_3}$.

The theorem can be expressed in the usual way as follows:

Under the assumptions of the theorem, if the double cover $h_{K'} : C_{K'} \to C_{D_4}$ of the trigonal curve C_{D_4} is étale or admits exactly two ramification points, the Prym variety $P(h_{K'})$ and the Jacobian of the tetragonal curve C_{S_3} are canonically isomorphic as principally polarized abelian varieties.

Proof By Proposition 5.8.9 the canonical principal polarization on $JC_{K'}$ induces twice the principal polarization λ_P on $P(h_{K'})$. Since $\mathrm{Ker}(2\lambda_P) = \mathrm{Ker}\,\gamma$, there is an isomorphism $\delta : \widehat{P(h_{K'})} \to JC_{S_3})$ such that the following diagram commutes

$$
\begin{array}{ccc}
P(h_{K'}) & \xrightarrow{\;\gamma\;} & JC_{S_3} \\[2pt]
{\scriptstyle 2\lambda_P}\downarrow & {\scriptstyle\simeq}\;\nearrow{\scriptstyle\delta} & \\[2pt]
\widehat{P(h_{K'})} & &
\end{array}
\quad.
$$

If we denote by λ the polarization on the dual abelian variety $\widehat{P(h_{K'})}$ induced via δ by the canonical principal polarization $\lambda_{JC_{S_3}}$, we may complete the above diagram to get the diagram

$$
\begin{array}{ccc}
P(h_{K'}) & \xrightarrow{\;\gamma\;} & JC_{S_3} \\[2pt]
{\scriptstyle 2\lambda_P}\downarrow & {\scriptstyle\simeq}\;\nearrow{\scriptstyle\delta} & \downarrow{\scriptstyle\lambda_{JC_{S_3}}} \\[2pt]
\widehat{P(h_{K'})} & & \widehat{JC_{S_3}} \\[2pt]
{\scriptstyle\lambda}\downarrow & {\scriptstyle\delta}\;\swarrow{\scriptstyle\simeq} & \\[2pt]
P(h_{K'}) & &
\end{array}
$$

The commutativity of the diagram implies that λ is a principal polarization with $\mathrm{Ker}(\lambda \circ 2\lambda_P) = P(h_{K'})[2]$. Hence λ is the dual polarization on $\widehat{P(h_{K'})}$ of the principal polarization λ_P. $\square$

5.8.3   The Degree of γ in the General Principally Polarized Case

Recall that $P(h_{K'})$ is principally polarized for any curve $\widetilde{C}$ with S_4 action, if and only if either $2u + 2v = 0$ or 2. The following proposition computes $\deg \gamma$ in the case $C \neq \mathbb{P}^1$.

Proposition 5.8.11 *Let $\widetilde{C}$ be a curve admitting an action of $\mathcal{S}_4$ with $g = g(C) > 0$ and either $2u + 2v = 0$ or 2. Then the degree of the isogeny $\gamma : P(h_{K'}) \to P(h_{\mathcal{S}_3})$ is*

$$
\deg \gamma = \begin{cases}
2^{4g-5+r+2s}, & \text{if } u = v = 0, \ r > 0 \text{ and } P(r_{D_4})[2] \subset \eta^{\perp}_{h_{K'}}; \\
2^{4g-6+2s}, & \text{if } u = v = 0 = r \text{ and } P(r_{D_4})[2] \subset \eta^{\perp}_{h_{K'}}; \\
2^{4g-6+r+2s}, & \text{if } u = v = 0, \ r > 0 \text{ and } P(r_{D_4})[2] \not\subset \eta^{\perp}_{h_{K'}}; \\
2^{4g-7+2s}, & \text{if } u = v = 0 = r \text{ and } P(r_{D_4})[2] \not\subset \eta^{\perp}_{h_{K'}}; \\
2^{4g-4+r+2s+v}, & \text{if } 2u + 2v = 2 \text{ and } r > 0; \\
2^{4g-5+2s+v}, & \text{if } 2u + 2v = 2 \text{ and } r = 0.
\end{cases}
$$

Proof Recall that according to Proposition 5.8.8 under our hypotheses,

$$
F = h^*_{K'}(P(r_{D_4})[2]) \cap P(h_{K'})[2].
$$

If $2u + 2v = 2$, then $h^*_{K'}$ is injective and $P(h_{K'})[2] = h^*_{K'}(JC_{D_4}[2])$ by Proposition 3.3.3. It follows that

$$
F = h^*_{K'}(P(r_{D_4})[2]) \quad \text{with} \quad |F| = 2^{4g-4+r+2s+v}.
$$

According to Lemma 5.8.5, we have

$$
\deg \gamma = \begin{cases} |F| & \text{if } r > 0; \\ \frac{|F|}{2} & \text{if } r = 0. \end{cases} \tag{5.54}
$$

Together this gives the assertion in the case $2u + 2v = 2$.

So suppose that $2u + 2v = 0$. From Corollary 5.8.7 we know that $\eta_{h_{K'}} \in P(r_{D_4})$ and also that

$$
P(h_{K'})[2] = h^*_{K'}(\eta^{\perp}_{h_{K'}}) \simeq \eta^{\perp}_{h_{K'}}/\{0, \eta_{h_{K'}}\}.
$$

So we have to analyze two cases:

(1) $P(r_{D_4})[2] \subset \eta^{\perp}_{h_{K'}}$
(2) $P(r_{D_4})[2] \not\subset \eta^{\perp}_{h_{K'}}$

In case (1) we obtain $F = h^*_{K'}(P(r_{D_4})[2])$, which is isomorphic to $P(r_{D_4})[2]/\{0, \eta_{H_{K'}}\}$. Therefore in this case, $|F| = 2^{4g-5+r+2s}$. This implies the assertion in this case by Eq. (5.54).

In case (2) we have

$$
F = h^*_{K'}(P(r_{D_4})[2] \cap \eta^{\perp}_{h_{K'}}) \simeq P(r_{D_4})[2] \cap \eta^{\perp}_{h_{K'}}/\{0, \eta_{h_{K'}}\},
$$

and therefore $|F| = 2^{4g-6+r+2s}$. Again by Eq. (5.54), we get the last two cases of the proposition.   □

5.8.4 The Non-Principally Polarized Case

In this subsection we compute $\deg \gamma$ in the case when $P(h_{K'})$ is not principally polarized, that is, $2u + 2v > 2$. For this we describe the set F by constructing the elements of $P(h_{K'})[2]$ coming from the ramification. Then we decide which of those lie in F.

We are interested in the ramification of the double cover $h_{K'} : C_{K'} \to C_{D_4}$. For this note first that there are four types of points in C over which the degree-four cover $h_{S_3} : C_{S_3} \to C$ ramifies: the r, s, u, and v points corresponding to the images of simple, triple, $(2, 2)$, or total type of ramification points, respectively.

According to diagram (5.46), the double cover $h_{K'} : C_{K'} \to C_{D_4}$ ramifies exactly over the preimages of the u and v points. We call them *points of type u,* respectively, v.

The points of type u: If $y \in C$ is of type u, then $r_{D_4} : C_{D_4} \to C$ is unramified over y. We denote

$$r_{D_4}^{-1}(y) = y_1' + y_2' + y_3'$$

with $y_i' \in C_{D_4}$ such that

$$h_{K'}^*(y_1') = 2x_1, \quad h_{K'}^*(y_2') = 2x_2 \quad \text{and} \quad h_{K'}^*(y_3') = x_5 + x_6, \quad x_i \in C_{K'}.$$

With respect to $h_K : C_K \to C_{D_4}$, we choose $n \in JC_{D_4}$ such that

$$n^2 = \mathcal{O}_{C_{D_4}}(y_1' - y_2') \quad \text{and} \quad h_K^*(n) \in \text{Ker}(1 + \kappa + \kappa^2).$$

Then

$$\mathcal{G} := \mathcal{O}_{C_{K'}}(x_2 - x_1) \otimes h_{K'}^*(n) \in P(h_{K'})[2] \quad \text{such that} \quad (1 + \kappa + \kappa^2)k_{\sigma\tau}^* \ell_{\sigma\tau}^*(\mathcal{G}) = 0.$$

In particular $\mathcal{G} \in F$.

Hence, if we label the u points of C as $y^1, \ldots, y^u$, we have that the corresponding points $x_1^1, x_2^1, \ldots, x_1^u, x_2^u$ of $C_{K'}$ are ramification points of $h_{K'} : C_{K'} \to C_{D_4}$. As above we construct for each $i = 1, \ldots, u$, elements $\mathcal{G}_{2i-1}$ in F given as follows:

$$\mathcal{G}_{2i-1} = \mathcal{O}_{C_{D_4}}(x_2^i - x_1^i) \otimes h_{K'}^*(n_i)$$

with $n_i \in JC_{D_4}$, $n_i^2 = \mathcal{O}_{h_{D_4}}(y'^i_1 - y'^i_2)$, and $(1 + \kappa + \kappa^2)h_K^*(n_i) = 0$.

Similarly, we construct for $i = 1, \ldots, u$ elements $\mathcal{G}_{2i}$ of $P(h_{K'})[2]$ as follows

$$\mathcal{G}_{2i} := \mathcal{O}_{C_{K'}}(x_1^{i+1} - x_2^i) \otimes h_{K'}^*(m_i)$$

with $m_i \in JC_{D_4}$, $m_i^2 = \mathcal{O}_{C_{D_4}}(\gamma_2'^i - \gamma_1'^i)$.

The points of type v: If $z \in C$ is of type v, then $r_{D_4} : C_{D_4} \to C$ is simply ramified over z. We denote

$$r_{D_4}^{-1}(z) = y_1' + 2y_2'$$

with $y_i' \in C_{D_4}$ such that

$$h_{K'}^*(y_1') = 2x_1', \quad h_{K'}^*(y_2') = 2x_2' \quad x_i' \in C_{K'}.$$

Choose $n \in JC_{D_4}$ such that

$$n^2 = \mathcal{O}_{C_{D_4}}(y_2' - y_1') \quad \text{and} \quad h_K^*(n) \in \mathrm{Ker}(1 + \kappa + \kappa^2).$$

Then

$$\mathcal{D} := \mathcal{O}_{C_{D_4}}(x_1' - x_2') \otimes h_{K'}^*(n) \in P(h_{K'})[2] \quad \text{such that} \quad (1 + \kappa + \kappa^2)k_{\sigma\tau}^* \ell_{\sigma\tau}^*(\mathcal{D}) = 0.$$

In particular $\mathcal{D} \in F$.

Hence, if we label the v points of C as $z^1, \ldots, z^v$, we have that the corresponding points $x_1'^1, x_2'^1, \ldots, x_1'^v, x_2'^v$ of $C_{K'}$ are ramification points of $h_{K'} : C_{K'} \to C_{D_4}$. As above we construct for each $i = 1, \ldots, v$, elements $\mathcal{D}_{2i-1}$ in F given as follows:

$$\mathcal{D}_{2i-1} = \mathcal{O}_{C_{D_4}}(x_2'^i - x_1'^i) \otimes h_K^*(n_i)$$

with $n_i \in JC_{D_4}$, $n_i^2 = \mathcal{O}_{h_{D_4}}(y_2'^i - y_1'^i)$ and $(1 + \kappa + \kappa^2)h_K^*(n_i) = 0$.

Similarly, we construct for $i = 1, \ldots, u$ elements $\mathcal{D}_{2i}$ of $P(h_{K'})[2]$ as follows

$$\mathcal{D}_{2i} := \mathcal{O}_{C_{K'}}(x_2'^{i+1} - x_1'^i) \otimes h_{K'}^*(m_i)$$

with $m_i \in JC_{D_4}$, $m_i^2 = \mathcal{O}_{C_{D_4}}(y_1'^i - y_2'^i)$.

Finally, if $uv > 0$, we consider one more sheaf which links both cases

$$\mathcal{L} := \mathcal{O}_{C_{K'}}(x_1'^1 - x_2^u) \otimes h_{K'}^*(m) \quad \text{with} \quad m \in JC_{D_4} \quad \text{and} \quad m^2 = \mathcal{O}_{C_{D_4}}(y_2^u - y_1'^1).$$

With these notations we get

Lemma 5.8.12 *Let $\widetilde{C}$ be a curve admitting an action of S_4 with $2u + 2v > 2$. Then the set F can be given as follows.*

$$F = \begin{cases} h_{K'}^*(P(r_{D_4})[2]) \oplus_{j=1}^{u} \mathcal{G}_{2j-1}\mathbb{Z}/2\mathbb{Z} \oplus_{j=1}^{v-1} \mathcal{D}_{2j-1}\mathbb{Z}/2\mathbb{Z} \ if \ uv > 0; \\ h_{K'}^*(P(r_{D_4})[2]) \oplus_{j=1}^{u-1} \mathcal{G}_{2j-1}\mathbb{Z}/2\mathbb{Z} & if \ v = 0; \\ h_{K'}^*(P(r_{D_4})[2]) \oplus_{j=1}^{v-1} \mathcal{D}_{2j-1}\mathbb{Z}/2\mathbb{Z} & if \ u = 0. \end{cases}$$

Proof With the above notation, Corollary 3.3.5 gives

$P(h_{K'})[2]$

$$= \begin{cases} h_{K'}^*(P(r_{D_4})[2]) \oplus_{j=1}^{2u-1} \mathcal{G}_j\mathbb{Z}/2\mathbb{Z} \oplus \mathcal{L}\mathbb{Z}/2\mathbb{Z} \oplus_{j=1}^{2v-2} \mathcal{D}_j\mathbb{Z}/2\mathbb{Z} \ if \ uv > 0; \\ h_{K'}^*(P(r_{D_4})[2]) \oplus_{j=1}^{2u-2} \mathcal{G}_j\mathbb{Z}/2\mathbb{Z} & if \ v = 0; \\ h_{K'}^*(P(r_{D_4})[2]) \oplus_{j=1}^{2v-2} \mathcal{D}_j\mathbb{Z}/2\mathbb{Z} & if \ u = 0. \end{cases}$$

Since we are assuming $2u + 2v > 2$, the map $h_{K'}^* : JC_{D_4} \to JC_{K'}$ is injective. Hence $h_{K'}^*(P(r_{D_4})[2]) \subset P(h_{K'})[2]$, and therefore the factor of F not coming from the ramification is $h_{K'}^*(P(r_{D_4})[2])$.

For $uv > 0$ let $\mathcal{F} := \{\mathcal{G}_{2i-1}, \mathcal{D}_{2j-1} \mid i = 1, \ldots, u, \ j = 1, \ldots, v-1\}$.

If $v = 0$, let $\mathcal{F} := \{\mathcal{G}_{2i-1}, \mid i = 1, \ldots, u-1\}$.

If $u = 0$, let $\mathcal{F} := \{\mathcal{D}_{2j-1} \mid j = 1, \ldots, v-1\}$.

Then note that in each case, the elements of $\mathcal{F}$ span the sheaves which come from the ramification and which are zero under the map $(1 + \kappa + \kappa^2)k_{\sigma\tau}^* \ell_{\sigma\tau}^*$, that is, those of the set F coming from the ramification. This gives the assertion on F and thus completes the proof of the lemma. □

Using the lemma, we compute $\deg \gamma$ in this case.

Proposition 5.8.13 *Let $\widetilde{C}$ be a curve admitting an action of S_4 with $2u + 2v > 2$. Then the isogeny $\gamma : P(h_{K'}) \to P(h_{S_3})$ is of degree*

$$\deg \gamma = \begin{cases} 2^{4g-5+r+2s+u+2v} \ if \ r > 0; \\ 2^{4g-6+2s+u+2v} & if \ r = 0. \end{cases}$$

Proof From Lemma 5.8.12 we conclude that

$$|F| = 2^{g(C_{D_4})-g(C))+u+v-1} = 2^{4g-5+r+2s+u+2v}$$

whenever $2u + 2v > 2$. By Lemma 5.8.5 we have $\deg \gamma = |F|$ for $r > 0$ and $\deg \gamma = \frac{1}{2}|F|$ if $r = 0$. This implies the assertion of the proposition. □

Proof of Theorem 5.8.1 The first and last equations follow directly from Propositions 5.8.9 and 5.8.13, respectively. The two middle equations follow from Proposition 5.8.11 using the definition of F and Lemma 5.8.5. □

Chapter 6
Some Series of Group Actions

In this chapter we study the decomposition of the Jacobians of curves with actions of three series of groups, namely, cyclic groups of order n, dihedral groups of order $2n$, and the semidirect product of the group of order 3 by an arbitrary self-product of the Klein group of order 4.

In the first section, we decompose the Jacobian of curves with cyclic groups of order p, p^2, and pq, where p and q are different prime numbers. We always assume that the corresponding cover $\widetilde{C} \to C$ is either étale or totally ramified. In each case we compute the isotypical decomposition as well as its degree and the dimensions of the isotypical components. In the case $n = pq$, we get the first example where not all factors are Jacobians or Prym varieties corresponding to subgroups. Here a component is only a Prym variety of a pair of covers.

In Sect. 6.2, following [6], we do the same for the dihedral groups D_n of order $2n$ in the cases $n = p$, $n = 2^\alpha$, an arbitrary power of 2, and $n = 2p$ where p is an odd prime number. The ramification in the first and second case can be determined, whereas in the second case, we do not use it, with the drawback of not being able to compute the dimensions of the corresponding factors. In each case there are some isogenies between isotypical factors of which we compute the degrees. Again in the case $n = 2p$, there occurs a factor which is only a Prym variety of a pair of covers.

In the last section, we consider for any positive integer n the group G_n, the semidirect product of the group of order 3 by the product of n copies of the Klein group of order 4. We show that for any curve C and any sufficiently large t, there is a Galois cover $\widetilde{C} \to C$ with group G_n with signature $(g; 3, \ldots, 3)$ and t numbers 3. The important case is $g = 0$ where we give a more precise estimate for t. The importance lies in the fact that this gives examples of curves whose Jacobian is isogenous to a product of an arbitrary number of Jacobians of the same genus. For the proof we determine the isotypical decomposition and deduce some specific isogenies between the factors. Here we follow [5].

H. Lange, R. E. Rodríguez, *Decomposition of Jacobians by Prym Varieties*, Lecture Notes in Mathematics 2310, https://doi.org/10.1007/978-3-031-10145-8_6

6.1 Cyclic Covers

In this section we study the decompositions of Jacobians of cyclic covers $f : \widetilde{C} \to C$ of degree n and their Prym varieties in the cases $n = p^2, 2^\alpha$, and pq, where p and q are different prime numbers. We restrict the discussion to the case where f is either totally ramified or étale.

6.1.1 Notation and First Results

Let $f : \widetilde{C} \to C$ be a cyclic cover of degree $n \geq 2$ with generating automorphism σ. Recall that we abbreviate $\widetilde{J}^\sigma := \widetilde{J}^{\langle \sigma \rangle}$ and similarly for every subset of $\widetilde{J}$.

Proposition 6.1.1

(a) $\widetilde{J}^\sigma = \mathrm{Ker}(1_{\widetilde{J}} - \sigma) = f^* J + P(f)_0$, with $P(f)_0 = P(f) \cap \widetilde{J}^\sigma \subseteq P(f)[n]$;
(b) *If moreover f is étale, then $\widetilde{J}^\alpha$ is connected; that is, $\widetilde{J}^\sigma = f^* J$.*

Proof (a) is a special case of Corollary 3.5.3.

For (b) we have to show, using (a), that $\mathrm{Ker}(1_{\widetilde{J}} - \sigma) \subseteq f^* J$. Let $\ell \in \mathrm{Ker}(1_{\widetilde{J}} - \sigma)$ considered as a line bundle on $\widetilde{J}$. So there is an isomorphism $\varphi : \ell \to \sigma(\ell)$. But then $\sigma^{n-1}(\varphi) \circ \cdots \circ \sigma(\varphi) \circ \varphi$ is an isomorphism of ℓ to itself and therefore equal to a non-zero constant c. Replacing φ by $\frac{1}{\sqrt[n]{c}}\varphi$, we obtain an automorphism of the line bundle ℓ of order n; that is, a $\langle \sigma \rangle$ action. Now, since f is étale, f^* gives an equivalence between the line bundle on C and the line bundles on $\widetilde{C}$ which admit a $\langle \sigma \rangle$-action (see [26]). Hence $\ell \in f^* J$. $\square$

Recall from Corollary 3.5.4 that we have for $n = 2$ and f étale: $\widetilde{J}^\sigma = f^* J + P(f)[2]$. For $n \geq 3$ an analogous statement is not valid by dimension reasons.

In order to determine the isotypical decomposition of $\widetilde{J}$, we have to know the irreducible rational representations of the cyclic group $\langle \sigma \rangle$. Recall that the irreducible complex representations are all of dimension 1; that is, are characters, given by

$$\chi_k(\sigma) = \xi_n^k \quad \text{for} \quad k = 1, \ldots, n \quad \text{where} \quad \xi_n := e^{2\pi i / n}.$$

Now consider the set of divisors of n,

$$\Omega_n := \{d \in \mathbb{N} \mid d \text{ divides } n\}.$$

For each $d \in \Omega_n$ let V_d denote the character given by $\chi_{n/d}$. It is of order d in the character group. Its character field is the cyclotomic field

$$K_d := \mathbb{Q}(\xi_n^{n/d}) = \mathbb{Q}(\xi_d).$$

Its Schur index is clearly $s = s(V_d) = 1$. K_d is a Galois extension of $\mathbb{Q}$ of degree $[K_d : \mathbb{Q}] = \varphi(d)$, where φ is the Euler function. For any $\gamma \in \mathrm{Gal}(K_d/\mathbb{Q})$, let V_d^γ denote the representation defined by the composition of $\chi_{n/d}$ with γ. Then the representation

$$W_d := \bigoplus_{\gamma \in \mathrm{Gal}(K_d/\mathbb{Q})} V_d^\gamma$$

is defined over the rationals for any $d \in \Omega_n$.

Proposition 6.1.2 *The representations W_d with $d \in \Omega_n$ are exactly the irreducible rational representations of the cyclic group $\langle \sigma \rangle$.*

Proof The irreducibility of W_d is obvious. That these are all irreducible rational representations follows from the well-known property of elementary number theory $\sum_{d|n} \varphi(d) = n$. $\square$

Let A_{W_d} denote the isotypical component of $\widetilde{J}$ corresponding to the representation W_d; that is,

$$A_{W_d} = \widetilde{J}^{e_d} \quad \text{with} \quad e_d = \frac{1}{n} \sum_{i=1}^{n} \mathrm{tr}_{K_d|\mathbb{Q}}(\chi_{n/d}(\sigma^{n-i}))\sigma^i$$

(see Eq. (2.23). Note that e_d is a symmetric idempotent here, not an integer). According to Theorem 2.9.1, the isotypical decomposition is given by the addition map

$$\mu : \prod_{d \in \Omega_n} A_{W_d} \to \widetilde{J}$$

which we want to determine in some cases.

Suppose $g = g_C$ is the genus of C. The Hurwitz formula or Eq. (3.2) gives

Lemma 6.1.3 *Let $f : \widetilde{C} \to C$ be a cyclic cover of degree n, either étale or totally ramified in r points. For any $r \geq 0$ such that $(n-1)r \equiv 0 \mod 2$ and $r \geq 2$ if $g = 0$, we have*

$$g_{\widetilde{C}} = ng + \frac{n-1}{2}(r-2) \quad \text{and} \quad \dim P(f) = \frac{n-1}{2}(2g+r-2).$$

Note that the formulas impose the condition $(n-1)r \equiv 0 \mod 2$. Moreover, if $g = 0$, we must have $r \geq 2$, to make sure that there are connected covers.

Proposition 6.1.4 *For any cyclic cover $f : \widetilde{C} \to C$ of degree n, the group algebra decomposition coincides with the isotypical decomposition.*

Proof This is an immediate consequence of Theorem 2.9.1 and Eq. (2.30), since all irreducible complex representations are characters. $\square$

6.1.2   Cyclic Covers of Degree p

Proposition 6.1.5 *Let* $f : \widetilde{C} \to C$ *be a cover as in Lemma 6.1.3 with* $n = p$ *a prime. Then the addition map*

$$\mu : f^*J \times P(f) \to \widetilde{J}$$

is an isogeny of degree

$$\deg \mu = \begin{cases} p^{2g} & \text{if } r > 0 \\ p^{2g-2} & \text{if } r = 0. \end{cases}$$

The map μ *is the isotypical decomposition of* $\widetilde{J}$ *with respect to the action of the group* $\langle \sigma \rangle$.

Proof The first assertion follows from Propositions 3.2.9 and 3.2.3. Now note that

$$\Omega_p = \{1, p\},$$

since p is a prime. So there are two irreducible rational representations, namely, the trivial representation W_1 and the non-trivial one W_p of dimension $p - 1$. Only W_1 acts on f^*J, whereas W_p acts on $P(f)$. This gives the last assertion.   □

6.1.3   Cyclic Covers of Degree p^2

First consider more generally a cyclic cover $f : \widetilde{C} \to C$ of degree n, either étale or totally ramified in r points. For each $d \in \Omega_n$ denote

$$C_d := \widetilde{C}/\langle \sigma^{n/d} \rangle.$$

Then the intermediate covers

$$k : \widetilde{C} \to C_d \text{ of degree } d \quad \text{and} \quad h : C_d \to C \text{ of degree } \frac{n}{d}$$

are also cyclic étale or totally ramified, respectively. From Lemma 6.1.3 we get

Lemma 6.1.6 *For any* $d \in \Omega_n$ *and* $r \geq 0$ *such that* $(n-1)r \equiv 0 \mod 2$ *and* $r \geq 2$ *if* $g = 0$, *we have*

$$g_{C_d} = \frac{n}{d}g + \frac{\frac{n}{d}-1}{2}(r-2) \quad \text{and} \quad \dim P(h) = \frac{\frac{n}{d}-1}{2}(2g+r-2).$$

Proposition 6.1.7 *The map*

$$\Psi : P(k) \times P(h) \times J \to \tilde{J}, \quad (x, y, z) \mapsto x + k^* y + f^* z$$

is a $\langle \sigma \rangle$-equivariant isogeny of degree

$$\deg \Psi = \begin{cases} d^{(\frac{n}{d}-1)(2g+r-2)} \cdot n^{2g} & \text{if } r > 0, \\ d^{(\frac{n}{d}-1)(2g-2)} \cdot n^{2g-1} & \text{if } r = 0. \end{cases}$$

Proof The $\langle \sigma \rangle$-invariance of μ is obvious. That Ψ is an isogeny and the formula for the degree of μ follow from Theorem 3.2.12(b) and Lemma 6.1.6. In fact,

$$\deg \Psi = \frac{|JC_d[d]| \cdot |JC[\frac{n}{d}]|}{|\operatorname{Ker} f^*|}$$

$$= \frac{1}{\deg f^*} \cdot d^{2g c_d} \cdot \left(\frac{n}{d}\right)^{2g}$$

$$= \frac{1}{\deg f^*} \cdot d^{2\frac{n}{d}g + (\frac{n}{d}-1)(r-2)} \cdot \left(\frac{n}{d}\right)^{2g} = \frac{1}{\deg f^*} \cdot d^{(\frac{n}{d}-1)(2g+r-2)} \cdot n^{2g}$$

which gives the assertion, since f^* is injective for $r > 0$ and of degree n onto its image in the étale case. $\qquad \square$

In the same way, using Corollary 3.2.14(b) instead of Theorem 3.2.12(b), one computes the degree of the addition map

$$\mu : P(k) \times k^* P(h) \times f^* J \to \tilde{J}.$$

Now consider the special case $n = p^2$ with a prime number p. Clearly

$$\Omega_n = \{1, p, p^2\}.$$

So Lemma 6.1.6 and Proposition 6.1.7 applied to $d = p$ give under the assumptions on r of Lemma 6.1.6

$$g_{C_p} = pg + \frac{1}{2}(p-1)(r-2) \quad \text{and} \quad \dim P(h) = \frac{1}{2}(p-1)(2g+r-2).$$

Furthermore, we have the following two propositions under these assumptions:

Proposition 6.1.8 *The addition map $\mu : P(k) \times k^* P(h) \times f^* J \to \tilde{J}$ gives the isotypical decomposition of $\tilde{J}$ with respect to the action of the group $\langle \sigma \rangle$.*

Proof The group $\langle \sigma \rangle$ has three irreducible rational representations, namely, W_1, W_p, and W_{p^2}. It acts on $f^* J$ by the trivial representation W_1, on $k^* P(h)$

by the representation W_{p^2}, and on $P(k)$ by the representation W_p. The uniqueness of the isotypical representation implies the assertion.   □

Proposition 6.1.9 *The canonical polarization Θ of $\widetilde{J}$ restricts to a polarization of type*

(1) $(p^2, \ldots, p^2)$ *(respectively $(1, p^2, \ldots, p^2)$) on f^*J, if $r > 0$ (respectively $r = 0$);*

(2) $(p, \ldots, p, p^2, \ldots, p^2)$ *on $k^*P(h)$ with g numbers p^2, if $r > 0$;*

(3) $(1, \ldots, 1, p, \ldots, p)$ *on $P(k)$ with g_{C_p} (respectively $g_{C_p} - 1$) numbers p, if $r > 0$ (respectively $r = 0$).*

In the same way as in (2), one computes the type of the restriction Θ on $k^*P(h)$ for $r = 0$. It depends then on whether or not $P(h)$ contains the p-division point defining the cover k_p.

Proof (1) and (3) follow from Propositions 3.2.1(a) and 3.2.2.

(2): The canonical polarization of JC_p induces a polarization of type $(1, \ldots, 1, p \ldots p)$ on $P(h)$ with g (respectively $g - 1$) numbers p. According to Proposition 3.2.1, the type of $k^*P(h)$ is the p-fold of the type of $P(h)$. This is clear in the case $r > 0$ and follows in the case $r = 0$, i.e., the étale case, since $P(h)$ does not contain the p-division point defining the cover k.   □

6.1.4   Cyclic Covers of Degree pq

In this section we consider the special case $n = pq$ with different primes p and q. We will see that in general there is an isotypical component, which is not a Prym variety, but only a Prym variety of a pair of covers.

Let $f : \widetilde{C} \to C$ be a cyclic cover with group generated by σ, and denote $C_p := \widetilde{C}/\langle \sigma^q \rangle$ and $C_q := \widetilde{C}/\langle \sigma^p \rangle$ so that $k_p : \widetilde{C} \to C_p$ is of degree p and $k_q : \widetilde{C} \to C_q$ is of degree q. Then we have the following commutative diagram

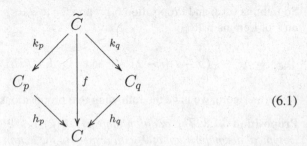

$$(6.1)$$

and let $P(k_p, k_q)$ denote the Prym variety of the pair (k_p, k_q).

Proposition 6.1.10 *If $n = pq$ with different primes p and q and r such that $(n - 1)r \equiv 0 \mod 2$ and $r \geq 2$ if $g = 0$, then the addition map*

$$\mu : f^*J \times k_p^*P(h_p) \times k_q^*P(h_q) \times P(h_p, h_q) \to \widetilde{J}$$

is a $\langle \sigma \rangle$-equivariant isogeny.

The addition map μ is the isotypical decomposition of $\widetilde{J}$ with respect to the action of the group $\langle \sigma \rangle$.

Proof According to Corollary 3.2.14, the addition map

$$\mu_1 : f^*J \times k_q^*P(h_q) \times P(k_q) \to \widetilde{J}$$

is a $\langle \sigma \rangle$-equivariant isogeny. According to the definition of $P(k_p, k_q)$, the addition map

$$\mu_2 : k_p^*P(h_p) \times P(k_p, k_q) \to P(k_q)$$

is an isogeny, which certainly is $\langle \sigma \rangle$-equivariant. Together these give the first assertion.

The set Ω_{pq} consists of four elements, namely,

$$\Omega_{pq} = \{1, p, q, pq\}.$$

For the corresponding irreducible rational representations, we have the group $\langle \sigma \rangle$ acts on f^*J by the trivial representation, acts on $k_p^*P(h_p)$ by the representation W_q, acts on $k_q^*P(k_q)$ by the representation W_p, and finally acts on $P(k_p, k_q)$ by the representation W_{pq}. The uniqueness of the isotypical representation implies the last assertion of the proposition.   □

For the dimensions of the components, we have

$$\dim f^*J = g, \quad \dim k_p^*P(h_p) = (q-1)(g + \frac{r-2}{2}),$$

$$\dim k_q^*P(h_q) = (p-1)(g + \frac{r-2}{2}),$$

and according to Proposition 3.4.4,

$$\dim P(k_p, k_q) = (p-1)(q-1)(g + \frac{r-2}{2}).$$

It is easy to check that the sum of the dimensions of the four abelian subvarieties equals the genus of $\widetilde{C}$.

Remark 6.1.11 If $f : \widetilde{C} \to C$ is a general cyclic cover of degree pq, then diagram (6.1) consists of all subcovers of $\widetilde{C}$, and it is easy to check that the isotypical component $P(k_p, k_q)$ of the representation W_{pq} is not a Prym variety of one of the subcovers.

In the special case

$$C = \mathbb{P}^1$$

we compute the types of the restricted polarizations. For the sake of simplicity of the formulas, we only do this in the ramified case $r > 0$. The étale case works in the same way by applying Proposition 3.2.3 (similarly as in the sentence after Proposition 6.1.9).

Proposition 6.1.12 *Let $n = pq$ as in Proposition 6.1.10 with $p < q$ and the additional assumptions $r > 0$ and $C = \mathbb{P}^1$. Then the canonical polarization Θ of $\tilde{J}$ restricts to a polarization of type*

(1) $(n, \ldots, n)$ *on $f^* J$ with g numbers n;*
(2) $(p, \ldots, p)$ *on $k_p^* P(h_p)$ with $\frac{1}{2}(q-1)(r-2)$ numbers p;*
(3) $(q, \ldots, q)$ *on $k_q^* P(h_q)$ with $\frac{1}{2}(p-1)(r-2)$ numbers q;*
(4) $(1, \ldots, 1, p, \ldots, p, n, \ldots, n)$ *on $P(k_p, k_q)$ with $(q-p)\frac{r-2}{2}$ numbers p and $(p-1)\frac{r-2}{2}$ numbers n.*

Proof (1) follows from Propositions 3.2.1(a) and 3.2.2 as does (2), since $P(h_p) = J_{C_p}$. The same argument works for (3).

As for (4), by Theorem 3.4.5 we have

$$K(\Theta|_{P(k_p, k_q)}) \simeq K(\Theta|_{P(k_q)}) \oplus K(\Theta|_{k_p^* P(h_p)})$$

$$\simeq [\mathbb{Z}/q\mathbb{Z} \oplus \cdots \oplus \mathbb{Z}/q\mathbb{Z})]^2 \oplus [\mathbb{Z}/p\mathbb{Z} \oplus \cdots \oplus \mathbb{Z}/p\mathbb{Z}]^2$$

with $g_{C_q} = (p-1)\frac{r-2}{2}$ numbers q and $\dim P(h_p) = g_{C_p} = (q-1)\frac{r-2}{2}$ numbers p. Since we assumed $p < q$, this gives

$$K(\Theta|P(k_p, k_q))) \simeq ([\mathbb{Z}/p\mathbb{Z} \oplus \cdots \oplus \mathbb{Z}/p\mathbb{Z}]^2 \oplus [\mathbb{Z}/n\mathbb{Z} \oplus \cdots \oplus \mathbb{Z}/n\mathbb{Z}]^2$$

with $(q-p)\frac{r-2}{2}$ numbers p and $(p-1)\frac{r-2}{2}$ numbers n. In terms of types, this gives assertion (4). $\qquad\square$

6.2   Covers with Dihedral Group Action

Let D_n denote the dihedral group of order $2n$ generated by σ of order n and τ of order 2; that is,

$$D_n = \langle \sigma, \tau \mid \sigma^n = \tau^2 = (\sigma\tau)^2 = 1 \rangle.$$

In this section we study covers of curves $f : \widetilde{C} \to C$ with an action of D_n on $\widetilde{C}$ such that $C = \widetilde{C}/D_n$. In particular, $\deg f = 2n$. Here we use [6]. We are interested mainly in the isotypical decomposition of $J\widetilde{C}$, which we abbreviate as always by $\widetilde{J}$, as well as in the Prym varieties of various intermediate covers. If we denote for every element $\alpha \in D_n$ by C_α the quotient curve $C_\alpha := \widetilde{C}/\langle\alpha\rangle$, we have for every $i = 0, \ldots, n-1$ the following diagram of curves

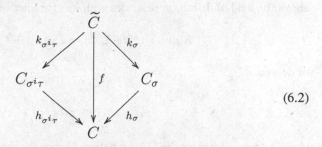

$$(6.2)$$

Note that $k_{\sigma^i \tau}$ and h_σ are of degree 2 and k_σ and $h_{\sigma^i \tau}$ are of degree n for every i.

6.2.1 The Irreducible Rational Representations of D_n

For any n we have the following two complex representations of degree 1 (see [34]), namely, the trivial representation χ_0 and

$$\text{the representation } \chi_1 \text{ with } \chi_1(\sigma^j) = 1, \ \chi_1(\sigma^j \tau) = -1$$

for all j. For even n, there are two more, namely,

$$\text{the representation } \chi_2 \text{ with } \chi_2(\sigma^j) = (-1)^j, \ \chi_2(\sigma^j \tau) = (-1)^j \quad \text{and}$$

$$\text{the representation } \chi_3 \text{ with } \chi_3(\sigma^j) = (-1)^j, \ \chi_3(\sigma^j \tau) = (-1)^{j+1}$$

for all j. For any n, the remaining irreducible complex representations are of degree two and given by

$$V_h(\sigma^j) = \begin{pmatrix} \xi_n^{hj} & 0 \\ 0 & \xi_n^{-hj} \end{pmatrix} \quad \text{and} \quad V_h(\sigma^j \tau) = \begin{pmatrix} 0 & \xi_n^{-hj} \\ \xi_n^{hj} & 0 \end{pmatrix}$$

for all j, where $\xi_n = e^{2\pi i/n}$ and $1 \le h < \frac{n}{2}$.

In order to define the irreducible rational representations of D_n, consider the following set

$$\Omega_{D_n} := \{d \in \mathbb{N} \mid d \text{ is a divisor of } n \text{ with } d < \frac{n}{2}\}.$$

For any $d \in \Omega_{D_n}$ consider the complex irreducible representation V_d of degree 2 of above. Its field of definition coincides with its character field

$$K_d := \mathbb{Q}(\xi_n^d + \xi_n^{-d}) = \mathbb{Q}(\xi_{n/d} + \xi_{n/d}^{-1})$$

of degree

$$[K_d : \mathbb{Q}] = \frac{1}{2}\varphi\left(\frac{n}{d}\right)$$

over $\mathbb{Q}$ where φ denotes the Euler function. Hence the representation

$$W_d := \bigoplus_{\gamma \in \mathrm{Gal}(K_d/\mathbb{Q})} V_d^\gamma$$

is clearly defined over the rationals and moreover irreducible. It is called the irreducible rational representation *associated to V_d*.

Proposition 6.2.1 *The representations χ_j, $j = 0, \ldots, 3$ if n is even respectively $j = 0, 1$ if n is odd and W_d with $d \in \Omega_{D_n}$ are exactly the irreducible rational representations of D_n.*

Proof We only have to show that these are all irreducible rational representations of D_n. But for odd n we have

$$2 + \sum_{d \in \Omega_{D_n}} \deg W_d = 2 + 2 \sum_{d \in \Omega_{D_n}} \varphi\left(\frac{n}{d}\right) = 2 + 2(n - \varphi(1)) = 2n.$$

A similar computation works for even n. $\square$

6.2.2   Curves with D_p-Action: Ramifications and Genera

Let p be an odd prime and $\widetilde{C}$ be a curve with an action of the dihedral group D_p. Some of the statements in this and the next subsections are direct generalizations of the corresponding statements in Sect. 4.2. Nevertheless, for convenience we give full proofs, since some of them are slightly more complicated.

Note that all involutions $\sigma^i \tau$ in D_p are conjugate to each other and hence all curves $C_{\sigma^i \tau}$ are isomorphic. It suffices to consider diagram (6.2) for $i = 0$, that is,

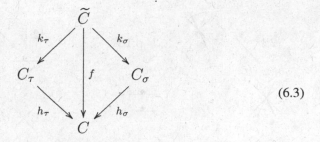

$$(6.3)$$

with k_τ and h_σ (cyclic) of degree 2, k_σ cyclic of degree p, and h_τ non-cyclic of degree p.

Lemma 6.2.2 *Let $c \in C$ be a branch point of f. There are exactly two possibilities for the ramification of the curves of the diagram:*

(i) *h_τ is totally ramified over c. Then h_σ and k_τ are unramified over c and $h_\tau^{-1}(c)$, respectively, and k_σ is totally ramified over each point of $h_\sigma^{-1}(c)$.*

(ii) *h_τ has $\frac{p-1}{2}$ simple ramification points and one unramified point over c. Then h_σ is ramified over c, k_τ is unramified over the $\frac{p-1}{2}$ points and ramified over the remaining point of $h_\tau^{-1}(c)$, and k_σ is unramified over the point $h_\sigma^{-1}(c)$.*

Proof Since f is of degree $2p$ with odd p, there are two possibilities for f over c: either $f^{-1}(c)$ consists of two points $\{P_1, P_2\}$, both fixed by σ and permuted by any involution, or of p points $\{Q_1, \ldots, Q_p\}$, each fixed by exactly one of the p involutions $\sigma^{-j} \tau \sigma^j$, $1 \le j \le p$, in D_p, and permuted by σ. To see this, note that if one involution would admit more than one fixed point in the same fiber, then there would be two involutions with the same fixed point, but any two involutions generate the group D_p, so the stabilizer group of this point would not be cyclic. We label these fixed points in such a way that Q_j is fixed by $\sigma^{-j} \tau \sigma^j$.

In the first case, h_σ and k_τ are necessarily unramified over c, and each point of $h_\tau^{-1}(c)$ and k_σ is totally ramified over each point of $h_\sigma^{-1}(c)$. Hence we are in case (i).

In the second case, observe that τ acts on each fiber $\{Q_1, \ldots, Q_p\}$ by fixing exactly one point, say Q_p, and identifying the other $p - 1$ points $\{Q_1, \ldots, Q_{p-1}\}$ in pairs: $\tau(Q_j) = Q_{p-j}$. Therefore k_τ ramifies at Q_p and is unramified over the other points of $h_\tau^{-1}(c)$. Then h_σ is ramified over c and k_σ is unramified over the point $h_\sigma^{-1}(c)$.   □

We call the corresponding branch points in C in Lemma 6.2.2 *of type* (i), respectively, *type* (ii).

(i)   (ii)

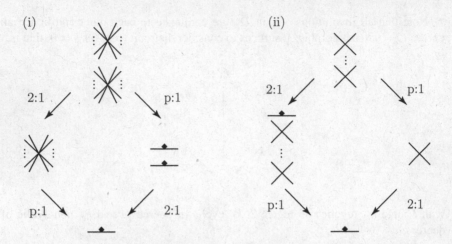

The number of fixed points of the involution τ is even. Since each involution acts non-trivially on the fixed points of σ, so is the number of fixed points of σ. We denote

- $2s := |\operatorname{Fix}\sigma|$
- $2t := |\operatorname{Fix}\tau|$

Corollary 6.2.3 *With these notations the ramification degrees* $|R(\cdot)|$ *of the maps in diagram* (6.3) *are as follows:*

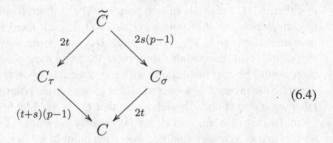

$$(6.4)$$

Proof Lemma 6.2.2 implies that there are s branch points of type (i) and $2t$ branch points of type (ii). Clearly the branch points of type (i) are disjoint from the branch points of type (ii). This gives the ramification degrees of k_τ and k_σ and then also of h_σ and h_τ. □

Note that $t \geq 1$ if $C = \mathbb{P}^1$, since $\mathbb{P}^1$ does not admit connected étale double covers. The Hurwitz formula then gives

Lemma 6.2.4 *Let p be an odd prime. Given a diagram* (6.3) *with $D_p \subset \operatorname{Aut}(\widetilde{C})$, s and t as above and $g = g_C$ the genus of C. Then we have for the genera of the other curves*

$$g_{C_\tau} = pg - p + 1 + \frac{p-1}{2}(s+t), \quad g_{C_\sigma} = 2g - 1 + t, \quad g_{\widetilde{C}} = 2pg - 2p + 1 + (p-1)s + pt.$$

Lemma 6.2.5 *Let p be an odd prime. Given a diagram (6.3) with $D_p \subset \mathrm{Aut}(\widetilde{C})$.*
Then

(i) $h_\tau^* : J \to JC_\tau$ *is injective.*
(ii) k_τ^* *is injective if and only if h_σ^* is injective.*
(iii) $\mathrm{Ker}\, h_\sigma^* = \mathrm{Ker}\, f^*$.
(iv) $\mathrm{Ker}\, k_\tau^* = h_\tau^*(\mathrm{Ker}\, h_\sigma^*)$.
(v) $h_\sigma^*(J) \cap \mathrm{Ker}\, k_\sigma^* = \{0\}$.

Note that Lemma 6.2.5 is valid for τ replaced by any involution $\sigma^i \tau$, since all involutions in D_p are conjugate.

Proof (i) and (ii) follow from Proposition 3.2.2, (i) since h_τ is non-cyclic and (ii) since k_τ is ramified if and only if h_σ is ramified.

As for (iii), clearly $\mathrm{Ker}\, h_\sigma^* \subseteq \mathrm{Ker}\, f^*$. So let $x \in \mathrm{Ker}\, f^*$. Then $h_\tau^*(x) \subset \mathrm{Ker}\, k_\tau^* \subseteq JC_\tau[2]$. By (i), h_τ^* is injective. Hence $x \in J[2]$. On the other hand, $h_\sigma^*(x) \in \mathrm{Ker}\, k_\sigma^* \subseteq JC_\sigma[p]$. Since $x \in J[2]$ and p is odd, we conclude $h_\sigma^*(x) = 0$; that is, $x \in \mathrm{Ker}\, h_\sigma^*$.

For (iv) just note that $\mathrm{Ker}\, h_\sigma^* = \mathrm{Ker}\, f^* = (h_\tau^*)^{-1}(\mathrm{Ker}\, k_\tau^*)$ with the first equality by (iii) and the second by the commutativity of (6.3).

For the proof of (v), suppose $y \in h_\sigma^*(J) \cap \mathrm{Ker}\, k_\sigma^*$. Then $y = h_\sigma^*(x)$ for some $x \in J$. Hence $f^*(x) = k_\sigma^*(y) = 0$. Using (iii), we get $x \in \mathrm{Ker}\, h_\sigma^*$ and therefore $y = 0$. □

Corollary 6.2.6 *With the notation of Lemma 6.2.5, we have for any involution $\sigma^i \tau \in D_p$: $k_{\sigma^i \tau}^* |_{P(h_{\sigma^i \tau})}$ is injective.*

Proof We may assume $i = 0$. Suppose k_τ^* is not injective. By Lemma 6.2.5(ii),

$$\mathrm{Ker}\, h_\sigma^* = \{0, \eta\} \subseteq J[2] \quad \text{with} \quad \eta \neq 0.$$

Then $h_\tau^*(\eta)$ is the non-zero element of $\mathrm{Ker}\, k_\tau^*$, since $\mathrm{Ker}\, h_\tau^*$ consists of points of odd order. But $h_\tau^*(J) \cap P(h_\tau)$ also consists of points of odd order. Hence $h_\tau^*(\eta) \notin P(h_\tau)$. □

6.2.3 Curves with D_p-Action: Decompositions of $\widetilde{J}$

Let p be an odd prime. Given a diagram of curves (6.3) with $D_p \subseteq \mathrm{Aut}(\widetilde{C})$, we want to determine the isotypical and group algebra decompositions of $\widetilde{J} = J\widetilde{C}$.

Note that Ω_{D_p} consists of the number 1 alone. Hence D_p admits three irreducible rational representations, χ_0, χ_1, and W_1. Let $A_{\chi_0}, A_{\chi_1}, A_{W_1}$ denote the corresponding isotypical components. The corresponding isotypical decomposition of $\tilde{J}$ is given by the addition map

$$\mu : A_{\chi_0} \times A_{\chi_1} \times A_{W_1} \to \tilde{J}.$$

The following Proposition identifies the isotypical components by the maps of diagram (6.3).

Proposition 6.2.7

$$A_{\chi_0} = f^* J, \quad A_{\chi_1} = k_\sigma^* P(h_\sigma), \quad A_{W_1} = P(k_\sigma).$$

Proof The first assertion is clear; D_p acts on $f^* J$ by the trivial representation χ_0.

Next we show that D_p acts on $k_\sigma^* P(h_\sigma)$ by the alternating representation χ_1. Let τ be any transposition of D_p. Since $\langle \sigma \rangle$ is a normal subgroup of D_p, τ induces an involution τ' on C_σ such that

$$\tau^* \circ k_\sigma^* = k_\sigma^* \circ \tau'^*.$$

By Corollary 3.5.2(b) we have $P(h_\sigma) = \mathrm{Ker}(1 + \tau')^0$. It follows that $\tau'(y) = -y$ for any $y \in P(h_\sigma)$. Hence

$$\tau^*(k_\sigma^*(y)) = -k_\sigma^*(y)$$

for any $y \in P(h_\sigma)$ and any involution $\tau \in D_p$. We also have $k_\sigma^* P(h_\sigma) \subset k_\sigma^* J C_\sigma \subset \tilde{J}^\sigma$; that is, σ acts trivially on $P(h_\sigma)$. So D_p acts on $k_\sigma^* P(h_\sigma)$ by the representation χ_1.

In order to complete the proof of the corollary, we have to show that D_p acts on $P(k_\sigma)$ by the representation W_1. But this is proved in Corollary 6.2.11 below. Now it suffices to check that $\dim f^* J + \dim k_\sigma^* P(h_\sigma) + \dim P(k_\sigma) = \dim \tilde{J}$, since then this implies that $A_{W_1} = P(k_\sigma)$. But this follows immediately using Lemma 6.2.4. □

Corollary 6.2.8 *With the notations of diagram* (6.3), *the addition map*

$$\mu : P(k_\sigma) \times k_\sigma^* P(h_\sigma) \times f^* J \longrightarrow \tilde{J}$$

gives the isotypical decomposition with respect to the action of D_p.

Proposition 6.2.9 *With the notations of diagram* (6.3) *and the above map*

$$\Psi : P(k_\sigma) \times P(h_\sigma) \times J \longrightarrow \tilde{J}, \qquad (x, y, z) \mapsto x + k_\sigma^* y + f^* z$$

is an isogeny of degree

$$\deg \Psi = \frac{2^{2g}\, p^{4g-2+2t}}{|\operatorname{Ker} k_\sigma^*| \cdot |\operatorname{Ker} h_\sigma^*|}$$

with

$$|\operatorname{Ker} k_\sigma^*| = \begin{cases} 1 \ if \ s > 0 \\ p \ if \ s = 0 \end{cases} \quad and \quad |\operatorname{Ker} h_\sigma^*| = \begin{cases} 1 \ if \ t > 0 \\ 2 \ if \ t = 0. \end{cases}$$

Note that Corollary 3.2.14(b) also gives the degree of the isotypical decomposition μ. This is only slightly more complicated to formulate, due to the fact that there are two possibilities for $|\operatorname{Ker} k_\sigma^*|_{P(h_\sigma)}|$.

Proof According to Theorem 3.2.12(b), we have

$$\deg \Psi = \frac{|P(h_\sigma)[p]|}{|\operatorname{Ker} k_\sigma^*|} \cdot \frac{|J[2p]|}{|\operatorname{Ker} h_\sigma^*|}.$$

According to Lemma 6.2.4 we have

$$|P(h_\sigma)[p]| \cdot |J[2p]| = p^{2g-2+2t} \cdot (2p)^{2g} = 2^{2g} \cdot p^{4g-2+2t}.$$

Finally, for the formula for the cardinalities of $\operatorname{Ker} k_\sigma^*$ and $\operatorname{Ker} h_\sigma^*$, we use Proposition 3.2.2 and Lemma 6.2.2. Together this completes the proof of the proposition.
□

The following lemma gives a particular abelian subvariety B_{W_1} for the group algebra decomposition of $\widetilde{J}$.

Lemma 6.2.10 *The map*

$$\varphi : P(h_\tau) \times P(h_\tau) \to P(k_\sigma), \quad (z_1, z_2) \mapsto k_\tau^*(z_1) + \sigma k_\tau^*(z_2)$$

is an isogeny with kernel

$$\widetilde{K} = \{(z, -z) \in P(h_\tau)[p] \times P(h_\tau)[p] \mid k_\tau^*(z) \ is \ D_p - invariant\}.$$

Proof Denote

$$B := k_\tau^* P(h_\tau).$$

According to Corollary 6.2.6, the map $k_\tau^*|_{P(h_\tau)} : P(h_\tau) \to B$ is an isomorphism. Hence we have to show that the composition

$$B \times B \xrightarrow{1 \times \sigma} B \times \sigma(B) \xrightarrow{\mu} P(k_\sigma),$$

where μ denotes the addition map, is an isogeny with kernel

$$K' := \{(x, -x) \in B[p] \times B[p] \mid x \text{ is } D_p - \text{invariant}\}.$$

Note first that by Corollary 3.5.2,

$$P(k_\sigma) = \{z \in \tilde{J} \mid \sum_{i=0}^{p-1} \sigma^i z = 0\}^0$$

and

$$B = \{x \in \tilde{J}^\tau \mid \sum_{i=0}^{p-1} \sigma^i x = 0\}^0.$$

Hence both B and $\sigma(B)$ are contained in $P(k_\sigma)$ and thus $\mu(B \times \sigma(B)) \subseteq P(k_\sigma)$. From Lemma 6.2.4 one checks that

$$2 \dim B = \dim P(k_\sigma).$$

Hence it suffices to show the assertion on the kernel of $\mu : B \times \sigma(B) \to P(k_\sigma)$.
    Suppose $(x_1, \sigma(x_2)) \in \text{Ker } \mu$, that is,

$$x_1 + \sigma(x_2) = 0.$$

Using the fact that both x_1 and x_2 are fixed under τ, we get

$$x_1 + \sigma^{p-1}(x_2) = x_1 + \tau\sigma(x_2) = \tau(x_1 + \sigma(x_2)) = 0.$$

Combining the two equations, we get

$$\sigma^2(x_1) = -\sigma^2(\sigma^{p-1}(x_2)) = -\sigma(x_2) = x_1.$$

In other words, x_1 is invariant under the action of $\langle \sigma^2 \rangle = \langle \sigma \rangle$. But then $B \subseteq P(k_\tau)$ implies

$$0 = (1 + \sigma + \cdots + \sigma^{p-1})(x_1) = px_1,$$

that is, $x_1 \in B[p]$. Moreover, since x_1 is invariant under D_p and $x_2 = -\sigma^{p-1}(x_1) = -x_1$, we get $\text{Ker } \mu \subset K'$. That $K' \subseteq \text{Ker } \mu$ is obvious.   □

Corollary 6.2.11 *With the notations of diagram (6.3), we have, the group D_p acts on $P(k_\sigma)$ by the representation W_1.*

Proof Clearly D_p acts on no non-trivial abelian subvariety of $P(k_\sigma)$ via χ_0, since any such subvariety is contained in f^*J. Since D_p only admits the three irreducible rational representations χ_0, χ_1, and W_1 and since D_p acts on $P(k_\sigma)$, it suffices to show that D_p acts on no non-trivial abelian subvariety of $P(k_\sigma)$ via χ_1. But this is clear, since $B \subset \tilde{J}^\tau$.   □

From Lemma 6.2.10 we get the following group algebra decomposition of $\tilde{J}$ with respect to the action of D_p.

Theorem 6.2.12 *Let the notations be as in diagram* (6.3). *Then with the notations of Lemma 6.2.10, the map*

$$P(h_\tau)^2 \times k_\sigma^* P(h_\sigma) \times f^* J \to \tilde{J}, \qquad (x_1, x_2, y, z) \mapsto \varphi(x_1, x_2) + y + z$$

gives a group algebra decomposition of $\tilde{J}$ with respect to the action of D_p.

Proof According to Corollary 6.2.8, the isotypical decomposition of $\tilde{J}$ is given by the addition map $\mu : P(k_\sigma) \times k_\sigma^* P(h_\sigma) \times f^* J \longrightarrow \tilde{J}$. Moreover we know that the rational group algebra of G decomposes as $\mathbb{Q}[G] \simeq 2W_1 \oplus \chi_1 \oplus \chi_0$. Hence $k_\sigma^* P(h_\sigma)$ and $f^* J$ are components of the group algebra decomposition with respect to χ_1 and χ_0, respectively, and, if we choose $B_{W_1} = P(h_\tau)$, then Lemma 6.2.10 implies the assertion.   □

6.2.4 The Degree of the Isogeny φ

In Lemma 6.2.10 we already saw that the map

$$\varphi : P(h_\tau)^2 \to P(k_\sigma), \quad (z_1, z_2) \mapsto k_\tau^*(z_1) + \sigma k_\tau^*(z_2)$$

is an isogeny with kernel given by $\tilde{K}$. Here we will compute the degree of φ.

Lemma 6.2.13 *The kernel of φ is isomorphic to*

$$K := \{ z \in P(h_\tau)[p] \mid k_\tau^*(z) \text{ is } D_p - invariant \}.$$

Proof This follows immediately from Lemma 6.2.10 by projection onto the first factor.   □

Consider the diagram (6.3) and denote the number of fixed points of σ by $2s$. If $s \geq 1$, we let $\{p_1, \ldots, p_s\}$ denote the set of total ramification points of $h_\tau : C_\tau \to C$. For each i, $2 \leq i \leq s$, we choose a line bundle $m_i \in \text{Pic}^0(C)$ such that

$$m_i^p = \mathcal{O}_C(h_\tau(p_1) - h_\tau(p_i))$$

and define

$$\mathcal{F}_i := \mathcal{O}_{C_\tau}(p_i - p_1) \otimes h_\tau^*(m_i) \in \text{Pic}^0(JC_\tau). \tag{6.5}$$

The group K will be described by the following lemma.

Lemma 6.2.14

(i) $h_\tau^* J[p] \subset K$.

(ii) *If* $s = 0$ *or* 1, *then*

$$K = h_\tau^* J[p].$$

(iii) *If* $s \geq 2$, *then*

$$K = h_\tau^* J[p] \oplus \bigoplus_{i=2}^{s} \mathcal{F}_i \mathbb{Z}/p\mathbb{Z}.$$

Proof

Step 1:   We claim $h_\tau^* J[p] \subseteq K$.

For the proof note that $\operatorname{Ker} \operatorname{Nm} h_\tau$ consists of only one connected component, since h_τ is not a Galois cover. So

$$P(h_\tau) = \operatorname{Ker} \operatorname{Nm} h_\tau.$$

Now $k_\tau^* \circ h_\tau^*(x)$ is D_p-invariant for every $x \in J$. This implies that $h_\tau^* J[p] \subseteq K$, since $\operatorname{Nm} h_\tau \circ h_\tau^* = $ multiplication by p in J.

Step 2:   If $s = 0$, we claim that $K = h_\tau^*(J)[p]$.

For the proof note first that, since the cover $k_\sigma : \widetilde{C} \to C_\sigma$ is cyclic étale, we have that $\widetilde{J}^\sigma = k_\sigma^*(JC_\sigma)$ which by Corollary 3.5.2 and Proposition 6.1.1 is equivalent to

$$\operatorname{Ker}(1 - \sigma) = \operatorname{Im}\left(\sum_{i=0}^{p-1} \sigma^i\right).$$

Since $K = \{z \in P(h_\tau)[p] \mid \sigma k_\tau^*(z) = k_\tau^*(z)\}$, it follows, using Corollary 6.2.6, that

$$K \simeq k_\tau^*(K) = k_\tau^*(P(C_\tau/C)) \cap \operatorname{Ker}(1 - \sigma)$$

$$= k_\tau^*(P(C_\tau/C)) \cap \operatorname{Im}\left(\sum_{i=1}^{p-1} \sigma^i\right).$$

Hence for any $z \in K$, there exists a $w \in \widetilde{J}$ such that

$$k_\tau^*(z) = \left(\sum_{i=0}^{p-1} \sigma^i \right)(w).$$

Now denote $f := h_\tau \circ k_\tau$ and let $x = \mathrm{Nm}\, f(w)$. Since $\mathrm{Nm}\, f(g(w)) = \mathrm{Nm}\, f(w)$ for every $g \in D_p$, we conclude that

$$px = p\,\mathrm{Nm}\, f(w) = \mathrm{Nm}\, f\left(\sum_{i=0}^{p-1} \sigma^i(w) \right)$$

$$= \mathrm{Nm}\, f(k_\tau^*(z)) = \mathrm{Nm}\, h_\tau \circ \mathrm{Nm}\, k_\tau(k_\tau^*(z))$$

$$= \mathrm{Nm}\, h_\tau(2z) = 2\,\mathrm{Nm}\, h_\tau(z) = 0.$$

So $x \in J[p]$. It follows that $h_\tau^*(\frac{p-1}{2}x) \in K$ and hence

$$z + h_\tau^*(\frac{p-1}{2}x) \in K.$$

But then

$$k_\tau^*\left(z + h_\tau^*(\frac{p-1}{2}x) \right) = k_\tau^*(z) + \frac{p-1}{2} f^*(x) = k_\tau^*(z) + \frac{p-1}{2} f^*(\mathrm{Nm}\, f(w))$$

$$= k_\tau^*(z) + \frac{p-1}{2} \sum_{g \in D_p} g(w)$$

$$= k_\tau^*(z) + \frac{p-1}{2} \left(\sum_{i=0}^{p-1} \sigma^i(w) + \tau \sum_{i=0}^{p-1} \sigma^i(w) \right)$$

$$= k_\tau^*(z) + \frac{p-1}{2}(k_\tau^*(z) + \tau k_\tau^*(z))$$

$$= k_\tau^*(z) + (p-1)k_\tau^*(z) = p k_\tau^*(z) = 0.$$

Therefore

$$z = -h_\tau^*(\frac{p-1}{2}x) \in h_\tau^*(J)[p]$$

which completes the proof of Step 2.

Step 3:   Proof of (ii) and part of (iii) for $s \geq 1$:

Let $\{p_1, \ldots, p_s\}$ denote the set of total ramification points of $h_\tau : C_\tau \to C$ (that is, such that the points $h_\tau(p_i)$ are the branch points of type (i)). If we denote for $i = 1, \ldots, s$,

$$k_\tau^*(p_i) = p_i' + p_i'',$$

then we have

$$\sigma(p_i') = p_i', \quad \sigma(p_i'') = p_i'' \quad \text{and} \quad \tau(p_i') = p_i''.$$

Choose an integer m such that $2m \geq g(\widetilde{C})$, and consider the Abel-Jacobi map given by

$$\alpha : \widetilde{C}^{(2m)} \to \mathrm{Pic}^0(\widetilde{C}) = \widetilde{J}, \qquad D \mapsto \mathcal{O}_{\widetilde{C}}(D - m(p_1' + p_2'')).$$

Now let $z \in K$. Then $k_\tau^*(z) \in \widetilde{J}$ and we have that the linear system $\alpha^{-1}(k_\tau^*(z)) = |k_\tau^*(z)|$ is non-empty, since $2m \geq g(\widetilde{C})$.

Since $\sigma(k_\tau^*(z)) = k_\tau^*(z)$, it follows that σ induces a projective transformation on $|k_\tau^*(z)|$, which has a fixed point $\mathcal{D}$ in $|k_\tau^*(z)|$, because σ generates a cyclic group. As a σ-invariant effective divisor, it must be of the form

$$\mathcal{D} = \sum_{i=1}^{s} (m_i p_i' + n_i p_i'') + \sum_{j=1}^{u} \left(\sum_{\ell=0}^{p-1} \sigma^\ell \right) q_j$$

for some points $q_j \in \widetilde{C}$. Then we have that

$$k_\tau^*(z) = \mathcal{O}_{\widetilde{C}}(\mathcal{D} - m(p_1' + p_1'')).$$

From $\tau(k_\tau^*(z)) = k_\tau^*(z)$ it follows that

$$k_\tau^*(2z) = \mathcal{O}_{\widetilde{C}} \left(\sum_{i=1}^{s} s_i(p_i' + p_i'') + \sum_{j=1}^{u} \left(\sum_{g \in D_p} g \right) q_j - 2m(p_1' + p_1'') \right)$$

with $s_j = m_j + n_j$. Applying $\mathrm{Nm}\, k_\tau$ to this equality, we obtain

$$4z = \mathcal{O}_{C_\tau} \left(\sum_{i=2}^{s} 2s_i(p_i - p_1) + \sum_{j=1}^{u} \left[2h_\tau^*(f(q_j)) - 2pp_1 \right] \right),$$

which may be written as

$$4z = \begin{cases} \mathcal{O}_{C_\tau} \left(\sum_{i=2}^{s} b_i(p_i - p_1) \right) \otimes h_\tau^*(\mathcal{L}) & \text{if } s \geq 2, \\ \mathcal{O}_{C_\tau} & \text{if } s = 1, \end{cases}$$

with $0 \leq b_j \leq p - 1$ and $\mathcal{L} \in J = \text{Pic}^0(C)$.

Since $4z$ is a point of order p in JC_τ, and since h_τ^* is injective, we conclude from the expression for $4z$ that

$$\mathcal{L}^p = \begin{cases} \mathcal{O}_C \left(\sum_{i=2}^{s} b_i(h_\tau(p_1) - h_\tau(p_i)) \right) & \text{if } s \geq 2. \\ \mathcal{O}_C & \text{if } s = 1. \end{cases}$$

In particular we have just shown that for $s = 1$ we have $K = h_\tau^* J[p]$ which completes the proof of (ii) of the lemma in this case.

So suppose $s \geq 2$. Then we have shown using the definition of $\mathcal{F}_i$ in Eq. (6.5)

$$4z \otimes \mathcal{F}_2^{-b_2} \otimes \cdots \otimes \mathcal{F}_s^{-b_s} = h_\tau^*(\mathcal{L} \otimes m_2^{-b_2} \otimes \cdots \otimes m_s^{-b_s}),$$

with

$$(\mathcal{L} \otimes m_2^{-b_2} \otimes \cdots \otimes m_s^{-b_s})^p = \mathcal{O}_C.$$

Hence

$$4z = \mathcal{F}_2^{b_2} \otimes \cdots \otimes \mathcal{F}_s^{b_s} \otimes h_\tau(\alpha) \quad \text{with} \quad \alpha \in J \quad \text{such that} \quad \alpha^p = \mathcal{O}_C.$$

Note that any multiple of $4z$, including z, has a similar expression, since the subgroups of K generated by $4z$ and z coincide.

Step 4: In order to complete the proof of (iii), it remains to show that the numbers b_i are uniquely determined.

Suppose that they are not, that there is a non-trivial linear combination of the $\mathcal{F}_i$ with coefficients in $\mathbb{Z}/p\mathbb{Z}$ which is trivial modulo $h_\tau^* J[p]$. In other words, after reordering the p_i if needed, there must exist an $\alpha \in J[p]$ such that

$$\mathcal{F}_2^{\nu_2} \otimes \mathcal{F}_3^{\nu_3} \otimes \cdots \otimes \mathcal{F}_s^{\nu_s} = h_\tau^*(\alpha)$$

with $0 \leq \nu_2 \leq \nu_3 \leq \cdots \leq \nu_s \leq p - 1$ and $\nu_s > 0$. The last condition just means that the linear combination is non-trivial. Inserting the definition of the $\mathcal{F}_i$, this gives

$$\mathcal{O}_{C_\tau} \left(\sum_{i=2}^{s} \nu_i p_i - (\sum_{j=2}^{s} \nu_j) p_1 \right) = h_\tau^* \left(\alpha \otimes \bigotimes_{i=2}^{s} m_i^{-\nu_i} \right).$$

Now, if

$$\sum_{j=2}^{s} v_j = mp - \epsilon \quad \text{with} \quad 0 \le \epsilon \le p - 1, \tag{6.6}$$

this can be written as

$$\mathcal{O}_{C_\tau}\left(\sum_{i=2}^{s} v_i p_i + \epsilon p_1\right) = h_\tau^*\left(\mathcal{O}_C(mh_\tau(p_1)) \otimes \alpha \otimes \bigotimes_{i=2}^{s} m_i^{-v_i}\right).$$

Consider the line bundle

$$\mathcal{L} := \mathcal{O}_C(mh_\tau(p_1)) \otimes \alpha \otimes \bigotimes_{i=2}^{s} m_i^{-v_i}.$$

Clearly, $\mathcal{L} \in \mathrm{Pic}^m J$ and we have just seen that

$$h_\tau^*(\mathcal{L}) = \mathcal{O}_{C_\tau}\left(\sum_{i=2}^{s} v_i p_i + \epsilon p_1\right).$$

Using the definition of the $\mathcal{F}_i$ and the properties of α, we get

$$\mathcal{L}^p = \mathcal{O}_C(pmh_\tau(p_1)) \otimes \alpha^p \otimes_{i=2}^{s} m_i^{-pv_i}$$

$$= \mathcal{O}_C(pmh_\tau(p_1)) \otimes \mathcal{O}_C\left(\sum_{i=2}^{s} v_i(h_\tau(p_i) - h_\tau(p_1))\right)$$

$$= \mathcal{O}_C\left(\sum_{i=2}^{s} v_i h_\tau(p_i) - [\sum_{j=2}^{s} v_j - pm]h_\tau(p_1)\right)$$

$$= \mathcal{O}_C\left(\sum_{i=2}^{s} v_i h_\tau(p_i) + \epsilon h_\tau(p_1)\right),$$

where for the last equation we used Eq. (6.6). But these equations mean just that $h_\tau : C_\tau \to C$ is a Galois cover with group $\mathbb{Z}/p\mathbb{Z}$ and branched over the divisor $\sum_{i=2}^{s} v_i h_\tau(p_i) + \epsilon h_\tau(p_1)$, which is absurd. This completes the proof of Lemma 6.2.14. $\square$

The following theorem is an immediate consequence of Lemma 6.2.14.

Theorem 6.2.15 *If g denotes the genus of C, the isogeny $\varphi : P(h_\tau) \times P(h_\tau) \to P(k_\sigma)$ is of degree*

$$\deg \varphi = |K| = \begin{cases} p^{2g} & if\ s = 0, \\ p^{2g+s-1} & otherwise. \end{cases}$$

6.2.5 Curves with D_{2^α}-Action, $\alpha \geq 2$

In this subsection we consider D_n with $n = 2^\alpha$, $\alpha \geq 2$. Note that for $\alpha = 1$, the group D_2 equals the Klein group V_4, and this was studied already in Chap. 5, as well as the group D_4, that is $\alpha = 2$.

As usual let

$$G := D_{2^\alpha} = \langle \sigma, \tau \mid \sigma^{2^\alpha} = \tau^2 = (\sigma\tau)^2 = 1 \rangle.$$

Apart from the cyclic subgroups $\langle \sigma^{2^{\alpha-j-1}} \rangle$, for each j, $0 \leq j \leq \alpha - 1$, there are two non-conjugate subgroups of G of index $2^{\alpha-j}$ in G, namely,

$$H_{2j} = \langle \sigma^{2^{\alpha-j}}, \tau \rangle \quad \text{and} \quad H'_{2j} = \langle \sigma^{2^{\alpha-j}}, \sigma\tau \rangle.$$

In particular H_{2j} and H'_{2j} are not normal in G for $j = 0, \ldots, \alpha - 2$. Note that $H_{2^0} = \langle \tau \rangle$ and $H'_{2^0} = \langle \sigma\tau \rangle$ and $H_{2j} \simeq H'_{2j} \simeq D_{2j}$, the dihedral group of order 2^{j+1}, for all $j = 1, \ldots \alpha - 1$.

Let now $\widetilde{C}$ be a curve with an action of G. The aim is to compute the isotypical and a group algebra decomposition of $\widetilde{J}$. For any subgroup H of G and any element $h \in G$, we denote by

$$C_H := \widetilde{C}/H \quad \text{and} \quad C_h := \widetilde{C}/\langle h \rangle$$

the quotient curve of $\widetilde{C}$ by H, respectively, by the subgroup generated by h. In particular we have

$$C_{H_{2^0}} = C_\tau \quad \text{and} \quad C_{H'_{2^0}} = C_{\sigma\tau}.$$

With these notations we have the following diagram of curves

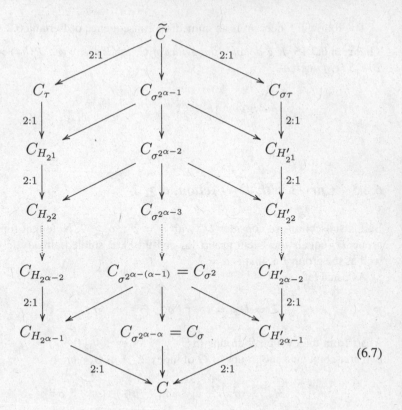

$$(6.7)$$

Moreover, we consider the following composed maps in the diagram:

$$f_{\alpha-j} : \widetilde{C} \longrightarrow C_{\sigma^{2\alpha-j}} \quad \text{for} \quad j = 0, \ldots, \alpha \quad \text{as well as}$$

$$\varphi_j : \widetilde{C} \longrightarrow C_{H_{2^j}} \quad \text{and} \quad \varphi'_j : \widetilde{C} \longrightarrow C_{H'_{2^j}} \quad \text{for} \quad j = 0 \ldots, \alpha - 1.$$

In particular, $\dot{f}_{\alpha-0} = \widetilde{C} \to \widetilde{C}$ is the identity map, and $\varphi_0 : \widetilde{C} \to C_\tau$ and $\varphi'_0 : \widetilde{C} \to C_{\sigma\tau}$ are the double covers given by the action of τ and $\sigma\tau$.

In order to determine the irreducible rational representations of G, recall from Sect. 6.2.1 that

$$\Omega_G = \{2^j \in \mathbb{N} \mid 0 \le j \le \alpha - 2\}.$$

According to Proposition 6.2.1 for every $2^j \in \Omega_G$, there is an irreducible rational representation W_{2^j} of degree $\varphi(\frac{2^\alpha}{2^j}) = 2^{\alpha-j-1}$ such that

$$\chi_0, \chi_1, \chi_2, \chi_3, W_{2^{\alpha-2}}, W_{2^{\alpha-3}}, \ldots, W_4 = W_{2^2}, W_2 = W_{2^1}, W_1 = W_{2^0}$$

are exactly all irreducible rational representations of $G = D_{2^\alpha}$. Note that the W's are ordered by increasing degree. In particular $W_{2^\alpha-2}$ is of degree 2, $W_{2^\alpha-3}$ of degree 4, etc. Moreover, the rational group algebra of G decomposes as

$$\mathbb{Q}[G] = \rho_{\{1\}} = \chi_0 \oplus \chi_1 \oplus \chi_2 \oplus \chi_3 \oplus 2W_{2^\alpha-2} \oplus 2W_{2^\alpha-3} \oplus \cdots, \oplus 2W_2 \oplus 2W_1. \quad (6.8)$$

The following theorem gives the isotypical decomposition of $\tilde{J}$ with respect to the action of $G = D_{2^\alpha}$. The main point of the proof uses the middle column of diagram (6.7).

Theorem 6.2.16 *Let $\tilde{C}$ be a curve with an action of the dihedral group $G = D_{2^\alpha}$ of order $2^{\alpha+1}$, and let the notation be as in diagram (6.7).*

The isotypical component of $\tilde{J}$ with respect to the action of G corresponding to the representation

(i) χ_0 *is* $A_{\chi_0} = f^* J$, χ_1 *is* $A_{\chi_1} = f_\sigma^* P(C_\sigma/C)$,
(ii) χ_2 *is* $A_{\chi_2} = \varphi_{\alpha-1}^* P(C_{H_{2^\alpha-1}}/C)$, χ_3 *is* $A_{\chi_3} = \varphi_{\alpha-1}'^* P(C_{H_{2^\alpha-1}'}/C)$,
(iii) W_{2j} *is* $A_{W_{2j}} = f_{\alpha-j}^* P(C_{\sigma^{2^\alpha-j}}/C_{\sigma^{2^\alpha-j-1}})$ *for* $j = \alpha - 2, \alpha - 3, \ldots, 1, 0$.

So the isotypical decomposition of $\tilde{J}$ is given by the addition map

$$\mu : \prod_{i=0}^{3} A_{\chi_i} \times \prod_{j=0}^{\alpha-2} A_{W_{2^\alpha-j}} \longrightarrow \tilde{J}.$$

Proof (i): Clearly, we have $A_{\chi_0} = f^* J$. The proof of $A_{\chi_1} = f_\sigma^* P(C_{\langle\sigma\rangle}/C)$ is the same as the analogous proof of Proposition 6.2.7.

The proof of (ii) is very similar to the analogous proof of Proposition 6.2.7, where we now use that $H_{2^\alpha-1}$ (respectively, $H_{2^\alpha-1}'$) is normal of index 2 in G. This time σ induces -1, and τ (respectively, $\sigma\tau$) acts trivially on $P(C_{2^\alpha-1}/C)$ (respectively, $P(C_{2^\alpha-1}'/C)$). This gives assertion (ii).

Alternatively, one can verify that

$$\rho_{\langle\sigma\rangle} = \chi_0 \oplus \chi_1, \quad \rho_{H_{2^\alpha-1}} = \chi_0 \oplus \chi_2 \text{ and } \rho_{H_{2^\alpha-1}'} = \chi_0 \oplus \chi_3$$

and apply Corollary 3.5.10.

(iii): Recall that for $j = 0, \ldots, \alpha$ we denote by $f_{\alpha-j} : \tilde{C} \to C_{\sigma^{2^\alpha-j}}$ the composition of the maps of the middle column of diagram (6.7). Moreover we denote for abbreviation

$$P_{\alpha-j} := P(C_{\sigma^{2^\alpha-j}}/C_{\sigma^{2^\alpha-j-1}})$$

for $j = 0, \ldots, \alpha - 1$. In particular, $P_{\alpha-0} = P(\tilde{C}/C_{\sigma^{2^\alpha-1}})$.

Applying successively Corollary 3.2.10 for $j = \alpha = \alpha - 0, \alpha - 1, \ldots, \alpha - (\alpha - 1) = 1$, we get the following isogenies

$$\tilde{J} \sim P(\tilde{C}/C_{\sigma^{2\alpha-1}}) \times f^*_{\alpha-1}JC_{\sigma^{2\alpha-1}} = P_{\alpha-0} \times f^*_{\alpha-1}JC_{\sigma^{2\alpha-1}}$$

$$\sim P_{\alpha-0} \times f^*_{\alpha-1}P_{\alpha-1} \times f^*_{\alpha-2}JC_{\sigma^{2\alpha-2}}$$

$$\sim P_{\alpha-0} \times f^*_{\alpha-1}P_{\alpha-1} \times f^*_{\alpha-2}P_{\alpha-2} \times f^*_{\alpha-3}JC_{\sigma^{2\alpha-3}}$$

$$\sim \cdots$$

$$\sim \prod_{j=0}^{\alpha-1} f^*_{\alpha-j}P_{\alpha-j} \times f^*_0 JC_\sigma.$$

$$\sim \prod_{j=0}^{\alpha-1} f^*_{\alpha-j}P_{\alpha-j} \times f^*_0 P(C_\sigma/J) \times f^*J.$$

Noting that the dihedral group of order 8 acts on C_{σ^4}, we may apply Proposition 5.3.4 to give that the addition map $\varphi^*_{\alpha-2}P(C_{H_{2\alpha-1}}/C) \times \varphi'^*_{\alpha-2}P(C_{H'_{2\alpha-1}}/C) \to f^*_1 P_1$ is an isogeny; we have that the addition map

$$\mu : f^*J \times f^*_0 P(C_0/C) \times \varphi^*_{\alpha-2}P(C_{H_{2\alpha-1}}/C) \times \varphi'^*_{\alpha-2}P(C_{H'_{2\alpha-1}}/C) \times \prod_{j=0}^{\alpha-2} f^*_{\alpha-j}P_{\alpha-j} \to \tilde{J}$$

is also an isogeny.

Now one shows by decreasing induction (i.e., starting with $\alpha - 2$ and going from from $\alpha - j$ to $\alpha - j - 1$) that $f^*_{\alpha-j}P_{\alpha-j}$ is the isotypical component of W_{2j}. For $j = \alpha - 2$ this follows just from Theorem 5.3.3, and the induction step is proved with the same argument as for the proof of Theorem 5.3.3. Also this shows that the addition map μ is equivariant. This completes the proof of the theorem. $\square$

By using the left-hand column of diagram (6.7), instead of the middle column, one can compute a group algebra decomposition of $\tilde{J}$ with respect to the action of D_{2^α}.

Proposition 6.2.17 *Let the notation be as in diagram (6.7). For $j = 0, \ldots, \alpha - 2$, a group algebra component of the decomposition of $\tilde{J}$ corresponding to the representation W_{2j} is*

$$B_{W_{2j}} = \varphi^*_j P(C_{H_{2j}}/C_{H_{2j+1}}).$$

A group algebra decomposition of $\tilde{J}$ is given by an isogeny

$$J \times P(C_\sigma/C) \times P(C_{H_{2\alpha-1}}/C) \times P(C_{H'_{2\alpha-1}}/C) \times \prod_{j=0}^{\alpha-2} P(C_{H_{2j}}/C_{H_{2j+1}})^2 \sim \tilde{J}$$

Proof Recall that for $j = 0, \ldots, \alpha - 1$, we denote by $\varphi_j : \widetilde{C} \to C_{H_{2j}}$ the composition of the corresponding maps of the left-hand column of diagram (6.7). Moreover we denote for abbreviation

$$P_{H_{2j}} := P(C_{H_{2j}}/C_{H_{2j+1}})$$

for $j = 0, \ldots, \alpha - 2$ and observe that

$$JC_{H_{2j}} \sim P_{H_{2j}} \times JC_{H_{2j+1}}.$$

Applying successively Corollary 3.2.10 to the left-hand column of diagram (6.7) for $j = 0 \ldots, \alpha - 2$, we get the following isogenies

$$\widetilde{J} \sim P(\widetilde{C}/C_{H_{20}}) \times \varphi_0^*(JC_{H_{20}}) = P(\widetilde{C}/C_\tau) \times \varphi_0^*(JC_\tau)$$

$$\sim P(\widetilde{C}/C_\tau) \times \varphi_0^* P_{H_{20}} \times \varphi_1^*(JC_{H_{21}})$$

$$\sim P(\widetilde{C}/C_\tau) \times \varphi_0^* P_{H_{20}} \times \varphi_1^* P_{H_{21}} \times \varphi_2^*(JC_{H_{22}})$$

$$\sim \cdots$$

$$\sim P(\widetilde{C}/C_\tau) \times \prod_{j-0}^{\alpha-2} \varphi_j^* P_{H_{2j}} \times \varphi_{\alpha-1}^*(JC_{H_{2\alpha-1}})$$

One checks that for $j = 0, \ldots, \alpha - 2$ we have

$$\rho_{H_{2j}} = \chi_0 \oplus \chi_2 \oplus W_{2j} \oplus W_{2j+1} \oplus \cdots \oplus W_{2\alpha-2}.$$

So we get

$$\rho_{H_{2j}} - \rho_{H_{2j+1}} = W_{2j}$$

and then Corollary 3.5.10 gives the first assertion.

Since the χ_i are characters and the irreducible complex representations V_{2i} Galois associated to W_{2i} are of dimension 2 for all i, the second assertion follows from Eq. (2.32). □

Similarly, using the right-hand column of diagram (6.7), we obtain

Proposition 6.2.18 *Let the notation be as in diagram* (6.7). *For $j = 0, \ldots, \alpha - 2$, the group algebra component of the decomposition of $\widetilde{J}$ corresponding to the representation W_{2j} is*

$$B_{W_{2j}} = \varphi_j'^* P(C_{H'_{2j}}/C_{H'_{2j+1}}).$$

Proof Denoting

$$P_{H'_{2j}} := P(C_{H'_{2j}}/C_{H'_{2j+1}})$$

for $j = 0, \ldots, \alpha - 2$, one checks in the same way as in the previous proof that

$$\tilde{J} \sim P(\tilde{C}/C_{H'_1}) \times \prod_{j-0}^{\alpha-2} \varphi_j'^* P_{H'_{2j}} \times \varphi_{\alpha-1}'^* (JC_{H'_{2\alpha-1}})$$

and

$$\rho_{H'_{2j}} = \chi_0 \oplus \chi_3 \oplus W_{2j} \oplus W_{2j+1} \oplus \cdots \oplus W_{2\alpha-2}.$$

So we get

$$\rho_{H'_{2j}} - \rho_{H'_{2j+1}} = W_{2j}$$

and again Corollary 3.5.10 gives the assertion.   □

Comparing the group algebra and the isotypical decomposition of $\tilde{J}$, we obtain the following proposition.

Proposition 6.2.19 *Let the notations be as in diagram (6.7). For every $j = 0, \ldots, \alpha - 2$, there is an isogeny*

$$P(C_{H_{2j}}/C_{H_{2j+1}}) \times P(C_{H'_{2j}}/C_{H'_{2j+1}}) \sim P(C_{\sigma^{2\alpha-j}}/C_{\sigma^{2\alpha-j-1}}).$$

There is also an isogeny

$$P(C_{H_{2\alpha-1}}/C) \times P(C_{H'_{2\alpha-1}}/C) \sim P(C_{\sigma^2}/C_\sigma).$$

Proof For the last isogeny, observe that

$$\rho_{\langle\sigma^2\rangle} = \chi_0 \oplus \chi_1 \oplus \chi_2 \oplus \chi_3$$
$$= \rho_{\langle\sigma\rangle} \oplus \chi_2 \oplus \chi_3.$$

Comparing the decompositions of $\tilde{J}$ and applying Theorem 6.2.16 and Propositions 6.2.17 and 6.2.18 gives the assertion.   □

6.2.6 Curves with a D_{2p}-Action: Ramification and Genera

In this section we consider the special case of the dihedral group of order $4p$ with an odd prime p. We will see that again in general there is an isotypical component, which is not a Prym variety, but only a Prym variety of a pair of covers.

So denote

$$G := D_{2p} = \langle \sigma, \tau \mid \sigma^{2p} = \tau^2 = (\sigma\tau)^2 = 1 \rangle.$$

G has three conjugacy classes of involutions represented by τ, $\sigma^p \tau$, and σ^p. There is one conjugay class of Klein subgroups, represented by $K := \langle \tau, \sigma^p \rangle$. There is one subgroup of order p, namely, the one generated by σ^2, and there are three subgroups of order $2p$, all normal in G: the dihedral groups of order $2p$ $D_p = \langle \sigma^2, \tau \rangle$ and $\widetilde{D}_p = \langle \sigma^2, \sigma^p \tau \rangle$ and the cyclic subgroup generated by σ. Finally, the cyclic subgroup of order p generated by σ^2 is normal in G.

If then $\widetilde{C}$ is a curve with a G-action, we have the following diagram of subcovers up to conjugacy, where on the right-hand side we denote the order of the corresponding subgroups.

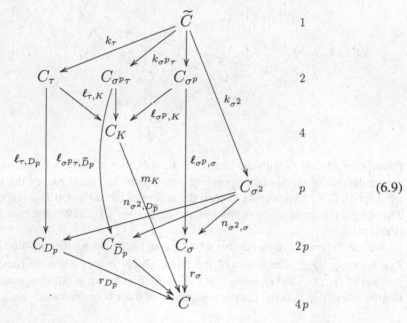

$$(6.9)$$

As in Lemma 5.3.1, one checks that the number of fixed points of σ is even, the number of fixed points of σ^2 not fixed by σ is divisible by 4, and the number of fixed points of σ^p not fixed by σ is divisible by $2p$. Hence we may denote

- $2s := |\operatorname{Fix}\sigma|$
- $2t_1 := |\operatorname{Fix}\tau|$

- $2t_2 := |\operatorname{Fix}(\sigma^p \tau)|$
- $4r := |\operatorname{Fix}\sigma^2 \setminus \operatorname{Fix}\sigma|$
- $2pu := |\operatorname{Fix}\sigma^p \setminus \operatorname{Fix}\sigma|$

Using these definitions, the Hurwitz formula, and the commutativity of diagram (6.9), we get the following lemma.

Lemma 6.2.20 *With these notations the ramification degrees $|R(f)|$ of the maps f in the diagram 6.9 are as follows:*

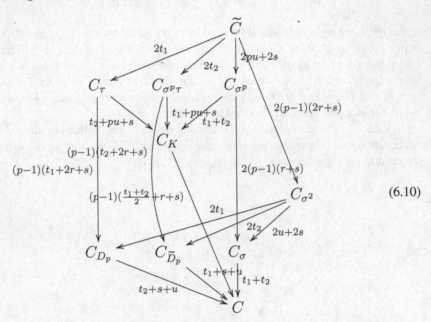

$$(6.10)$$

Proof Considering $\widetilde{C}$ with the action of the subgroup K only, we get the ramification degrees of the six corresponding maps at the left-hand part of the diagram by diagram (5.8). Considering $\widetilde{C}$ with the action of the subgroup D_p, respectively, $\widetilde{D}_p$, we get the ramification degrees of the maps over C_{D_p}, respectively, $C_{\widetilde{D}_p}$ from diagram (6.4).

We get the ramification degrees of $\ell_{\sigma^p,\sigma}$ and $n_{\sigma^2,\sigma}$ using the equation $\ell_{\sigma^p,\sigma} \circ k_{\sigma^p} = n_{\sigma^2,\sigma} \circ k_{\sigma^2}$. The action of the Klein group $\langle \sigma^2, \tau \rangle$ gives the ramification degrees of $r_{D_p}, r_{\widetilde{D}_p}$ and r_σ using diagram (5.8). Finally we get the last ramification degree, namely, m_K, using the commutativity of the whole diagram. $\square$

Remark 6.2.21 There are restrictions on the numbers $4r, s, t_1, t_2,$ and u. For example, the Hurwitz formula implies that $t_1 + t_2$ and $t_i + s + u$ must be even for $i = 1$ and 2. If $C = \mathbb{P}^1$, the numbers $t_1 + t_2$ and $t_i + s + u$ must be positive.

The following corollary is an immediate consequence of the Hurwitz formula.

Corollary 6.2.22 *Let $\widetilde{C}$ be a curve with and action of the group D_{2p} with C of genus $g \geq 0$ and the numbers $r, s, t_1, t_2,$ and u as defined above. Then the curves in diagram (6.9) are of genus*

$$g(C_{D_p}) = 2g - 1 + \frac{1}{2}(t_2 + s + u), \quad g(C_{\widetilde{D}_p}) = 2g - 1 + \frac{1}{2}(t_1 + s + u),$$

$$g(C_\sigma) = 2g - 1 + \frac{1}{2}(t_1 + t_2),$$

$$g(C_{\sigma^2}) = 4g - 3 + t_1 + t_2 + s + u, \quad g(C_K) = pg + \frac{p-1}{2}(\frac{t_1 + t_2}{2} + r + s - 2),$$

$$g(C_\tau) = 2pg - 1 + (p-1)(\frac{t_1 + t_2}{2} + r + s - 2) + \frac{1}{2}(t_2 + s + pu),$$

$$g(C_{\sigma^p\tau}) = 2pg - 1 + (p-1)(\frac{t_1 + t_2}{2} + r + s - 2) + \frac{1}{2}(t_1 + s + pu),$$

$$g(C_{\sigma^p}) = 2pg - 2p + 1 + p\frac{t_1 + t_2}{2} + (p-1)(r + s),$$

$$g(\widetilde{C}) = 4pg - 3 + p(t_1 + t_2) + 2(p-1)(r + s - 2) + s + pu.$$

6.2.7 Curves with D_{2p}-Action: Decomposition of $\widetilde{J}$

Let p be an odd prime. According to Proposition 6.2.1, we have the following lemma, where the statement follows by comparing dimensions with a little additional argument.

Lemma 6.2.23 *The six representations $\chi_j, \cdot j = 0, \ldots, 3$ of degree 1 and*

$$W_1 = \bigoplus_{i=1}^{(p-1)/2} V_{2i-1} \quad and \quad W_2 = \bigoplus_{i=1}^{(p-1)/2} V_{2i}$$

of degree $p-1$ are exactly the irreducible rational representations of D_{2p}. Moreover the rational group algebra of D_{2p} decomposes as

$$\mathbb{Q}[D_{2p}] = \rho_{\{1\}} = \chi_0 \oplus \chi_1 \oplus \chi_2 \oplus \chi_3 \oplus 2W_1 \oplus 2W_2.$$

Let $\widetilde{C}$ be a curve with D_{2p} action with the notation of the preceeding subsection. The following theorem gives the isotypical decomposition of $\widetilde{J}$ with respect to the action of the group D_{2p}. For this we denote for any subgroup H of D_{2p} by f_H the natural quotient map $f_H : \widetilde{C} \to C_H$. If H is generated by an element h, we write $f_h = f_{\langle h \rangle}$ and finally $f = f_{D_{2p}} : \widetilde{C} \to C$.

Theorem 6.2.24 *Let $\widetilde{C}$ be a curve with an action of the dihedral group D_{2p} of order $4p$, and let the notation be as in diagram (6.9).*

The isotypical component of $\widetilde{J}$ with respect to the action corresponding to the representation

(i) χ_0 *is* $A_{\chi_0} = f^* J$; χ_1 *is* $A_{\chi_1} = f_\sigma^* P(r_\sigma)$.

(ii) χ_2 *is* $A_{\chi_2} = f_{D_p}^* P(r_{D_p})$; χ_3 *is* $A_{\chi_3} = f_{\widetilde{D}_p}^* P(r_{\widetilde{D}_p})$.

(iii) W_1 *is* $A_{W_1} = P(k_{\sigma^p}, k_{\sigma^2})$. *A group algebra component for W_1 is*

$$B_{W_1} = [k_\tau^* P(C_\tau/C_{D_p}) \cap k_\tau^* P(C_\tau/C_K)]^0 = k_\tau^* P(\ell_{\tau, D_p}, \ell_{\tau, K}).$$

(iv) W_2 *is* $A_{W_2} = k_{\sigma^p}^* P(\ell_{\sigma^p, \sigma})$. *A group algebra component for W_2 is*

$$B_{W_2} = P(C_K/C).$$

So the isotypical decomposition of $\widetilde{J}$ is given by the addition map

$$\mu : \prod_{i=0}^{3} A_{\chi_i} \times A_{W_1} \times A_{W_2} \to \widetilde{J}.$$

A group algebra decomposition of $\widetilde{J}$ is given by an isogeny

$$\widetilde{J} \sim \prod_{i=0}^{3} A_{\chi_i} \times B_{W_1}^2 \times B_{W_2}^2.$$

Proof The assertion $A_{\chi_0} = f^* J$ is clear, since D_{2p} acts on $f^* J$ by χ_0 as always. Moreover one checks

$$\rho_{\langle \sigma \rangle} = \chi_0 + \chi_1 = \rho_G + \chi_1, \quad \rho_{D_p} = \rho_G + \chi_2 \quad \text{and} \quad \rho_{\widetilde{D}_p} = \rho_G + \chi_3.$$

So Proposition 3.5.10 gives the rest of (i) and assertion (ii).

(iii): For the first assertion, one verifies that

$$\rho_1 = \rho_{\sigma^p} + \chi_2 + \chi_3 + 2W_1 = \rho_{\sigma^2} + 2W_2 + 2W_1$$

and applies Corollary 3.5.12.

Then one checks

$$\rho_{\langle\tau\rangle} = \rho_{D_p} \oplus W_1 \oplus W_2 = \rho_K \oplus W_1 \oplus \chi_3.$$

So Corollary 3.5.14 implies

$$B_{W_1} = [P(C_\tau/C_{D_p}) \cap P(C_\tau/C_K)]^0.$$

According to Corollary 3.4.3, we have for the Prym variety of the pair of covers $\ell_{\tau,D_p} : C_\tau \to C_{C_p}$ and $\ell_{\tau,K} : C_\tau \to C_K$,

$$P(\ell_{\tau,D_p}, \ell_{\tau,K}) \subseteq [P(C_\tau/C_{D_p}) \cap P(C_\tau/C_K)]^0.$$

But using Corollary 6.2.22, one immediately checks that both abelian varieties are of the same dimension. This implies the assertion on B_{W_1}.

(iv): One checks

$$\rho_{\langle\sigma^p\rangle} = \rho_{\langle\sigma\rangle} + 2W_2 \quad \text{and} \quad \rho_K = \chi_0 + W_2.$$

Let V_2 denote a complex representation associated to W_2. Then we have

$$t_{W_2} := \dim V_2^{\sigma^p} - \dim V_2^\sigma = 2.$$

(For the definition of t_{W_2}, see Corollary 3.5.9.) Then the second assertion of Corollary 3.5.10 gives for the isotypical component of W_2 in $\widetilde{J}$

$$A_{W_2} = k_{\sigma^p}^* P(C_{\sigma^p}/C_\sigma) = k_{\sigma^p}^* P(\ell_{\sigma^p,\sigma}).$$

The first assertion of the proposition gives a componen B_{W_2} as

$$B_{W_2} = P(C_K/C).$$

This completes the proof of the theorem.   □

Remark 6.2.25 If $f : \widetilde{C} \to C$ is a general Galois cover with group D_{2p}, then diagram (6.9) consists of all subcovers of $\widetilde{C}$, and it is easy to check using Corollary 6.2.22 that the isotypical component $A_{W_1} = P(k_{\sigma^p}, k_{\sigma^2})$ is not a Prym variety of one of the subcovers.

6.2.8 Some Isogenies Between Prym Subvarieties

Let $\widetilde{C}$ be a curve with a D_{2p} action and diagram (6.9) of subcovers. Considering $\widetilde{C}$ with an action of the subgroups $D_p = \langle\tau, \sigma^2\rangle$ and $\widetilde{D}_p = \langle\sigma^p\tau, \sigma^2\rangle$ as well as considering the curve C_{σ^p} with an action of $D_p \simeq \langle \tau, \sigma \rangle$, we get from Lemma 6.2.10 the following isogenies.

Proposition 6.2.26 *The maps*

(i) $P(C_\tau/C_{D_p}) \times P(C_\tau/D_{D_p}) \to P(\widetilde{C}/C_{\sigma^2}), \quad (x, y) \mapsto k_\tau^*(x) + \sigma k_\tau^*(y),$

(ii) $P(C_{\sigma^p\tau}/C_{\widetilde{D}_p}) \times P(C_{\sigma^p\tau}/D_{\widetilde{D}_p}) \to P(\widetilde{C}/C_{\sigma^2}), \quad (x, y) \mapsto k_{\sigma^p\tau}^*(x) + \sigma k_{\sigma^p\tau}^*(y),$

(iii) $P(C_K/C) \times P(C_K/C) \to P(C_{\sigma^p}/C_\sigma), \quad (x, y) \mapsto \ell_{\sigma^p, K}^*(x) + \sigma \ell_{\sigma^p, K}^*(y)$

are isogenies of degrees given by Theorem 6.2.15.

The Klein group of order 4 acts on the curve C_{σ^2}. In fact, the group $\langle \sigma^2 \rangle$ is normal of index 4 in G. So σ and τ induce automorphisms on C_{σ^2} which we denote by the same letter. Hence we get from Proposition 5.2.3 the following isogenies.

Proposition 6.2.27 *The maps*

(i) $P(C_{D_p}/C) \times P(C_{\widetilde{D}_p}/C) \to P(C_{\sigma^2}/C_\sigma), \quad (x, y) \mapsto n_{\sigma^2, D_p}^*(x) + n_{\sigma^2, \widetilde{D}_p}^*(y),$

(ii) $P(C_{D_p}/C) \times P(C_\sigma/C) \to P(C_{\sigma^2}/C_{\widetilde{D}_p}), \quad (x, y) \mapsto n_{\sigma^2, D_p}^*(x) + n_{\sigma^2, \sigma}^*(y),$

(iii) $P(C_{\widetilde{D}_p}/C) \times P(C_\sigma/C) \to P(C_{\sigma^2}/C_{D_p}), \quad (x, y) \mapsto n_{\sigma^2, \widetilde{D}_p}^*(x) + n_{\sigma^2, \sigma}^*(y)$

are isogenies with kernel in the products of the 2-division points.

A special result for the case of a D_{2p} action is the following theorem. Let the notation be as in Sect. 6.2.6. In particular the ramifications are given by the numbers $s, t_1, t_2, r,$ and u. Then note that

$$\dim P(C_\tau/C_{D_p}) = \dim P(C_{\sigma^p\tau}/C_{\widetilde{D}_p})$$

which suggest that there might be an isogeny between them.

In fact, we want to show that $P(C_\tau/C_{D_p})$ is isogenous to $P(C_{\sigma^p\tau}/C_{\widetilde{D}_p})$. For this it turns out to be convenient to take a conjugate Prym variety: the curve $C_{\sigma\tau}$ is conjugate to $C_{\sigma^p\tau}$. Since $\sigma\tau \in \widetilde{D}_p = \langle \sigma^2, \sigma^p\tau \rangle = \langle \sigma^2, \sigma\tau \rangle$, this implies that $P(C_{\sigma^p\tau}/C_{\widetilde{D}_p})$ is isomorphic to $P(C_{\sigma\tau}/C_{\widetilde{D}_p})$. So instead of the above mentioned isogeny, we will show the following theorem, where $k_{\sigma\tau} : C_{\sigma\tau} \to C_{\widetilde{D}_p})$ denotes the natural projection.

Theorem 6.2.28 *The map*

$$\varphi = \mathrm{Nm}\, k_{\sigma\tau} \circ k_\tau^*|_{P(C_\tau/C_{D_p})} : P(C_\tau/C_{D_p}) \to P(C_{\sigma\tau}/C_{\widetilde{D}_p})$$

is an isogeny of degree

$$\deg \varphi = \begin{cases} p^{2g-2+s+t_2+u} & \text{if } r = 0; \\ p^{2g-2+r+t_2+u} & \text{if } s = 0; \\ p^{2g-1+r+s+t_2+u} & \text{if } rs > 0. \end{cases}$$

Proof Let $A := k_\tau^*(P(C_\tau/C_{D_p})) \subset \tilde{J}$ and $B := k_{\sigma\tau}^*(P(C_{\sigma\tau}/C_{\tilde{D}_p})) \subset \tilde{J}$. Then we have the following commutative diagram

$$
\begin{array}{ccc}
 & A \xrightarrow{\;\;1+\sigma\tau\;\;} B & \\
{\scriptstyle k_\tau^*}\nearrow & {\scriptstyle \mathrm{Nm}\,k_{\sigma\tau}^*}\searrow & \nearrow{\scriptstyle k_{\sigma\tau}^*} \\
P(C_\tau/C_{D_p}) \xrightarrow{\;\;\varphi\;\;} & P(C_{\sigma\tau}/C_{\tilde{D}_p}) &
\end{array}
\tag{6.11}
$$

According to Proposition 3.5.6, we have

$$
A = \{x \in \tilde{J}^\tau \mid (1 + \sigma^2 + \sigma^4 + \cdots + \sigma^{2(p-1)})(x) = 0\}^0
$$

and

$$
B = \{y \in \tilde{J}^{\sigma\tau} \mid (1 + \sigma^2 + \sigma^4 + \cdots + \sigma^{2(p-1)})(y) = 0\}^0.
$$

Note that $(1 + \sigma\tau)|_A = (1 + \sigma)|_A$.

Step 1: φ is an isogeny.

To see this, recall the irreducible complex representations $\chi_0, \ldots, \chi_3$ of degree 1 and $V_1, \ldots, V_{p-1}$ of degree 2 of D_{2p}, and let

$$
T_0\tilde{J} = m_0\chi_0 \oplus \cdots \oplus m_3\chi_3 \oplus n_1 V_1 \oplus \cdots \oplus n_{p-1}V_{p-1}
$$

denote the complex isotypical decomposition of the tangent space $T_0\tilde{J}$. Comparing with the corresponding actions, we have

$$
T_0 A = dk_\tau^*(T_0 P(C_\tau/C_{\tilde{D}_p})) = \bigoplus_{i=1}^{p-1}(n_i V_i)^\tau \quad \text{and}
$$

$$
T_0 B = dk_{\sigma\tau}^*(T_0 P(C_{\sigma\tau}/C_{D_p})) = \bigoplus_{i=1}^{p-1}(n_i V_i)^{\sigma\tau}.
$$

Furthermore, clearly we have for each i

$$
(1 + \sigma\tau)(n_i V_i^\tau) = n_i V_i^{\sigma\tau}.
$$

It follows that

$$
(1 + \sigma\tau)\Big(\bigoplus_{i=1}^{p-1} n_i V_i^\tau\Big) = \bigoplus_{i=1}^{p-1} n_i V_i^{\sigma\tau}.
$$

It follows from diagram (6.11) that φ is an isogeny.

Step 2: $\mathrm{Ker}(1 + \sigma\tau)|_A \subseteq A[p]$.

To see this, note first that

$$\mathrm{Ker}(1 + \sigma\tau)|_A = \mathrm{Ker}(1 + \sigma)|_A = \{x \in A \mid x = -\sigma(x)\}.$$

So, if $x \in \mathrm{Ker}(1 + \sigma)|_A$, it follows that

$$0 = (1 + \sigma^2 + \sigma^4 + \cdots + \sigma^{2(p-1)})(x) = px,$$

where the last equation holds, since $\sigma^2(x) = x$.

Step 3:   We claim that

$$\mathrm{Ker}\,\varphi = \{x \in P(C_\tau/C_{D_p})[p] \mid \sigma k_\tau^*(x) + k_\tau^*(x) = 0\} \simeq K/L \qquad (6.12)$$

where

$$K := \{x \in P(C_\tau/C_{D_p})[p] \mid \sigma^2 k_\tau^*(x) = k_\tau^*(x)\} \quad \text{and}$$

$$L := \{x \in P(C_\tau/C_{D_p})[p] \mid \sigma k_\tau^*(x) = k_\tau^*(x)\}.$$

For the proof note first that

$$k_\tau^{*-1}(A[p]) = P(C_\tau/C_{D_p})[p] + \mathrm{Ker}\,k_\tau^*$$

and

$$\mathrm{Ker}\,k_\tau^* \subset JC_\tau[2].$$

Since p and 2 are relatively prime, this gives

$$k_\tau^{*-1}(A[p]) = P(C_\tau/C_{D_p})[p]$$

and hence

$$\mathrm{Ker}\,\varphi \subseteq P(C_\tau/C_{D_p})[p].$$

Analogously we get $\mathrm{Ker}\,k_{\sigma\tau}^* \subseteq JC_{\sigma\tau}[2]$ which implies finally

$$\mathrm{Ker}(k_{\sigma\tau} \circ \varphi) = \mathrm{Ker}\,\varphi.$$

By the commutativity of diagram (6.11), this gives the first equation of (6.12). The isomorphism with K/L is a triviality.

Step 4:   We claim that the cardinality of K is

$$|K| = \begin{cases} p^{4g-2+t_2+u} & \text{if } r = s = 0, \\ p^{4g-3+t_2+u+2r+2s} & \text{otherwise.} \end{cases} \tag{6.13}$$

For the proof we consider the curve $\widetilde{C}$ with the action of the subgroup $D_p := \langle \sigma^2, \tau \rangle$ and the following subdiagram of diagram (6.9):

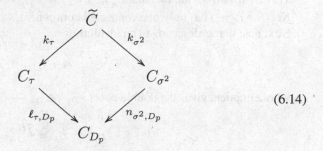

$$\tag{6.14}$$

Now note that K of Step 3 coincides with K of Lemma 6.2.13 applied to diagram (6.14). According to diagram (6.10),

$$\alpha := 2r + s$$

denotes the set of total ramification points of the map ℓ_{τ, D_p}. (Note that t_1 counts the number of simple ramification points). Hence we can apply Theorem 6.2.15 to give

$$\deg \varphi = \begin{cases} p^{2g(C_{D_p})} & \text{if } \alpha = 0, \\ p^{2g(C_{D_p})+\alpha-1} & \text{if } \alpha \geq 1. \end{cases}$$

Inserting the genus of C_{D_p} according to Corollary 6.2.22 gives the assertion of Step 4.

Step 5:   We claim that the cardinality of L is

$$|L| = \begin{cases} p^{2g-2+r+s} & \text{if } r \geq 2 \text{ and } s \geq 2, \\ p^{2g-1+s} & \text{if } r \leq 1 \text{ and } s \geq 2, \\ p^{2g-1+r} & \text{if } r \geq 2 \text{ and } s \leq 1, \\ p^{2g} & \text{if } r \leq 1 \text{ and } s \leq 1. \end{cases} \tag{6.15}$$

For the proof note first that

$$L = \{x \in P(C_\tau/C_{D_p})[p] \mid \sigma k_\tau^*(x) = k_\tau^*(x)\}$$
$$= \{x \in P(C_\tau/C_{D_p})[p] \mid k_\tau^*(x) \text{ is } D_{2p} - invariant\}.$$

Again we consider the curve $\widetilde{C}$ with the action of the subgroup $D_p \subset D_{2p}$ with corresponding diagram (6.14). We want to apply Theorem 6.2.15 to this situation, which is a bit different from the situation in Step 4.
Observe that

$$P(C_\tau / C_{D_p}) = \mathrm{Ker}(\mathrm{Nm}\, \ell_{\tau, D_p}). \tag{6.16}$$

This follows from the fact that $\ell^*_{\tau, D_p} : C_\tau \to C_{D_p}$ is not a Galois cover. Hence $\mathrm{Ker}(\mathrm{Nm}\, \ell_{\tau, D_p})$ has only one connected component, which was to be shown. Next note that σ descends to an involution

$$\widetilde{\sigma} : C_{D_p} \to C_{D_p}$$

whose quotient gives the double cover $r_{D_p} : C_{D_p} \to C$. It follows that

$$(k^*_\tau \circ \ell^*_{\tau, D_p}) J C^{\widetilde{\sigma}}_{D_p} \subseteq \widetilde{J}^{D_{2p}}.$$

On the other hand, we have

$$\ell^*_{\tau, D_p}(J C^{\widetilde{\sigma}}_{D_p}[p]) \subseteq \mathrm{Ker}(\mathrm{Nm}\, \ell_{\tau, D_p}) = P(C_\tau / C_{D_p})$$

since $\mathrm{Nm}\, \ell_{\tau, D_p} \circ \ell^*_{\tau, D_p} = p J C_{D_p}$ and by (6.16). It follows that

$$L = \ell^*_{\tau, D_p}(J C^{\widetilde{\sigma}}_{D_p}[p]) + \text{part coming from the ramification of } \ell_{\tau, D_p}. \tag{6.17}$$

Since p is an odd prime, it follows from Corollary 3.5.4 that

$$J C^{\widetilde{\sigma}}_{D_p}[p] = r^*_{D_p}(J[p]).$$

We have to compute the part coming from the ramification. For this note that total ramification points of the cover $\ell_{\tau, D_p} : C_\tau \to C_{D_p}$ appear only from points of types r and s, that is, coming from points of $\mathrm{Fix}\,\sigma^2 \setminus \mathrm{Fix}\,\sigma$ and of $\mathrm{Fix}\,\sigma$.
Let $\{p_1, \ldots, p_s\}$ in C_τ denote the points of type s, and let $\{q_1, \ldots q_{2r}\}$ in C_τ denote the points of type r. So, each p_i is a total ramification point of ℓ_{τ, D_p} such that $\widetilde{\sigma}\ell_{\tau, D_p}(p_i) = \ell_{\tau, D_p}(p_i)$ for all i, and, labelling the q_j appropriately, each q_j is a total ramification point of ℓ_{τ, D_p} such that $\widetilde{\sigma}\ell_{\tau, D_p}(q_j) = \ell_{\tau, D_p}(q_{j+1})$ for $j = 1, 3, \ldots .2r - 1$.

In order to apply Theorem 6.2.15, for each $i = 2, 3, \ldots, s$ and each $j = 3, 5, \ldots, 2r - 1$, choose m_i and n_j in $J C^{\widetilde{\sigma}}_{D_p}$ such that

$$m_i^2 = \mathcal{O}_{C_{D_p}}(\ell_{\tau, D_p}(p_1) - \ell_{\tau, D_p}(p_i))$$

and

$$n_j^2 = \mathcal{O}_{C_{D_p}}(\ell_{\tau,D_p}(q_1) + \ell_{\tau,D_p}(q_2) - \ell_{\tau,D_p}(q_j) - \ell_{\tau,D_p}(q_{j+1}))$$

and define for $i = 2, 3, \ldots, s$,

$$\mathcal{F}_i = \mathcal{O}_{C_\tau}(p_i - p_1) \otimes \ell^*_{\tau,D_p}(m_i)$$

and for $j = 3, 5, \ldots, 2r - 1$,

$$\mathcal{G}_j = \mathcal{O}_{C_\tau}(q_j + q_{j+1} - q_1 - q_2) \otimes \ell^*_{\tau,D_p}(n_j).$$

Then Theorem 6.2.15 gives

$$L = \ell^*_{\tau,D_p} \mathrm{or}^*_{D_p} J[p] \oplus \bigoplus_{i=2}^{s} \mathcal{F}_i \mathbb{Z}/p\mathbb{Z} \oplus \mathcal{G}_3 \mathbb{Z}/p\mathbb{Z} \oplus \mathcal{G}_5 \mathbb{Z}/p\mathbb{Z} \oplus \cdots \oplus \mathcal{G}_{2r-1} \mathbb{Z}/p\mathbb{Z}.$$

Note that there is no relation between the $\mathcal{F}_i$ and $\mathcal{G}_j$, because we have already taken away the relevant elements of $P(\ell_{\tau,D_p})[p]$ coming from the ramification. This gives the assertion of Step 5.

Step 6: Since $\mathrm{Ker}\,\varphi \simeq K/L$ by Step 3, we conclude the proof of Theorem 6.2.28 by combining Step 4 and Step 5.
$\square$

6.3 Semidirect Products of a Group of Order 3 by Powers of K_4

In this section we will show that for any positive integer N, there exists a smooth projective curve X whose Jacobian is isogenous to the product of $m \geq N$ Jacobian varieties of the same dimension. For this we use the action of a semidirect product of a cyclic group of order 3 by a self-product with a high number of factors of the Klein group K_4 of order 4 on a smooth curve. Here we follow [5].

6.3.1 The Group G_n

Consider the alternating group $\mathcal{A}_4$ and its presentation as a semidirect product $\mathcal{A}_4 = K_4 \rtimes B_3$ of the Klein group K_4 and a cyclic group $B_3 = \langle b \rangle \simeq \mathbb{Z}_3$. For any positive integer n, let

$$G_n := N_n \rtimes B_3 := (\langle a_1, a_2 \rangle \times \langle a_3, a_4 \rangle \times \cdots \times \langle a_{2n-1}, a_{2n} \rangle) \rtimes \langle b \rangle$$

where the subgroups $\langle a_j, a_{j+1} \rangle \simeq K_4$ for each odd j and $\langle a_j, a_{j+1} \rangle \rtimes \langle b \rangle \simeq \mathcal{A}_4$ for $j = 1, 3, \ldots, 2n - 1$. Obviously G_n is of order

$$|G_n| = 2^{2n}3$$

and B_3 is a 3-Sylow subgroup of G_n with $B_3 = N_{G_n}(B_3) = C_{G_n}(B_3)$, where N_{G_n} and C_{G_n} denote the normalizer and centralizer in G_n, respectively.

Lemma 6.3.1 G_n *does not admit any subgroup of index* 2.

Proof Suppose R is a subgroup of G_n of index 2. Then R is normal in G_n and by the Sylow Theorem, $B_3 \subseteq R$. This implies $G_n = N_n B_3 = N_n R$ and hence

$$Q := N_n \cap R \lhd G_n \quad \text{with} \quad [N_n : Q] = 2.$$

Hence $N_n = \langle n_1 \rangle Q$ with some $n_1 \in N_n$. Since $Q \lhd G_n$, we have

$$(n_1)^b := bn_1 b^{-1} = n_2 \notin Q \quad \text{and} \quad (n_1)^{b^2} = n_1 n_2 \notin Q.$$

The last equation follows from the fact that $\langle n_1, n_2 \rangle \simeq K_4$ and $(n_1)^{b^2} \neq n_1, n_2$. Then $\langle n_1, n_2 \rangle \cap Q = \{1\}$ and hence

$$2 = [N_n : Q] = \frac{|\langle n_1, n_2 \rangle|}{|\langle n_1, n_2 \rangle \cap Q|} = 4,$$

a contradiction.   □

There are however subgroups of index 4. For example,

$$M_1 := (\langle a_3, a_4 \rangle \times \cdots \times \langle a_{2n-1}, a_{2n} \rangle) \rtimes \langle b \rangle$$

is such a subgroup. According to Lemma 6.3.1, any such subgroup M is maximal in G_n; that is, there is no subgroup strictly between M and G_n. In each conjugacy class of subgroups of index 4, there is exactly one which contains B_3. In this section we will see that if we consider only such subgroups containing B_3, they correspond bijectively to the subgroups U of index 4 in N_n that are normal in G_n and furthermore correspond bijectively to the irreducible representations of degree 3 of G_n.

Lemma 6.3.2 *Let* $M \subseteq G_n$ *be any subgroup of index* 4 *containing* B_3. *Then*

$$G_n = N_n M \quad \text{and} \quad U := N_n \cap M \lhd G, \ [N_n : U] = 4.$$

Proof Since M is maximal of index 4 in G_n and $N_n \lhd G_n$, we get $G_n = N_n M$. Since $N_n \lhd G_n$, we have $U = N_n \cap M \lhd M$. Also $U \lhd N_n$ since N_n is abelian. Therefore, $U \lhd N_n M = G_n$. Also $4 = [G_n : M] = [N_n : U]$.   □

It is well known by elementary number theory that for any positive integer n, the number

$$m := \frac{2^{2n} - 1}{3}$$

is an integer.

In order to determine the complex irreducible representations of G_n, we follow [34, Proposition 25]. The subindex 0 will denote the trivial representation for any group in what follows. The group G_n acts on the character group $\widehat{N}_n :=$ Hom$(N_n, \mathbb{C}^*)$ by

$$(g\psi)(h) := \psi(g^{-1}hg) \quad \text{for any} \quad g \in G_n, \; \psi \in \widehat{N}_n, \; h \in N_n.$$

In particular the subgroup B_3 acts on $\widehat{N}_n$ with stabilizer B_3 of the trivial character χ_0, whereas the stabilizer of any non-trivial character of N_n is trivial. Hence there are $1 + m$ orbits for the action of B_3 on $\widehat{N}_n$. Let $\psi_0, \ldots, \psi_m$ be a set of representatives. Then if ν_0, ν_1, ν_2 are the irreducible characters of B_3, consider the following representations of G_n:

- $\chi_0 = \psi_0 \otimes \nu_0$, $\chi_1 = \psi_0 \otimes \nu_1$, $\chi_2 = \psi_0 \otimes \nu_2$ of degree 1
- the induced representations

$$\theta_i := \text{Ind}_{N_n}^{G_n}(\psi_i) \quad \text{for} \quad i = 1, \ldots, m$$

of degree 3.

Then [34, Proposition 9] gives the following proposition.

Proposition 6.3.3 *The irreducible complex representations of G_n are exactly given by χ_0, χ_1, χ_2 of degree 1 and $\theta_1, \ldots, \theta_m$ of degree 3.*

Since the representations θ_i are all rational, this implies immediately

Corollary 6.3.4 *The irreducible rational representations of G_n are exactly the trivial representation χ_0, the representations θ_i of degree 3 for $i = 1, \ldots, m$, and the representation $\psi := \chi_1 \oplus \chi_2$ of degree 2.*

Lemma 6.3.5 *Let ψ be a non-trivial irreducible character of N_n and $\theta :=$ Ind$_{N_n}^{G_n}(\psi)$ the corresponding irreducible representation of G_n. If b denotes a generator of B_3, then*

(i) $[N_n : \text{Ker}(\psi)] = 2$.
(ii) $U := \text{Ker}(\psi) \cap \text{Ker}(\psi^b) = \text{Ker}(\psi) \cap \text{Ker}(\psi \psi^b)$; $\quad [N_n : U] = 4$, $\text{Ker}(\theta) = U \lhd G_n$.

Proof

(i): Since $N_n / \mathrm{Ker}(\psi)$ is isomorphic to a finite cyclic subgroup of $\mathbb{C}^*$ and $N_n \simeq (\mathbb{Z}_2)^{2n}$, we have $[N_n : \mathrm{Ker}(\psi)] = 2$.

(ii): Since $[N_n : \mathrm{Ker}(\psi)] = 2$ and $\psi \neq \psi^b$, we have $N_n = \mathrm{Ker}(\psi)\,\mathrm{Ker}(\psi^b)$. Hence $2 = [N_n : \mathrm{Ker}(\psi)] = [\mathrm{Ker}(\psi^b) : \mathrm{Ker}(\psi) \cap \mathrm{Ker}(\psi^b)]$. Therefore $[N_n : U] = 4$.

We have $\mathrm{Ker}(\psi) \cap \mathrm{Ker}(\psi^b) \subseteq \mathrm{Ker}(\psi\psi^b)$. So

$$ U := \mathrm{Ker}(\psi) \cap \mathrm{Ker}(\psi^b) = \mathrm{Ker}(\psi) \cap \mathrm{Ker}(\psi^b) \cap (\mathrm{Ker}(\psi\psi^b)) = \bigcap_{i=0}^{2} \mathrm{Ker}(\psi^i). $$

In particular U is normal in G_n. It is well known that $\mathrm{Ker}(\theta) = \bigcap_{g \in G_n} \mathrm{Ker}(\psi^g)$. Hence

$$ \mathrm{Ker}(\theta) = \bigcap_{n \in N_n, i=0,1,2} \mathrm{Ker}(\psi^{nb^i}) = \bigcap_{i=0}^{2} \mathrm{Ker}(\psi^{b^i}) = U. $$

$\square$

Conversely we have

Lemma 6.3.6 *Let* $U \lhd G_n$ *such that* $[N_n : U] = 4$. *Then*

(i) *There is a non-trivial character* ψ *of* N_n *such that* $\theta = \mathrm{Ind}_{N_n}^{G_n}(\psi)$ *is an irreducible representation of* G_n *with* $U = \mathrm{Ker}(\theta)$.

(ii) $M := UB_3 \subset G_n$ *with* $[G_n : M] = 4$.

Proof

(i): The quotient group N_n / U is isomorphic to the Klein group of order 4. Consider $L \subset N_n$ such that $U \subset L$ and $[L : U] = 2$. Then $U \subset L^{b^i}$ for all i, since $U \lhd G_n$.

There is a non-trivial character ψ of N_n such that $L = \mathrm{Ker}(\psi)$. Then by the previous lemma, we have for the induced representation $\theta = \mathrm{Ind}_{N_n}^{G_n}(\psi)$ that $U = \mathrm{Ker}(\theta)$.

(ii): Since $U \lhd G_n$ with $U \subset N_n$ and $[N_n : U] = 4$, we have $M = UB_3 \subset G_n$ and $[G_n : M] = 4$. $\square$

Combining the two lemmas, we obtain

Proposition 6.3.7 *There are canonical bijections between the following sets:*

- $\{U \lhd G_n \mid U \subset N_n,\ [N_n : U] = 4\}$.
- $\{\text{ subgroups } M \subset G_n \mid [G_n : M] = 4 \text{ and } B_3 \subset M\}$.
- $\{\theta \in \mathrm{Irr}(G_n) \mid \deg(\theta) = 3\}$.

In particular there are exactly m subgroups of index 4 in N_n which are normal in G_n.

Proof The bijections are given by

$$U = M \cap N \leftrightarrow M = U B_3 \leftrightarrow U = \mathrm{Ker}(\theta).$$

where $U \mapsto \theta$ is given by Lemma 6.3.6 and $\theta \mapsto U$ by Lemma 6.3.5. The last assertion follows from the first part of the proposition and Corollary 6.3.4. $\qquad\square$

Remark 6.3.8 Let $U \lhd G_n$ with $U \subset N_n$ of index $[N_n : U] = 4$. Let $M = U B_3$ the corresponding maximal subgroup and $U = \mathrm{Ker}\,\theta$ as in Proposition 6.3.7.

Consider $L \subset N_n$ with $U \subset L$ and $[N_n : L] = 2$. Observe that there are three such L, forming a conjugacy class of subgroups of G_n. Also consider ψ, the character of N_n with $\mathrm{Ker}(\psi) = L$. Then one checks that if b denotes a generator of B_3, then

- $U \subset \mathrm{Ker}(\psi^{b^i}) =: L_i$ for $i = 0, 1, 2$. So $L_0 = L$.
- $\theta = \mathrm{Ind}_{N_n}^{G_n}(\psi^{b^i})$ for $i = 0, 1, 2$.
- $\rho_{N_n} \simeq \chi_0 \oplus \chi_1 \oplus \chi_2$.
- $\rho_{L_i} \simeq \rho_{N_n} \oplus \theta$.
- $\rho_M \simeq \chi_0 \oplus \theta$.

6.3.2 The Diagram of Subcovers of a Curve with G_n-Action

Let

$$G_n = N_n \rtimes B_3 = (\langle a_1, a_2\rangle \times \langle a_3, a_4\rangle \times \cdots \times \langle a_{2n-1}, a_{2n}\rangle) \rtimes \langle b\rangle \simeq \mathbb{Z}_2^{2n} \rtimes \mathbb{Z}_3$$

be the group of the previous section, and let $\widetilde{C}$ denote a smooth projective curve with an action of G_n. In this subsection we study the diagram of subcovers and compute the genera of the occurring curves in a special case.

For any subgroup H of G_n, let C_H denote as usual the quotient curve

$$C_H := \widetilde{C}/H \quad \text{and} \quad C = \widetilde{C}/G_n.$$

According to Proposition 6.3.7, the subgroup N_n has exactly $m = \frac{2^{2n}-1}{3}$ subgroups of index 4 which are normal in G_n. Let $U_1, \ldots, U_m$ be these subgroups and $M_1, \ldots, M_m$ denote the corresponding maximal subgroups of G_n. For any U_j we choose a subgroup $L_j \subseteq N_n$ containing U_j with $[N_n : L_j] = 2$. According to Remark 6.3.8, the conjugacy class of L_j is uniquely determined by U_j. With these notations we have the following diagram of covers of curves.

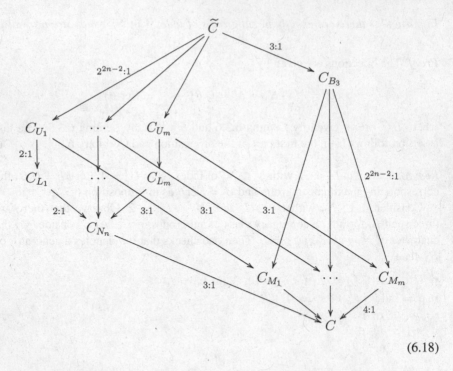

$$(6.18)$$

In the special case $n = 1$; that is, $G_1 = \mathcal{A}_4$, we have $U_1 = \{1\}$ and $M_1 = B_3$. Hence in this case the diagram coincides with diagram (5.36).

In diagram (5.37) we saw that already for $n = 1$, that is, $G_1 = \mathcal{A}_4$; there are two types of ramification for $\widetilde{C} \to C$. This implies that for high n there are many types. Here we consider only the case that the ramification over any branch point of $\widetilde{C} \to C$ is of type $(3, 3, \ldots, 3)$ with 2^{2n} numbers 3. In other words the cover $\widetilde{C} \to C$ is of signature $(g; 3, \ldots, 3)$ with say t numbers 3.

From Theorem 3.1.1 one gets immediately

Proposition 6.3.9 *For any C and sufficiently large t, there is a Galois cover $\widetilde{C}$ with group G_n and signature $(g; 3, \ldots, 3)$ with t numbers 3.*

In the special case we are interested in, that is, $g = 0$, the following proposition is more precise.

Proposition 6.3.10 *For any $n \geq 2$ and $t \geq 2n$, there is a curve $\widetilde{C}$ with an action of the group G_n with signature $(0; 3, \ldots, 3)$ with t numbers 3.*

Proof For simplicity we give a proof only for $t = 2n$ and leave the easy generalization to arbitrary $t \geq 2n$ to the reader. Recall that b is a generator of B_3 and $N_n = \langle a_1, a_2 \rangle \times \cdots \times \langle a_{2n-1}, a_{2n} \rangle$.

For $n = 2$ consider the elements $c_1 = a_1 b$, $c_2 = b$, $c_3 = a_3 b^2$, and $c_4 = b^2 a_4 a_1$. Now $c_1 c_2 c_3 c_4 = a_1 b b a_3 b^2 b^2 a_4 a_1 = a_1 b^2 a_3 b a_4 a_1 = 1$, since $b^2 a_3 b = a_4$. So there is a generating vector of type $(0; 3, 3, 3, 3)$ satisfying the conditions of Theorem 3.1.1.

For $n = 3$, it is clear that $(c_1 = a_1 b, c_2 = b, c_3 = a_3 b^2, c_4 = b^2 a_4 a_1, c_5 = a_5 b, c_6 = b^2 a_5)$ is a generating vector of type $(0; 3, 3, 3, 3, 3, 3)$ satifying the conditions of Theorem 3.1.1.

For $n \geq 4$ it is easy to generalize the proceedure described above successively.

$\square$

Lemma 6.3.11 *Let* $\widetilde{C} \to C = \mathbb{P}^1$ *be a Galois cover with group* G_n *and action with signature* $(0; 3, \ldots, 3)$ *with* t *numbers* 3. *Then*

(i) $C_{N_n} \to \mathbb{P}^1$ *is totally ramified and* $\widetilde{C} \to C_{N_n}$ *is étale.*
(ii) *Over each of the* t *branch points* p_i *of* $\widetilde{C} \to \mathbb{P}^1$, *the map* $C_{B_3} \to \mathbb{P}^1$ *admits* m *ramification points of multiplicity* 3 *and* 1 *point which is unramified over* $\mathbb{P}^1$.

Proof

(i): Since $C_{N_n} \to \mathbb{P}^1$ is cyclic of degree 3, it is either étale over p_i or totally ramified. Suppose it is étale. Since p_i is a branch point of $\widetilde{C} \to \mathbb{P}^1$, at least one of its preimages in C_{N_n} is also a branch point of the Galois cover $\widetilde{C} \to C_{N_n}$ and hence of type $(3, \ldots, 3)$. This gives a contradiction, since its degree is 2^{2n} which is not divisible by 3. Hence $C_{N_n} \to \mathbb{P}^1$ is totally ramified, and consequently $\widetilde{C} \to C_{N_n}$ is étale, since otherwise the signature of $\widetilde{C} \to \mathbb{P}^1$ would not be $(0; 3, \ldots, 3)$.

(ii): For $i = 1, \ldots, t$ let $q_i \in C_{B_3}$ be a point in the fibre over p_i. Then either q_i is of multiplicicty 3 over p_i or unramified over p_i. In the first case, the cover $\widetilde{C} \to C_{B_3}$ is étale over q_i and in the second case totally ramified over q_i.

Let r_i be the number of points in the fibre in C_{B_3} over p_i at which the map $C_{B_3} \to C$ is ramified. According to the Hurwitz formula, we get

$$g(C_{B_3}) = \sum_{i=1}^{t} r_i + 2^{2n}(-1) + 1 \leq m(t-3)$$

where the inequality follows from the inequalities $r_i \leq m$. So we have

$$g(C_{B_3}) = m(t-3) \Leftrightarrow r_i = m \text{ for all } i. \tag{6.19}$$

On the other hand, Theorem 3.1.6 gives a method to compute the genus of C_{B_3}. Let $G_{p_i} \simeq \mathbb{Z}_3$ be the stabilizer of G_n at the branch point p_i for $i = 1, \ldots, t$. Then

$$g(C_{B_3}) = [G_n : B_3](-1) + 1 + \frac{1}{2} \sum_{i=1}^{t} ([G_n : B_3] - |B_3 \backslash G_n / G_{p_i}|)$$

$$= 2^{2n}(-1) + 1 + \frac{1}{2} \sum_{i=1}^{t} (2^{2n} - m - 1) = m(t - 3).$$

Comparing this with (6.19) gives the assertion. $\square$

Using the Hurwitz formula, we immediately get the following Proposition.

Proposition 6.3.12 *Under the assumptions of the previous lemma, we have*

- $g(\widetilde{C}) = 2^{2n}(t - 3) + 1$
- $g(C_{N_n}) = t - 2$
- $g(C_{L_i}) = 2t - 5$
- $g(C_{U_i}) = 4t - 11$
- $g(C_{M_i}) = t - 3$
- $g(C_{B_3}) = m(t - 3)$

An immediate consequence is the following corollary,

Corollary 6.3.13 *Let the assumptions be as in Lemma 6.3.11. Then we have*

$$g(C_{B_3}) = \sum_{i=1}^{m} (P(C_{L_i} \to C_{N_n})) = \sum_{i=1}^{m} g(C_{M_i}).$$

The corollary suggests that there is a non-trivial relation between these Prym varieties and the Jacobian JC_{B_3} and moreover for $i = 1, \ldots, m$, a relation between the Prym variety $P(C_{L_i} \to C_{N_i})$ and the Jacobian JC_{M_i}. In the next subsection, we will see that this is the case for the first relation. The relation between $P(C_{L_i} \to C_{N_n})$ and JC_{M_i} is explained by the following corollary.

Corollary 6.3.14 *Let the assumptions be as in the previous corollary. For $i = 1, \ldots, m$, there is a canonical isomorphism of principally polarized abelian varieties*

$$P(C_{L_i} \to C_{N_n}) \xrightarrow{\simeq} JC_{M_i}.$$

Proof According to Proposition 6.3.7, the cover $C_{U_i} \to \mathbb{P}^1$ is Galois with group $\mathcal{A}_4$. On the other hand, by Lemma 6.3.11 the double cover $C_{U_i} \to C_{L_i}$ is étale. Hence we may apply the trigonal construction of Theorem 5.6.2 to give the asserted isomorphism, taking into account the canonical isomorphism of principally polarized abelian varieties $\widehat{JC_{M_i}} \simeq JC_{M_i}$. $\square$

6.3.3 A Jacobian Isogenous to the Product a Big Number of Jacobians

Let $\widetilde{C}$ be a curve with an action of the group G_n, and let the notation be as in diagram (6.18) with the additional assumption

$$\widetilde{C} \to C \text{ is a cover with signature } (0; 3, \ldots, 3) \text{ with } t \text{ numbers } 3.$$

In particular $C = \mathbb{P}^1$. For $i = 1, \ldots, m = \frac{1}{3}(2^{2n} - 1)$, we denote $\mu_i : C_{B_3} \to C_{M_i}$ the cover given by the inclusion $B_3 \subset M_i$. The pullback homomorphisms

$$\mu_i^* : JC_{M_i} \to JC_{B_3}$$

are isogenies onto their images. Considering the composition of these isogenies with the addition map, we get a canonical homomorphism

$$a : \prod_{i=1}^{m} JC_{M_i} \to JC_{B_3}.$$

The main result of this subsection is the following theorem.

Theorem 6.3.15 *Under the above assumptions, the homomorphim a : $\prod_{i=1}^{m} JC_{M_i} \to JC_{B_3}$ is an isogeny.*

Proof Recall that for any subgroup H of G_n, ρ_H denotes the representation of G_n induced by the trivial representation of H. Moreover, χ_0 denotes the trivial representation of G_n and θ_i, $i = 1, \ldots, m$ the irreducible rational representations of degree 3 of G_n. It is easy to see that

$$\rho_{B_3} = \chi_0 \oplus \bigoplus_{i=1}^{m} \theta_i. \tag{6.20}$$

Let $f_{B_3, \widetilde{\chi}_0}$ and $f_{B_3, \widetilde{\theta}_i}$ denote the corresponding central idempotents of the Hecke algebra $\mathcal{H}_{B_3}$ for the subgroup B_3 of G_n. Then according to Equation (2.37), the idempotent p_{B_3} of the subgroup B_3 in G_n decomposes as

$$p_{B_3} = f_{B_3, \widetilde{\chi}_0} + \sum_{i=1}^{m} f_{B_3, \widetilde{\theta}_i}.$$

Considering the idempotents as elements of $\mathrm{End}_{\mathbb{Q}}(JC_{B_3})$,

$$JC_{B_3, \widetilde{\chi}_0} = \mathrm{Im}(f_{B_3, \widetilde{\chi}_0}) = \mathrm{Im}(p_{G_n}) = 0,$$

since $g(C_{G_n}) = g(C) = 0$ and

$$JC_{B_3, \widetilde{\theta}_i} = \mathrm{Im}(f_{B_3, \widetilde{\theta}_i}) = \mathrm{Im}(\mu_i^*),$$

since $\theta_i = \rho_{M_i} - \chi_0$ according to Remark 6.3.8. Hence according to Eq. (2.38), the addition map gives an isogeny

$$+ : \prod_{i=1}^{m} JC_{B_3, \widetilde{\theta}_i} \longrightarrow JC_{B_3}.$$

Combining this with the isogenies (onto their images) μ_i^*, we get the isogeny a : $\prod_{i=1}^{m} JC_{M_i} \to JC_{B_3}$ as claimed.   □

Corollary 6.3.16 *For any integer $n \geq 2$, consider the integer $m = \frac{1}{3}(2^{2n} - 1)$. Then for any $t \geq 2n$ there exists a curve of genus $m(t - 3)$ whose Jacobian is isogenous to the product of m Jacobians of the same genus.*

Proof According to Lemma 6.3.10, there is a curve $\widetilde{C}$ of genus $2^{2n}(t - 3) + 1$ with an action of the group G_n and $C_{G_n} = \mathbb{P}^1$. By Proposition 6.3.12, $g(C_{B_3}) = m(t-3)$. So Theorem 6.3.15 gives the assertion.   □

Corollary 6.3.17 *Given any positive integer N, there exists a curve whose Jacobian is isogenous to the product of $m \geq N$ Jacobian varieties of the same dimension.*

Proof Choose an integer n such that $m = \frac{1}{3}(2^{2n} - 1) \geq N$. This is equivalent to $2^{2n} > 3N$. According to the previous corollary, there exists a curve whose Jacobian is isogenous to the product of $m \geq N$ Jacobians.   □

Remark 6.3.18

(a): The isogeny $+ : \prod_{i=1}^{m} JC_{B_3, \widetilde{\theta}_i} \to JC_{B_3}$ is the isotypical decomposition of JC_{B_3} with respect to the Hecke algebra action of $\mathcal{H}_{B_3}$ on JC_{B_3}.
(b): The paper [5] also gives an estimate of the degree of the isogeny a . In fact, $\deg a \leq 2^{2g(C_{B_3})}$.

Chapter 7
Some Special Groups and Complete Decomposability

In this last chapter, we will study some particular groups and questions which we think are worthwhile. The first section contains the decomposition of the Jacobian $\widetilde{J}$ of a curve with an action of the group $\mathcal{A}_5$ of even permutations of degree 5, which is the first interesting simple group. This turns out to be easy due to the fact that we know already the decomposition if we consider the decomposition of $\widetilde{J}$ with respect to the action of some subgroups of $\mathcal{A}_5$, namely, $\mathcal{A}_4$, D_5, K, and D_3.

All Schur indices of the groups occurring up to now were equal to 1. In Sect. 7.2 we study the action of two groups admitting a representation of Schur index 2. The first such group is the group Q_8 of Hamiltonian quaternions. Here a group algebra decomposition is given by the isotypical decomposition and to every irreducible rational representation there corresponds exactly one Prym variety.

More difficult is the group of Sect. 7.2.2, which is a group of order 24, given as the semidirect product of the group Q_8 by the group $\mathbb{Z}_3$. This is the first group occurring in this book admitting an isotypical component, which is equal to a group algebra component, but which is not equal not even isogenous to the Prym variety of a subcover nor a pair of such subcovers. This is not a surprise; in fact for more complicated groups, one should expect this to occur more often.

In the last section, we give an application of our methods, namely, the proof of a result, which is related to the results of Sect. 6.3, namely, the criterion of Ekedahl-Serre[12]. In Sect. 6.3 we gave an example of a curve whose Jacobian is isogenous to the product of an arbitrary number of Jacobians. Ekedahl and Serre asked whether there are Jacobians which are isogenous to a product of a large number of elliptic curves. They gave a criterion for those decompositions coming from a group action. We give several examples.

7.1   The Alternating Group $\mathcal{A}_5$

The alternating group $\mathcal{A}_5$ of permutations of degree 5 is the first interesting simple group. On the other hand, according to Corollary 3.1.2, there are lots of curves $\widetilde{C}$ with an action of this group. Hence it seems worthwhile to study the decompositions of the Jacobian $\widetilde{J} = J\widetilde{C}$. Moreover we work out some isogenies between Prym varieties of subcovers.

7.1.1   The Irreducible Rational Representations of $\mathcal{A}_5$

Let $\mathcal{A}_5$ denote the alternating group of degree 5, that is, the group of permutations of the elements 1,2,3,4,5:

$$\mathcal{A}_5 = \langle (1\ 2\ 3\ 4\ 5), (1\ 2\ 3), (1\ 2)(3\ 4)\rangle.$$

It consists of five classes of conjugacy:

- The unit element 1.
- 15 pairs of transpositions: $(i\ j)(k\ l)$ with representative $\tau := (1\ 2)(3\ 4)$.
- 20 elements of order 3: $(i\ j\ l)$ with representative $\sigma := (1\ 2\ 5)$.
- 12 elements of order 5: $(j\ k\ l\ m\ n)$ with representative $\kappa := (1\ 2\ 3\ 4\ 5)$.
- 12 elements of order 5 with representative $\kappa^2 = (1\ 3\ 5\ 2\ 4)$.

In the next section, we need the following proposition of the proof of which we omit.

Proposition 7.1.1 *The group $\mathcal{A}_5$ admits exactly one conjugacy class of proper subgroups for each appropriate order as given by representatives below.*

- *Order 2: $\langle\tau\rangle$; 15 subgroups in the class.*
- *Order 3: $\langle\sigma\rangle$; ten subgroups.*
- *Order 4: $K = \langle\tau, (1\ 4)(2\ 3)\rangle$; five subgroups.*
- *Order 5: $\langle\kappa\rangle$; six subgroups.*
- *Order 6: $D_3 = \langle\tau, \sigma\rangle$; ten subgroups.*
- *Order 10: $D_5 = \langle\tau, \kappa\rangle$; six subgroups.*
- *Order 12: $A_4 = \langle\tau, \sigma'\rangle$ with $\sigma' = (1\ 2\ 3)$; five subgroups.*

From the fact that there are five conjugacy classes of elements, it follows that there are exactly five irreducible complex representations. Their degrees d_i satisfy

$$\sum_{i=0}^{4} d_i^2 = |\mathcal{A}_5| = 60.$$

Up to permutation, this equation has a unique solution, namely, $d_0 = 1, d_1 = d_2 = 3, d_3 = 4$, and $d_4 = 5$. This proves the first part of the following lemma. For the assertion on the characters, we refer to [14, Section 3.1].

Lemma 7.1.2 *The group* $\mathcal{A}_5$ *has exactly five irreducible complex representations, the trivial representation* χ_0, *two representations* U_3 *and* U_3' *of degree* 3, *a representation* U_4 *of degree* 4, *and a representation* U_5 *of degree* 5 *with character table*

$\mathcal{A}_5$	1	15	20	12	12
	1	(12)(34)	(123)	(12345)	(21345)
χ_0	1	1	1	1	1
U_3	3	-1	0	$\frac{1}{2}(1+\sqrt{5})$	$\frac{1}{2}(1-\sqrt{5})$
U_3'	3	-1	0	$\frac{1}{2}(1-\sqrt{5})$	$\frac{1}{2}(1+\sqrt{5})$
U_4	4	0	1	-1	-1
U_5	5	1	-1	0	0

Since all Schur indices are 1, the representations U_4 and U_5 come from rational representations which we denote by the same letter. On the other hand, there is an irreducible rational representation W_6 of degree 6, defined by

$$W_6 \otimes \mathbb{C} = U_3 \oplus U_3'.$$

This gives

Corollary 7.1.3 *The group* $\mathcal{A}_5$ *admits exactly the following four irreducible rational representations:*

- *The trivial representation* χ_0.
- *The representation* U_4 *of degree* 4.
- *The representation* U_5 *of degree* 5.
- *The representation* W_6 *of degree* 6.

7.1.2 Ramification and Genera

Now let $\widetilde{C}$ be a curve with an action of the alternating group $\mathcal{A}_5$ with corresponding quotient map $f : \widetilde{C} \to C$. As always we denote the quotient of $\widetilde{C}$ by a cyclic group $\langle \alpha \rangle$ by C_α and by a subgroup $U \subset \mathcal{A}_5$ by C_U.

Consider the following diagram of subcovers of $\widetilde{C}$:

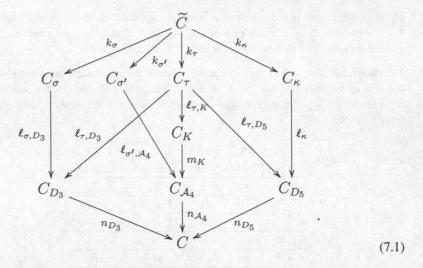

$$(7.1)$$

The number of fixed points of the involution τ is even. Since σ and κ are contained in D_3 and D_5, respectively, we know that each of them has an even number of fixed points. We denote these numbers by

- $2s := |\operatorname{Fix}\sigma| = |\operatorname{Fix}\sigma'|$, since σ and σ' are conjugate,
- $2t := |\operatorname{Fix}\tau|$,
- $2r := |\operatorname{Fix}\kappa|$.

With these notations we have the following lemmas.

Lemma 7.1.4 *The Galois cover $f : \widetilde{C} \to C$ is of signature $(g; 2, \ldots, 2, 3, \ldots, 3, 5, \ldots, 5)$ with t numbers 2, s numbers 3, and r numbers 5.*

Proof We have to count the numbers of fixed points which have the same image under f. As we saw in Sect. 3.1, if $p \in \widetilde{C}$ with non-trivial stabilizer $\mathcal{A}_{5,p}$, then the points in its orbit have stabilizers running through the complete conjugacy class of $\mathcal{A}_{5,p}$ in $\mathcal{A}_5$.

As we saw in Proposition 7.1.1, $\mathcal{A}_5$ admits exactly 15 conjugate subgroups of order 2. Since the fibre passing through a fixed point of τ consists of 30 points, exactly 2 of the fixed points of each involution belong to that fibre. This implies that the number 2 occurs exactly t times in the signature.

Similarly, $\mathcal{A}_5$ admits exactly ten conjugate subgroups of order 3. Since the fibre passing through a fixed point of σ consists of 20 points, exactly 2 of the fixed points of each such element belong to that fibre. So 3 occurs exactly s times in the signature.

Since $\mathcal{A}_5$ admits exactly six conjugate subgroups of order 5, the same argument gives that 5 occurs exactly r times in the signature. □

Lemma 7.1.5 *With the above numbers of fixed points, the ramification degrees $r(g)$ of the maps g in the diagram (7.1) are as follows:*

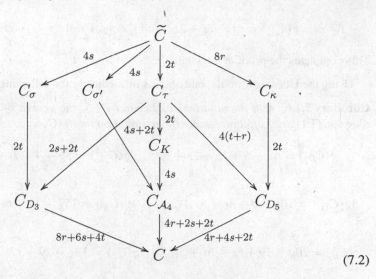

$$(7.2)$$

Proof Considering $\widetilde{C}$ with an action of the subgroups $\mathcal{A}_4$ and D_5 of $\mathcal{A}_5$, we can apply Lemma 5.5.1 to the upper middle triangle and Corollary 6.2.3 to the right-hand square to give all ramification degrees apart from those of $n_{\mathcal{A}_4}$, n_{D_3}, n_{D_5} and ℓ_{τ,D_3}, ℓ_{σ,D_3}.

Considering $\widetilde{C}$ with the action of the subgroup D_3 of $\mathcal{A}_5$, we can apply Corollary 6.2.3 to the upper left-hand square to give the ramification degrees of ℓ_{σ,D_3} and ℓ_{τ,D_3}.

For the computation of $r(n_{D_3})$, $r(n_{\mathcal{A}_4})$, and $r(n_{D_5})$, we apply Corollary 3.1.7. According to Lemma 7.1.4, the cover $f : \widetilde{C} \to C$ is of geometric signature

$$(g; G_1, \ldots, G_t, H_1, \ldots, H_s, K_1, \ldots, K_r)$$

where the G_i, respectively H_i, and respectively K_i are subgroups of order 2, respectively 3, and respectively 5 of $\mathcal{A}_5$. Then we have according to Corollary 3.1.7

$$r(n_{D_3}) = 10(r + s + t) - \sum_{i=1}^{t} |D_3 \backslash \mathcal{A}_5 / G_i| - \sum_{i=1}^{s} |D_3 \backslash \mathcal{A}_5 / H_i| - \sum_{i=1}^{r} |D_3 \backslash \mathcal{A}_5 / K_i|$$

Now we compute (using the computer program Magma) that for all i

$$|D_3 \backslash \mathcal{A}_5 / G_i| = 6, \quad |D_3 \backslash \mathcal{A}_5 / H_i| = 4 \quad \text{and} \quad |D_3 \backslash \mathcal{A}_5 / K_i| = 2.$$

Hence we get

$$r(n_{D_3}) = 10(r + s + t) - 2r - 4s - 6t = 8r + 6s + 4t.$$

In the same way, we compute

$$r(n_{\mathcal{A}_4}) = 4r + 2s + 2t \quad \text{and} \quad r(n_{D_5}) = 4r + 4s + 2t.$$

This completes the proof of the lemma.   □

Using the Hurwitz formula, one obtains immediately the following corollary.

Corollary 7.1.6 *With the notations of Lemma 7.1.5, the genera of the curves in diagram (7.1) are as follows, under the assumption that $g(C) = g$:*

$$g(C_{D_5}) = 6g - 5 + 2r + 2s + t, \qquad g(C_{\mathcal{A}_4}) = 5g - 4 + 2r + s + t,$$

$$3g(C_{D_3}) = 10g - 9 + 4r + 3s + 2t, \qquad g(C_K) = 15g - 14 + 6r + 5s + 3t,$$

$$g(C_\sigma) = 20g - 19 + 8r + 6s + 5t, \qquad g(C_\tau) = 30g - 29 + 12r + 10s + 7t,$$

$$C_\kappa = 12g - 11 + 4r + 4s + 3t, \qquad g(\widetilde{C}) = 60g - 59 + 24r + 20s + 15t.$$

7.1.3 Decompositions of $\widetilde{J}$

Let $\widetilde{C}$ be a curve with an action of the group $\mathcal{A}_5$ with quotient map $f : \widetilde{C} \to C$. We want to determine the isotypical and the group algebra decomposition of $\widetilde{J} = J\widetilde{C}$.

According to Corollary 7.1.3, the group $\mathcal{A}_5$ admits exactly the four irreducible rational representations χ_0, U_4, U_5 of degree 4 and 5, respectively, and W_6 of degree 6. Let A_{χ_0}, A_{U_4}, A_{U_5}, and A_{W_6} denote the corresponding isotypical components of $\widetilde{J}$. Since as always $A_{\chi_0} = f^*J$, the isotypical decomposition is given by the addition map

$$\mu : f^*J \times A_{U_4} \times A_{U_5} \times A_{W_6} \to \widetilde{J}.$$

It remains to determine A_{U_4}, A_{U_5}, and A_{W_6}. For this we first derive a group algebra decomposition.

Theorem 7.1.7 *Let $\widetilde{C}$ be a curve with an action of the alternating group $\mathcal{A}_5$, and let the notation be as in diagram (7.1). A group algebra decomposition is given by an isogeny*

$$\psi : J \times P(n_{\mathcal{A}_4})^4 \times P(n_{D_5})^5 \times P(\ell_\kappa)^3 \to \widetilde{J}.$$

Proof One checks

$$\rho_{\mathcal{A}_4} \simeq \chi_0 \oplus U_4, \qquad \rho_{D_5} \simeq \chi_0 \oplus U_5 \qquad \text{and} \qquad \rho_{\langle \kappa \rangle} \simeq \rho_{D_5} \oplus W_6.$$

Hence according to Corollary 3.5.10, there are abelian subvarieties B_{U_4}, B_{U_5}, and B_{W_6}, which are the images of primitive idempotents associated to the representations U_4, U_5, and W_6, respectively, such that

$$B_{U_4} \sim P(n_{\mathcal{A}_4}), \qquad B_{U_5} \sim P(n_{D_5}) \qquad \text{and} \qquad B_{W_6} \sim P(\ell_\kappa).$$

According to Eq. (3.14), we have

$$\tilde{J} \sim J \times B_{U_4}^4 \times B_{U_5}^5 \times B_{W_6}^3,$$

since $\dim U_i = i$ for $i = 3, 4, 5$, $W_6 \otimes \mathbb{C} \simeq U_3 \oplus U_3'$, and all Schur indices of $\mathcal{A}_5$ are 1. Together this implies the assertion of the theorem. $\qquad \square$

Let $\iota_{\mathcal{A}_4}, \iota_{D_5}$ respectively ι_κ denote the canonical embedding of $P(n_{\mathcal{A}_4})^4$, $P(n_{D_5})^5$ respectively $P(\ell_\kappa)^3$ into $J \times P(n_{\mathcal{A}_4})^4 \times P(n_{D_5})^5 \times P(\ell_\kappa)^3$. Then we have

Corollary 7.1.8 *Under the assumptions of Theorem 7.1.7, the isotypical components of $\tilde{J}$ are given by*

$$A_{U_4} = \mathrm{Im}(\psi \circ \iota_{U_4}), \qquad A_{U_5} = \mathrm{Im}(\psi \circ \iota_{D_5}) \qquad \text{and} \qquad A_{W_6} = \mathrm{Im}(\psi \circ \iota_\kappa).$$

Proof This follows immediately from the relation between the isotypical components and any group algebra components associated to them (see Theorem 2.9.2). From the uniqueness of the isotypical components of $\tilde{J}$, we deduce that the images of the maps occurring in Corollary 7.1.8 do not depend on the chosen isogeny ψ. $\qquad \square$

7.1.4 Some Isogenies Between Prym Varieties of Subcovers

Let $\tilde{C}$ be a curve with an $\mathcal{A}_5$ action and diagram (7.1) of subcovers. Considering $\tilde{C}$ with an action of the subgroup $\mathcal{A}_4 = \langle \tau, \sigma' \rangle$, we get from Theorem 5.5.6 the degree of the following isogeny.

Proposition 7.1.9 *The map*

$$\psi := \mathrm{Nm}\, k_{\sigma'} \circ h_{k_\tau}^* |_{P(\ell_{\tau,K})} : P(\ell_{\tau,K}) \to P(\ell_{\sigma'})$$

is an isogeny of degree

$$\deg \psi = \begin{cases} 2^{4g(C_{\mathcal{A}_4})-6+4s} & \text{if } t = 0 \text{ and } P(m_K)[2] \not\subset (\mathrm{Ker}(\ell_{\tau,K}^*))^\perp \\ 2^{4g(C_{\mathcal{A}_4})-5+4s+t} & \text{otherwise} \end{cases}$$

and kernel contained in $P(\ell_{\tau,K})[2]$.

Considering $\widetilde{C}$ with an action of the subgroup $D_5 = \langle \tau, \kappa \rangle$, we get from Theorem 6.2.15 the following isogeny.

Proposition 7.1.10 *The map*

$$\varphi := P(\ell_{\tau,D_5}) \times P(\ell_{\tau,D_5}) \to P(k_\kappa), \qquad (z_1, z_2) \mapsto k_\tau^*(z_1) + \kappa k_\tau^*(z_2)$$

is an isogeny of degree

$$\deg \varphi = \begin{cases} 5^{2g(C_{D_5})} & \text{if } r = 0, \\ 5^{2g(C_{D_5}+r-1)} & \text{otherwise.} \end{cases}$$

Finally, considering $\widetilde{C}$ with an action of the subgroup $D_3 = \langle \tau, \sigma \rangle$ with $\sigma = (1\ 2\ 5)$, we have the commutative diagram.

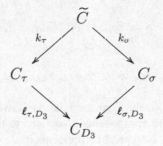

Applying Theorem 6.2.15 to it, we get the following isogeny.

Proposition 7.1.11 *The map*

$$\varphi' := P(\ell_{\tau,D_3}) \times P(\ell_{\tau,D_3}) \to P(k_\sigma), \qquad (z_1, z_2) \mapsto k_\tau^*(z_1) + \sigma k_\tau^*(z_2)$$

is an isogeny of degree

$$\deg \varphi = \begin{cases} 3^{2g(C_{D_3})} & \text{if } s = 0, \\ 3^{2g(C_{D_3})+s-1} & \text{otherwise.} \end{cases}$$

7.2 Groups with Schur Index Larger Than One

In this section we give two examples of groups admitting an irreducible representation of Schur index > 1. The first group is the group of Hamiltonian quaternions and the second one a degree-3 extension of it. The last group admits an isotypical component which is not a Prym variety of a subcover nor a Prym variety of a pair of such covers.

7.2.1 The Group of Hamiltonian Quaternions

The group of Hamiltonian quaternions Q_8 is the group consisting of the eight symbols

$$\{\pm 1, \pm i, \pm j, \pm k\}$$

with the presentation

$$Q_8 = \langle i, j \mid i^2 = j^2 = -1, k = ij = -ji \rangle.$$

There is an exact sequence

$$1 \to \{\pm 1\} \to Q_8 \xrightarrow{\pi} K_4 \to 1$$

where $K_4 = Q_8/\{\pm 1\}$ is the Klein group of order 4. Then for each one-dimensional representation ψ of K_4, the composition $\psi\pi$ gives a one-dimensional representation of Q_8. So we get four one-dimensional representations of Q_8, namely, the trivial representation χ_0 and the representations χ_1, χ_2, and χ_3 which give $\chi_1(i) = 1, \chi_1(j) = -1, \chi_2(i) = -1, \chi_2(j) = 1$, and $\chi_3 = \chi_1\chi_2$. There is also a two-dimensional irreducible complex representation V_2 given by the Pauly spin matrices:

$$i \mapsto \begin{pmatrix} \sqrt{-1} & 0 \\ 0 & -\sqrt{-1} \end{pmatrix}, \quad j \mapsto \begin{pmatrix} 0 & 1 \\ -1 & 0 \end{pmatrix} \quad \text{and} \quad k \mapsto \begin{pmatrix} 0 & \sqrt{-1} \\ \sqrt{-1} & 0 \end{pmatrix}$$

Since $4 \cdot 1 + 2^2 = 8$, these five representations are all the irreducible complex representations of $\mathbb{Q}_8$.

Note that the character table of Q_8 coincides with the character table of the dihedral group D_4 of order 8. Nevertheless there is an essential difference between their representations. Whereas the Schur indices of all representations of D_4 are equal to 1, the representation V_2 is of Schur index 2. In fact, the character field of V_2 is $K = \mathbb{Q}$, and its field of definition is $L = \mathbb{Q}(\sqrt{-1})$. By Eq. (2.20) this implies that the irreducible rational representation associated to V_2 is W_4 with

$$W_4 \otimes \mathbb{C} \simeq 2V_2$$

which is of dimension 4 over $\mathbb{Q}$. Hence Q_8 admits exactly five irreducible rational representations, $\chi_0, \ldots, \chi_3$ and W_4.

Now let $f : \widetilde{C} \to C$ be a Galois cover with group Q_8, where C as usual is of genus g. The group Q_8 admits three cyclic subgroups of order 4, namely, $\langle i \rangle$, $\langle j \rangle$, and $\langle k \rangle$, and one subgroup of order 2, namely, the normal subgroup $\{\pm 1\}$. Hence we get the following diagram of curves, where all covers are of degree 2:

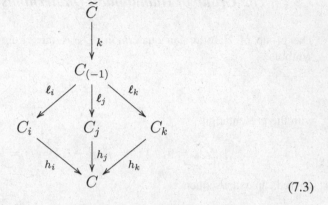

$$(7.3)$$

Since the ramification degree of a cover of even degree is always even, we may denote

$$2s := |\operatorname{Fix} i|, \quad 2t := |\operatorname{Fix} j|, \quad 2r := |\operatorname{Fix} k| \quad \text{and} \quad 2u := |\operatorname{Fix}(-1)|.$$

Note that u is not independent of i, j, and k, since it satisfies the equations

$$u = s + u_i = t + u_j = r + u_k$$

with integers u_i, u_j, and u_k given by $2u_i = |\operatorname{Fix}(-1) \setminus \operatorname{Fix} i|$ and similarly for u_j and u_k.

Lemma 7.2.1 *Let $\widetilde{C} \to C$ be a Galois cover with group Q_8 with the above notations. Then the ramification degree of the covers of diagram (7.3) is as follows:*

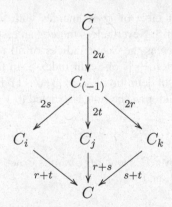

Proof Since $\{\pm 1\}$ is normal in Q_8, the Klein group K_4 acts on the curve $C_{(-1)}$, and the assertion follows from diagram (5.4). □

The Hurwitz formula immediately implies

Corollary 7.2.2 *With the assumptions of the lemma, the genera of the curves of diagram (7.3) are*

$$g(C_i) = 2g - 1 + \frac{r+t}{2}, \quad g(C_j) = 2g - 1 + \frac{r+s}{2}, \quad g(C_k) = 2g - 1 + \frac{s+t}{2},$$

$$g(C_{(-1)}) = 4g - 3 + r + s + t, \quad and \quad g(\widetilde{C}) = 8g - 7 + 2r + 2s + 2t + u.$$

As usual denote for every irreducible rational representation W of Q_8 by A_W the corresponding isotypical component in $\widetilde{J}$ given by W. For any element $\alpha \in Q_8$, let $f_\alpha : \widetilde{C} \to C_\alpha$ denote the corresponding cover. The main result of this section is the following theorem.

Theorem 7.2.3 *Let $\widetilde{C}$ be a curve with Q_8 action and notation as in diagram (7.3). Then the isotypical components of $\widetilde{J}$ with respect to this action are*

$$A_{\chi_0} = f^* J, \ A_{\chi_1} = f_i^* P(C_i/C), \ A_{\chi_2} = f_j^* P(C_j/C),$$

$$A_{\chi_3} = f_k^* P(C_k/C), \ A_{W_4} = P(\widetilde{C}/C_{(-1)}).$$

Or we could say that the isotypical decomposition and a group algebra decomposition are given by the addition map

$$\mu : f^* J \times f_i^* P(C_i/C) \times f_j^* P(C_j/C) \times f_k^* P(C_k/C) \times P(\widetilde{C}/C_{(-1)}) \to \widetilde{J}.$$

Proof As usual we have $A_{\chi_0} = f^* J$. According to Lemma 2.8.1, we compute the induced representation ρ_α of the trivial representation of a subgroup of Q_8 generated by the element $\alpha \in Q_8$ as

$$\rho_i = \chi_0 + \chi_1, \quad \rho_j = \chi_0 + \chi_2 \quad and \quad \rho_k = \chi_0 + \chi_3.$$

By Corollary 3.5.10 this implies $A_{\chi_1} = f_i^* P(C_i/C)$ and similarly for A_{χ_2} and A_{χ_3}. Similarly we have

$$\rho_{\{1\}} = \chi_0 + \chi_1 + \chi_2 + \chi_3 + W_4 \quad and \quad \rho_{(-1)} = \chi_0 + \chi_1 + \chi_2 + \chi_3.$$

So Corollary 3.5.10 implies $A_{W_4} = P(\widetilde{C}/C_{(-1)})$. Together this implies the assertion on the isotypical decomposition. The last assertion follows from Eq. (3.14), since the Schur index of V_2 is 2. □

7.2.2   A Group of Order 24

Here we give an example of a Galois cover admitting an isotypical component which is not a Prym variety and not a Prym variety of a pair of subcovers. Of course for more complicated groups, this should be the case more often.

Let G_{24} be the semidirect product of the group $Q_8 = \langle i, j \rangle$ of the previous subsection and the group $Z_3 = \langle b \rangle$ of order 3:

$$G_{24} := Q_8 \rtimes Z_3.$$

So G_{24} has the presentation

$$G_{24} = \langle i, j, b \mid i^2 = j^2 = -1, b^3 = 1, ij = -ji, b^{-1}ib = -j, b^{-1}jb = ij \rangle.$$

Let $\widetilde{C}$ be a curve with an action of G_{24}. In this case there are three subgroups of order 4, namely

$$\langle i \rangle, \qquad \langle j \rangle \qquad \text{and} \qquad \langle ij \rangle$$

and four subgroups of order 6, namely

$$H_0 = \langle -b \rangle, \quad H_1 = \langle ibi \rangle, \quad H_2 = \langle jbj \rangle \quad \text{and} \quad H_3 = \langle ijbij \rangle.$$

The subgroups of order 4 as well as the subgroups of order 6 are pairwise conjugate. Hence up to conjugacy we have the following diagram of subcurves (note that $\langle b \rangle \subset \langle -b \rangle$).

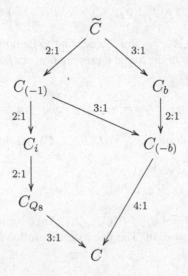

$$(7.4)$$

For the computation of the ramification degrees of the subcovers, we need some preliminaries. Consider first $\tilde{C}$ with the action of the subgroup $H_0 = \langle -b \rangle$ with corresponding subdiagram the upper square of diagram (7.4).

Since a cyclic cover of degree 2 or 3 is either unramified or totally ramified over each point, for any branch point p of the composed cyclic map $\tilde{C} \to C_{(-b)}$, there are three possibilities:

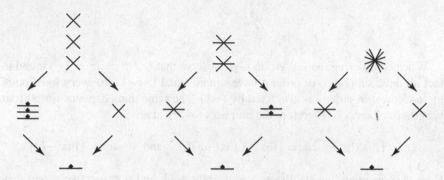

The ramification of the left-hand diagram corresponds to points of $\mathrm{Fix}(-1) \setminus \mathrm{Fix}(-b)$, of the middle diagram to points of $\mathrm{Fix}\, b \setminus \mathrm{Fix}(-b)$, and of the right-hand diagram to points of $\mathrm{Fix}\, b \cap \mathrm{Fix}(-1) = \mathrm{Fix}(-b)$. Therefore the numbers to look are

$$3\tilde{u} := |\mathrm{Fix}(-1) \setminus \mathrm{Fix}(-b)|, \qquad 2v := |\mathrm{Fix}(b) \setminus \mathrm{Fix}(-b)|, \qquad w := |\mathrm{Fix}(-b)|$$

and the upper square has the following ramification degrees:

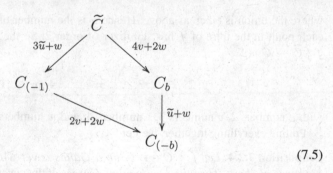

$$(7.5)$$

On the other hand, consider the action of the cyclic subgroup $\langle i \rangle$ of order 4. We may define

$$2\widehat{u} := |\mathrm{Fix}(-1) \setminus \mathrm{Fix}(i)| \quad \text{and} \quad 2s := |\mathrm{Fix}\, i|.$$

So the ramification degrees of the corresponding subdiagram of (7.4) are

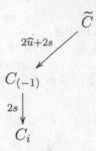

When considering the full group G_{24}, observe that $(-b)^3 = i^2 = (-1)$, and in fact all three subgroups of order 4 have square equal to (-1), so every fixed point of an element of order 4 is also fixed by (-1). The same thing happens for all four subgroups of order 6. Therefore the numbers to look at are

$$2s := |\operatorname{Fix} i|, \qquad 2v := |\operatorname{Fix} b \setminus \operatorname{Fix}(-b)| \qquad \text{and} \qquad w := |\operatorname{Fix}(-b)|.$$

Here the corresponding stabilizers are of order 4, 3, and 6, respectively, and they are exactly the fixed points in $\widetilde{C}$ with these stabilizers.

We remain with the fixed points of -1 with stabilizer of order 2. They are given by the set $\operatorname{Fix}(-1) \setminus \cup_\alpha \operatorname{Fix} \alpha$ where the union is to be taken over all $\alpha \in G_{24}$ of order 4 and 6. Since the cover $\widetilde{C} \to C$ is of degree 24, their number is divisible by 12. So we may define

$$12u := |\operatorname{Fix}(-1) \setminus \cup_\alpha \operatorname{Fix} \alpha|$$

where the union is taken as above. Hence u is the number of points in C such that each point in the fibre of f has stabilizer of order 2. So the signature of the cover $f : \widetilde{C} \to C$ is

$$(g; 2, \ldots, 2, 3, \ldots, 3, 4, \ldots, 4, 6, \ldots, 6)$$

with u numbers 2, v numbers 3, s numbers 4, and w numbers 6.

Putting everything together, we get

Proposition 7.2.4 *Let $f : \widetilde{C} \to C$ be a Galois cover with group G_{24} with the notations of above. Then the ramification degrees of the covers of diagram (7.4) are*

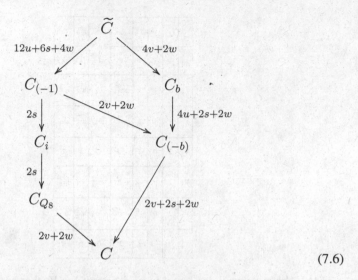

$$(7.6)$$

Proof Consider first the ramification degree of $\widetilde{C} \to C_{(-1)}$. The number $12u$ is clear, the number $6s$ comes from the three subgroups of order 4, and the number $4w$ comes from the four subgroups of order 6. We get the ramification degrees of $\widetilde{C} \to C_b$ and $C_{(-1)} \to C_{(-b)}$ from diagram (7.5). The commutativity of the upper square gives the ramification degree of $C_b \to C_{(-b)}$.

Considering $\widetilde{C}$ with the action of the subgroup Q_8, we may apply Lemma 7.2.1. Noting that the elements i, j, and ij are conjugate in G_{24}, we get the ramification degrees of $C_{(-1)} \to C_i$ and $C_i \to C_{Q_8}$ to be both $2s$.

Finally, since the subgroup generated by (-1) is normal in G_{24} with quotient $\simeq \mathcal{A}_4$, we may apply Lemma 5.5.1, to give the last two ramification degrees. $\qquad \square$

Using the Hurwitz formula, this immediately implies

Corollary 7.2.5 Let $f : \widetilde{C} \to C$ be a Galois cover with group G_{24} and C of genus $g \geq 0$ with the notation of above. Then the curves in diagram (7.4) are of genus

$$g(C_{Q_8}) = 3g - 2 + v + w, \quad g(C_{(-b)}) = 4g - 3 + v + s + w,$$

$$g(C_i) = 6g - 5 + 2v + s + 2w, \quad g(C_{(-1)}) = 12g - 11 + 4v + 3s + 4w,$$

$$g(C_b) = 8g - 7 + 2u + 2v + 3s + 3w \quad g(\widetilde{C}) = 24g - 23 + 6u + 8v + 9s + 10w.$$

The group has seven conjugacy classes of elements and seven complex irreducible representations, with character table given below, where w_3 denotes a primitive cubic root of unity.

	1_G	i	-1	b	b^2	$-b$	$-b^2$
χ_0	1	1	1	1	1	1	1
χ_1	1	1	1	w_3	w_3^2	w_3	w_3^2
χ_1'	1	1	1	w_3^2	w_3	w_3^2	w_3
V_2	2	0	-2	-1	-1	1	1
V_3	2	0	-2	$-w_3$	$-w_3^2$	w_3	w_3^2
V_3'	2	0	-2	$-w_3^2$	$-w_3$	w_3^2	w_3
V_4	3	-1	3	0	0	0	0

All V_j have Schur index equal to one, except for V_2 that has $K = \mathbb{Q}$ and $L = \mathbb{Q}(w_3)$ and hence Schur index 2. From this one immediately checks

Proposition 7.2.6 *The group G_{24} has exactly five irreducible rational representations, namely, the trivial character χ_0 and the representations $W_1, \ldots, W_4$ with*

- $W_1 \otimes_{\mathbb{Q}} \mathbb{C} \simeq \chi_1 \oplus \chi_1'$ *of dimension 2,*
- $W_2 \otimes_{\mathbb{Q}} \mathbb{C} \simeq 2V_2$ *of dimension 4,*
- $W_3 \otimes_{\mathbb{Q}} \mathbb{C} \simeq V_3 \oplus V_3'$ *of dimension 4,*
- $W_4 \otimes_{\mathbb{Q}} \mathbb{C} \simeq V_4$ *of dimension 3.*

As usual let A_{W_i} denote the isotypical component corresponding to W_i. So the isotypical decomposition of $\widetilde{J}$ is given by the addition map

$$f^* J \times A_{W_1} \times A_{W_2} \times A_{W_3} \times A_{W_4} \to \widetilde{J}$$

and it suffices to compute the isotypical components A_{W_i}.

Let B_{W_i} denote a suitable group algebra component corresponding to the representation W_i. Moreover for any element $g \in G_{24}$, respectively, for the subgroup Q_8, let $f_g : \widetilde{C} \to C_g$, respectively, $f_{Q_8} : \widetilde{C} \to C_{Q_8}$, denote the corresponding cover.

Proposition 7.2.7 *Let $f : \widetilde{C} \to C$ be a Galois cover with group G_{24} and notation as in diagram (7.6). Then suitable group algebra components of $\widetilde{J}$ with respect to this action are $B_{\chi_0} = J$,*

$$B_{W_1} = f_{Q_8}^* P(C_{Q_8}/C), \quad B_{W_3} = f_b^* P(C_b/C_{(-b)}),$$
$$B_{W_4} = f_{(-b)}^* P(C_{(-b)}/C) \sim f_i^* P(C_i/C_{Q_8}).$$

Proof The assertion on B_{χ_0} is clear. Using Lemma 2.8.1 one checks for the induced representation of the trivial representations of the subgroups

$$\rho_{Q_8} = \rho_G \oplus W_1, \qquad \rho_b = \rho_{-b} \oplus W_3, \qquad \rho_{-b} = \rho_G \oplus W_4, \qquad \rho_i = \rho_{Q_8} \oplus W_4.$$
$$(7.7)$$

The assertion follows from Corollary 3.5.10.   □

The following proposition gives the isotypical components of $\widetilde{C}$. In particular it gives B_{W_2} which is missing in Proposition 7.2.7. In fact, we have $B_{W_2} = A_{W_2}$.

Proposition 7.2.8 *Let $f : \widetilde{C} \to C$ be a Galois cover with group G_{24} and notation as in diagram (7.6). The isotypical components of $\widetilde{J}$ with respect to this action are given as follows.*

(i)

$$A_{\chi_0} = f^* J, \quad A_{W_1} = f^*_{Q_8} P(C_{Q_8}/C) \quad and \quad A_{W_4} = f^*_{(-1)} P(C_{(-1)}/C_{Q_8}).$$

There are suitable homomorphisms with finite kernel, $\alpha_{W_3} : P(C_b/C_{(-b)})^2 \to P(\widetilde{C}/C_{(-1)})$ such that $A_{W_3} = \alpha_{W_3}(P(C_b/C_{(-b)})^2)$. Furthermore, the addition map $A_{W_2} \times A_{W_3} \to P(\widetilde{C}/C_{(-1)})$ is an isogeny.

(ii) *Let $\mu : f^* J \times A_{W_1} \times A_{W_3} \times A_{W_4} \to \widetilde{J}$ denote the addition map. Then $A_{W_2} = B_{W_2}$ is the complement of Im μ in $\widetilde{J}$ with respect to the canonical polarization. It is of dimension*

$$\dim A_{W_2} = 4g - 4 + 2s + 2u + 2v + 2w.$$

Proof

(i): From Eq. (2.30) we get

$$A_{\chi_0} = B_{\chi_0}, \quad A_{W_1} = B_{W_1}, \quad A_{W_3} \sim B_{W_3}^2 \quad and \quad A_{W_4} \sim B_{W_4}^3,$$

since the χ_i are characters for $i = 1, 2, 3$ and $\dim V_3 = 2$ and $\dim V_4 = 3$ with Schur index 1.

Now we claim that $A_{W_4} = f^*_{-1} P(C_{(-1)}/C_{Q_8})$. To see this, note first that

$$\rho_{-1} - \rho_{Q_8} = 3 W_4.$$

Since $A_{W_4} \sim B_{W_4}^3$, the uniqueness of the isotypical component gives the assertion.

Now note that the addition map

$$A_{\chi_0} \times A_{W_1} \times A_{W_4} = f^* J \times f^*_{Q_8} P(C_{Q_8}/C) \times f^*_{(-1)} P(C_{(-1)}/C_{Q_8}) \to \widetilde{J}$$

has image equal to $f_{(-1)}^* JC_{-1}$. Together with the uniqueness of the isotypical decomposition and the previous proposition, this gives assertion (i).

(ii): The first assertion follows from Eq. (2.30) and the uniqueness of the isotypical decomposition. For the dimension of A_{W_2}, we compute

$$\dim A_{W_2} = g(\widetilde{C}) - g(C) - \dim A_{W_1} - \dim A_{W_3} - \dim A_{W_4}$$
$$= 4g - 4 + 2s + 2u + 2v + 2w$$

using Corollary 7.2.5.   □

Remark 7.2.9 Checking the dimensions of all Prym varieties of the covers occurring in diagram (7.4), we conclude that $B_{W_2} = A_{W_2}$ is not one of them. From the previous proposition, one can also say that it is the complement of A_{W_3} in $P(\widetilde{C}/C_{(-1)})$.

Similarly, one can verify that for any non-trivial $N \leq G$, $\langle \rho_N, W_2 \rangle = 0$. Therefore, according to diagram (7.4), the only possibility for B_{W_2} to be a Prym variety for a pair of covers would be to be the Prym variety of the pair of covers $(f_{(-1)}, f_b)$. We will see next that this is not the case.

Moreover, we can also say that $B_{W_2} = A_{W_2}$ is the complement of an explicitly given abelian subvariety in the Prym variety of the pair of covers $(f_{(-1)}, f_b)$. For this we need the following lemma.

Lemma 7.2.10 *With the notations of above, we have for the Prym variety of the pair* $(f_{(-1)}, f_b)$

$$P(f_{(-1)}, f_b) = (P(\widetilde{C}/C_{(-1)}) \cap P(\widetilde{C}/C_b))^0.$$

Proof One checks the following equations for corresponding induced representations

$$\rho_1 = \rho_{-1} \oplus W_2 \oplus 2W_3 \qquad \text{and}$$

$$\rho_1 = \rho_b \oplus W_2 \oplus W_1 \oplus W_3 \oplus 2W_4.$$

By Corollary 3.5.9 we get isogenies

$$P(\widetilde{C}/C_{(-1)}) \sim B_{W_2} \times B_{W_3}^2$$

and

$$P(\widetilde{C}/C_b) \sim B_{W_1} \times B_{W_2} \times B_{W_3} \times B_{W_4}^2,$$

from where we conclude

$$(P(\widetilde{C}/C_{(-1)}) \cap P(\widetilde{C}/C_b))^0 \sim B_{W_2} \times B_{W_3} = B_{W_2} \times P(C_b/C_{(-b)}). \tag{7.8}$$

On the other hand, we have by Corollary 3.4.3

$$P(f_{(-1)}, f_b) \subseteq (P(\widetilde{C}/C_{(-1)}) \cap P(\widetilde{C}/C_b))^0.$$

and by Proposition 3.4.4

$$\dim P(f_{(-1)}, f_b) = 8g - 8 + 4s + 4u + 3v + 4w.$$

Using Corollary 7.2.5 we get

$$8g - 8 + 4s + 4u + 3v + 4w = \dim P(f_{(-1)}, f_b) \leq \dim(P(\widetilde{C}/C_{(-1)}) \cap P(\widetilde{C}/C_b))^0$$
$$= \dim B_{W_2} + \dim B_{W_3} = 8g - 8 + 4s + 4u + 3v + 4w$$

where the last equality follows from Propositions 7.2.7 and 7.2.8. Together this implies $P(f_{(-1)}, f_b) = (P(\widetilde{C}/C_{(-1)}) \cap P(\widetilde{C}/C_b))^0$. □

Corollary 7.2.11 *There is an isogeny* $\beta : B_{W_3} = P(C_b/C_{(-b)}) \to P(f_{(-1)}, f_b)$ *onto its image, such that* $B_{W_2} = A_{W_2}$ *is the complement of* $\beta(B_{W_3})$ *in* $P(f_{(-1)}, f_b)$ *with respect to the canonical polarization.*

Proof Combining Eq. (7.8) with Lemma 7.2.10, we see that there is a homomorphism with finite kernel $\beta : P(C_b/C_{(-b)}) \to P(f_{(-1)}, f_b)$. Then the assertion follows with Eq. (7.8). □

Finally Proposition 7.2.7 implies an isogeny between Prym varieties of subcovers:

Proposition 7.2.12 *Let* $f : \widetilde{C} \to C$ *be a Galois cover with group* G_{24} *and notation as in diagram (7.6). Then there is an isogeny*

$$P(C_i/C_{Q_8}) \to P(C_{(-b)}/C).$$

Proof By Proposition 7.2.7 we have that $f_i^* P(C_i/C_{Q_8})$ and $f_{(-b)}^* P(C_{(-b)}/C)$ are isogenous, which implies the assertion. □

Since $G_{24}/\pm 1 \simeq \mathcal{A}_4$, we can apply Theorem 5.5.6 to give even the degree of such an isogeny. We leave this to the reader.

7.3   Completely Decomposable Jacobians

An abelian variety is called *completely decomposable* if it is isogenous to a product of elliptic curves. In this section we study these varieties in the case of a Jacobian with a group action.

7.3.1   The Theorem of Ekedahl-Serre

Let $f : \widetilde{C} \to C$ be a Galois cover with Galois group G. If W is an irreducible rational representation with corresponding isotypical component A_W in $\widetilde{J}$, we know that A_W is isogenous with B_W^n, where B_W is a group algebra component associate to W and n is given by Eq. (2.30). So, if $\dim B_W = 1$, then A_W is completely decomposable. However, as the following example shows, A_W might be completely decomposable although $\dim B_W > 1$.

Example 7.3.1 Let E be an elliptic curve, and let $f : \widetilde{C} \to E$ a cover of degree $n \geq 3$ simply ramified at exactly two points. It is well known that such curves exist (see, e.g., [17, Corollary 1.2]). Then $\widetilde{C}$ is of genus 2, and it is easy to see that for a general such curve the automorphism group $G = \operatorname{Aut} \widetilde{C}$ is generated by the hyperelliptic involution. So $C = \widetilde{C}/G = \mathbb{P}^1$. For the non-trivial character $W = \chi_1$ of G, we have

$$A_W = B_W = P(\widetilde{C}/\mathbb{P}^1) = \widetilde{J},$$

the Jacobian of $\widetilde{C}$. On the other hand, $\widetilde{J} \sim E \times E'$ where E' denotes the complement of f^*E in $\widetilde{J}$. So $\widetilde{J}$ is completely decomposable although B_W is not an elliptic curve. It is not difficult to construct examples of higher genus by taking covers of $\widetilde{C}$.

We say that A_W is *completely decomposable under the action of* G, if $\dim B_W = 1$, that is, if B_W is an elliptic curve. We say that the Jacobian $\widetilde{J}$ of $\widetilde{C}$ is *completely decomposable under the action of* G if all its non-zero isotypical components have this property. The following theorem gives a characterization for this. We call it the *criterion of Ekedahl-Serre*, because the main implication $(iii) \Rightarrow (i)$ is due to Ekedahl and Serre [12]. For this we need some notation.

Let $V_0, \ldots, V_s$ denote the irreducible complex representations and $W_0, \ldots, W_r$ the irreducible rational representations. For the irreducible rational representation W, let $m(W_i)$ denote the coefficient of W_i in the rational representation of G

$$\rho_r = \sum_{i=0}^{r} m(W_i) W_i.$$

Similarly let $n(V_j)$ denote the coefficient of V_j in the analytic representation of G

$$\rho_a = \sum_{J=0}^{s} n(V_j)V_j.$$

Theorem 7.3.2 *Let* $f : \widetilde{C} \to C$ *be a Galois cover of geometric signature* $(g; G_1, \dots G_t)$. *For any irreducible rational representation* W *with associate irreducible complex representation* V *and Schur index* $s(V)$, *the following statements are equivalent:*

(i) *The isotypical component* A_W *is completely decomposable under the action of* G.

(ii)

$$\frac{1}{2}m(W)s_V^2[K_V : \mathbb{Q}] = 1.$$

(iii) *The following 4 conditions are satisfied:*

   (a) $n(V) = 1$.
   (b) *The character field* K_V *of* V *is either* $\mathbb{Q}$ *or imaginary quadratic.*
   (c) $s(V) = 1$.
   (d) *if* K_V *is imaginary quadratic, then* $n(\overline{V}) = 0$, *where* $\overline{V}$ *is the complex conjugate representation.*

Proof (i) ⇔ (ii):   According to Proposition 2.9.3,

$$\dim B_W = \frac{1}{2}m(W)s_V[L_V : \mathbb{Q}] = \frac{1}{2}m(W)s_V^2[K_V : \mathbb{Q}]$$

where $m(W)$ denotes the coefficient of W in the rational representation of G and L_V denotes the field of definition of V. This gives the equivalence of (i) and (ii).
(iii) ⇒ (ii):   By what we just said, we have using condition (c),

$$\dim B_W = \frac{1}{2}m(W)[K_V : \mathbb{Q}].$$

We have to see how $m(W)$ and $n(V)$ are related. For this recall

$$\rho_r \simeq \rho_a \oplus \overline{\rho_a}. \tag{7.9}$$

According to (b), K_V is either $\mathbb{Q}$ or imaginary quadratic. If $K_V = \mathbb{Q}$, then (7.9) implies that $m(W) = 2n(V)$ which is 2 according to (a). So

$$\dim B_W = \frac{1}{2} \cdot 2 = 1.$$

If K_V is imaginary quadratic, then (d) and (a) imply that $m(W) = n(V) = 1$. So

$$\dim B_W = \frac{1}{2} \cdot 1 \cdot 2 = 1.$$

So in both cases we obtain (ii).

(ii) $\Rightarrow$ (iii):   Assume the equation (ii). We have to show conditions (a) to (d). Clearly (c) is valid and also K_V is either $\mathbb{Q}$ or imaginary quadratic. If $K_V = \mathbb{Q}$, then as we saw above, $m(W) = 2n(V) = 2$. So (a) is valid.

If K_V is imaginary quadratic, then $m(W)$ has to be 1. But if (d) is not valid, then by (7.9) we have $m(W) = 2n(V)$ contradicting (ii). So (d) and also (a) have to be fulfilled.   □

The following remark gives another statement which we do not use. It is applied often (see [29]) by using computer programs on groups.

Proposition 7.3.3 *With the notations of Theorem 7.3.2, the isotypical component A_W is completely decomposable under the action of G if and only if*

$$[L_V : \mathbb{Q}] \left(\dim V(g - 1) + \frac{1}{2} \sum_{j=0}^{r} \left(\dim V - \dim V^{G_j} \right) \right) = 1.$$

Proof This is just an application of Corollary 3.5.17.   □

7.3.2 Examples

We give a few examples. For more examples see [29] and the papers quoted there.

Example 1 (Curves with a Klein Group Action) Let us start with an easy example and construct curves with the action of the Klein group $K_4 = \langle \sigma, \tau \rangle$ of order 4 with completely decomposable Jacobian.

We use the notation of Sect. 5.2.1. So if $\widetilde{C}$ is a curve with the action of K_4, let $2s = |\text{Fix}\,\sigma|, 2t = |\text{Fix}\,\tau|$, and $2r = |\text{Fix}\,\sigma\tau|$, and let χ_i, $i = 0, \ldots, 3$ be the characters of K_4 as in that section. According to Proposition 5.2.2, the corresponding group algebra components are

$$B_{\chi_0} = J, \quad B_{\chi_1} = P(h_\sigma), \quad B_{\chi_2} = P(h_\tau) \quad \text{and} \quad B_{\chi_3} = P(h_{\sigma\tau}).$$

So $\widetilde{C}$ is completely decomposable under the action of K_4 if and only if J, $P(h_\sigma)$, $P(h_\tau)$, and $P(h_{\sigma\tau})$ all are of dimension ≤ 1.

To construct such a curve, let C be any elliptic curve and let $r = s = t = 1$. According to Theorem 3.1.6, there is a curve $\widetilde{C}$ with the action of K_4 with two fixed points of σ, τ, and $\sigma\tau$, respectively. According to Proposition 5.2.1, the curve $\widetilde{C}$ is of genus 4 with $\dim P(h_\sigma) = \dim P(h_\tau) = \dim P(h_{\sigma\tau}) = 1$. So JC is completely decomposable.

Example 2 (Curves of Genus 5 with S_4 Action) According to [25, Section 7.2] there is exactly one type of action of S_4 on a curve of genus 5. We assume the notation of Sect. 5.7. In particular let $r = s = 2$ and $u = v = 0$ be as in Sect. 5.7.1. Then the Galois cover $f : \widetilde{C} \to C$ has signature $(0; 2, 2, 3, 3)$.

According to Corollary 5.7.2, we have $\dim P(C_{\mathcal{A}_4}/C) = \dim P(C_{K'}/C_{D_4}) = 0$ and $\dim P(C_{D_4}/C) = \dim P(C_\sigma/C_{D_4}) = 1$. So Theorem 5.7.6 implies the following proposition.

Proposition 7.3.4 *Let $\widetilde{C}$ be a curve of genus 5 with S_4 action. Then $\widetilde{J}$ is completely decomposable. To be more precise,*

$$\widetilde{J} \sim E_1^2 \times E_2^3$$

with elliptic curves $E_2 = JC_{D_4}$ and $E_2 = P(C_\sigma/C_{D_4})$. The group S_4 acts on the isotypical component given by E_1^2 respectively E_2^3 by the representation W_1 respectively W_3.

Example 3 (Curves with a $PSL(2, 7)$-Action) The group $PSL(2, 7)$ is a simple group of order 168. It admits a presentation as a subgroup of S_8 given by

$$PSL(2, 7) = \langle a := (1, 2, 8)(4, 5, 6), \ b := (2, 4, 3)(5, 7, 8) \rangle.$$

That a and b generate the group $PSL(2, 7)$ was checked with the computer program GAP.

The group admits one conjugacy class of subgroups of order 3, represented by $G_3 = \langle a \rangle$, and one conjugacy class of cyclic subgroups of order 4, represented by $G_4 = \langle ab = (1, 2, 5, 7)(3, 8, 6, 4) \rangle$.

According to Theorem 3.1.4, there exists a curve $\widetilde{C}$ of genus 8 with action of $PSL(2, 7)$ with geometric signature $(0; G_3, G_3, G_4)$. In fact, a generating vector is $(c_1, c_2, c_3) = (a, b, (ab)^{-1})$.

There is exactly one irreducible complex representation W of degree 8; it is also a rational representation. One computes that

$$\langle W, \rho_{G_3} \rangle = \langle W, \rho_{G_4} \rangle = 2.$$

Applying Theorem 3.5.15 we obtain that W appears in the rational representation with multiplicity 2. Hence Corollary 3.5.17 implies

$$\dim B_W = 1 \qquad \text{and} \qquad \dim A_W = 8.$$

There are exactly two conjugacy classes of subgroups H of $PSL(2, 7)$ such that $\langle \rho_H, W \rangle = 1$, namely, D_4 and S_3. Hence we obtain

Proposition 7.3.5 *The Jacobian of the above curve $\widetilde{C}$ of genus 8 with an action of*
$PSL(2, 7)$ *is completely decomposable with isogenies*

$$\widetilde{J} \sim C_{D_4}^8 \sim C_{\mathcal{S}_3}^8.$$

Example 4 (A Curve of Genus 26 with an Action of $PSL(2, 11)$*)* Again we checked
with GAP that the group $PSL(2, 11)$ admits a presentation as a subgroup of $\mathcal{S}_{11}$
given by

$$PSL(2, 11) = \langle a := (3, 4)(5, 9)(7, 8)(10, 11), \ b := (1, 6, 9)(2, 11, 7)(4, 10, 5)\rangle.$$

The group admits one conjugacy class of subgroup of order 2, represented by
$G_2 = \langle a \rangle$; one conjugacy class of subgroups of order 3, represented by $G_3 = \langle b \rangle$;
and one conjugacy class of subgroups G_{11} of order 11, represented by $\langle ab = $
$(1, 6, 5, 3, 4, 11, 8, 7, 2, 10, 9)\rangle$.

According to Theorem 3.1.4 there exists a curve $\widetilde{C}$ of genus 26 and action of
$PSL(2, 11)$ with geometric signature $(0; G_2, G_3, G_{11})$. In fact, a generating vector
is $(c_1, c_2, c_3) = (a, b, (ab)^{-1})$.

This curve was much studied in the literature. In fact, it is the modular curve
$X(11)$ of level 11. It is well known that $X(11)$ has some other incarnations (for
these see [1] and the references given there): Klein showed that it is isomorphic
to the singular locus of the Hessian hypersurface of the threefold K_3 given by the
Klein cubic with equation $x_0^2 x_1 + x_1^2 x_2 + x_2^2 x_3 + x_3^2 x_4 + x_4^2 x_0 = 0$. Moreover,
it is isomorphic to the singular locus of the Palatini surface of the Fano threefold
of degree 14, dual to K_3 [1, Theorem 51.1]. It seems reasonable to compute the
decompositions of the Jacobian of this curve. The results are well known and can
also be proved also by different methods.

So let $\widetilde{C}$ be as above. According to Theorem 3.5.15, one checks that in this case
the rational representation is given by

$$\rho_r = W_5 \oplus W_{10}^2 \oplus W_{11}^2 \tag{7.10}$$

where W_5 is a rational irreducible representation of degree 10 with $W_5 \otimes \mathbb{C} \simeq$
$V_5 \oplus \overline{V_5}$ with V_5 complex irreducible of degree 5. For $i = 10$ and 11, W_i is a
rational irreducible of degree i with associated complex irreducible representation
also of degree i.

On the other hand, there are four conjugacy classes of subgroups H such that C_H
is an elliptic curve, namely, $H_1 = D_6$, $H_2 = G_{11}$, $H_3 = \mathbb{Z}_{11} \rtimes \mathbb{Z}_5$, and $H_4 = \mathcal{A}_4$.

Proposition 7.3.6 *The Jacobian of the above curve $\widetilde{C}$ a curve of genus 26 with*
$PSL(2, 11)$-*action is completely decomposable. To be more precise,*

$$\widetilde{J} \sim B_{W_5}^5 \times B_{W_{10}}^{10} \times B_{W_{11}}^{11}$$

with elliptic curves $B_{W_5} \sim C_{D_6}$, $B_{W_{10}} \sim C_{\mathcal{A}_4}$ and $B_{W_{11}} \sim C_{G_{11}} \sim C_{\mathbb{Z}_{11} \rtimes \mathbb{Z}_5}$.

Proof From Corollary 3.5.17 we get that

$$\dim B_{W_5} = \dim B_{W_{10}} = \dim B_{W_{11}} = 1 \qquad \text{as well as}$$

$$\dim A_{W_5} = 5, \ \dim A_{W_{10}} = 10, \ \dim A_{W_{11}} = 11.$$

Now one checks that $\langle \rho_{D_6}, V_5 \rangle = 1$ from where $B_{W_5} \sim C_{D_6}$ and similarly $\langle \rho_{\mathcal{A}_4}, W_{10} \rangle = 1$ from where $B_{W_{10}} \sim C_{\mathcal{A}_4}$ and finally $\langle \rho_{G_{11}}, W_{11} \rangle = \langle \rho_{C_{\mathbb{Z}_{11} \rtimes \mathbb{Z}_5}}, W_{11} \rangle = 1$ from where we get $B_{W_{11}} \sim C_{G_{11}} \sim C_{\mathbb{Z}_{11} \rtimes \mathbb{Z}_5}$. $\qquad \square$

Example 5 (A Curve of Genus 29 with an Action of $\mathrm{PGL}(2, 7) \times \mathbb{Z}_2$*)*
There exists a curve $\widetilde{C}$ with an action of the group $G = \mathrm{PGL}(2, 7) \times \mathbb{Z}_2$ (see [29, p. 433]). $\mathrm{PGL}(2, 7)$ admits a presentation as a subgroup of S_8 given by

$$\mathrm{PGL}(2, 7) = \langle (1, 6, 7, 2)(3, 4, 5, 8), (1, 3, 7, 8)(2, 4, 5, 6), (3, 7)(4, 8)(5, 6) \rangle.$$

Hence G admits the following presentation

$$G = \langle (1, 6, 7, 2)(3, 4, 5, 8), (1, 3, 7, 8)(2, 4, 5, 6), (3, 7)(4, 8)(5, 6) \rangle \times \langle (9, 10) \rangle$$

G acts on $\widetilde{C}$ with geometric signature

$$(0; G_1, G_2, G_3) \quad \text{with} \quad \begin{cases} G_1 = \langle (3, 7)(4, 8)(5, 6) \rangle, \\ G_2 = \langle (1, 3, 6, 5)(2, 4, 8, 7)(9, 10) \rangle, \\ G_3 = \langle (1, 6, 7, 4, 2, 3)(9, 10) \rangle. \end{cases}$$

In particular $\widetilde{C}$ is of genus 29 (by Theorem 3.1.6 using Magma).

G admits exactly 18 complex irreducible representations, all with Schur index 1; four of degree one, six of degree 6, four of degree 7, and four of degree 8. Almost all are rational irreducible, except for four of degree six, which combine in pairs into two rational irreducible ones.

Applying Theorem 3.5.15 with the given geometric signature, the rational representation of G is given by

$$\rho_r = 2W_1 \oplus 2W_2 \oplus 2W_3 \oplus 2W_4$$

with $W_1 = V_1$ of degree 6, $W_2 = V_2$ of degree 7, and $W_3 = V_3$ and $W_4 = V_4$ of degree 8.

Furthermore, Proposition 2.9.3 (or Corollary 3.5.17) gives

$$\dim B_{W_i} = 1 \text{ for } 1 \le i \le 4.$$

Hence we obtain the following proposition.

Proposition 7.3.7 *The Jacobian of the above curve $\widetilde{C}$ a curve of genus 29 with $PSL(2, 7) \times \mathbb{Z}_2$-action is completely decomposable. To be more precise,*

$$\widetilde{J} \sim E_1^6 \times E_2^7 \times E_3^8 \times E_4^8.$$

with elliptic curves $E_1, \ldots, E_4$.

7.3.3 Intermediate Covers

With the notation of Corollary 3.5.8, if a group algebra decomposition of the Jacobian of $\widetilde{C}$ with an action of a group G is given by

$$\widetilde{J} \sim J \times B_2^{\frac{\dim V_2}{s_2}} \times \cdots \times B_r^{\frac{\dim V_r}{s_r}},$$

then an induced decomposition for the Jacobian variety of the quotient curve $C_H = \widetilde{C}/H$ for every subgroup H of G is given by

$$JC_H \sim J \times B_2^{\frac{\dim V_2^H}{s_2}} \times \cdots \times B_r^{\frac{\dim V_r^H}{s_r}}.$$

Hence, if $\widetilde{J}$ is completely decomposable, so is JC_H. However, in this way we get more examples of completely decomposable Jacobians.

Example (A Completely Decomposable Jacobian of Dimension 12) In [29] it is verified computationally that there is no curve of genus 12 with a completely decomposable Jacobian under the action of a group. Here we give an example of a completely decomposable Jacobian of dimension 12 given by an intermediate cover.

According to Example 5 above, there is a curve $\widetilde{C}$ of genus 29 with an action of the group $G = PGL(2, 7) \times \mathbb{Z}_2$ whose Jacobian is completely decomposable under the action of G,

$$\widetilde{J} \sim E_1^6 \times E_2^7 \times E_3^8 \times E_4^8.$$

Moreover, there is a subgroup H of G of order 2, such that C_H is of genus 12. By what we said above, also JC_H is completely decomposable.

Indeed: G is of order 672 and $H := G_1 = \langle (3, 7)(4, 8)(5, 6) \rangle$ satisfies $|H \backslash G / G_1| = 174$, $|H \backslash G / G_2| = 84$, $|H \backslash G / G_3| = 56$, and then it follows from Theorem 3.1.6 that

$$g(C_H) = 12.$$

Furthermore, one verifies that $\dim V_j^H = 3$ for $1 \leq j \leq 4$ and therefore

$$JC_H \sim E_1^3 \times E_2^3 \times E_3^3 \times E_4^3.$$

Bibliography

1. A. Adler, S. Ramanan, *Moduli of Abelian Varieties* (Springer LNM. 1644, 1996)
2. Ya. G. Berkovich, E.M. Zhmud', *Characters of finite groups*. Part 2. Translations of Mathematical Monographs, vol. 181 (American Mathematical Society, Providence, RI, 1999)
3. Ch. Birkenhake, H. Lange, *Complex Abelian Varieties*, 2nd edn. Grundl. der Math. Wiss. 302 (Springer, 2004)
4. A. Carocca, H. Lange, R. E. Rodríguez, A. M. Rojas, Prym-Tyurin varieties via Hecke algebras. J. Reine Angew. Math. **634**, 209–234 (2009)
5. A. Carocca, H. Lange, R. E. Rodríguez, Decomposable Jacobians. Ann. Sc. Norm. Super. Pisa Cl Sci. (5) **22**(4), 1673–1690 (2021)
6. A. Carocca, S. Recillas, R. E. Rodríguez, Dihedral groups acting on Jacobians. Contemp. Math. **311**, 41–77 (2002)
7. A. Carocca, R. E. Rodríguez, Jacobians with group actions and rational idempotents. J. Alg. **306**, 322–343 (2006)
8. A. Carocca, R. E. Rodríguez, Hecke algebras acting on abelian varieties. J. Pure Appl. Algebra **222**, 2626–2647 (2018)
9. C. W. Curtis, I. Reiner, *Representation Theory of Finite Groups and Associated Algebras* (Wiley, 1962)
10. C. W. Curtis, I. Reiner, *Methods of Representation Theory, Vol 1* (Wiley, 1981)
11. R. Donagi, *The fibres of the Prym map*. Cont. Math., 136 (Am. Math. Soc., 1992)
12. T. Ekedahl, J.-P. Serre, Exemples de courbes algébriques à jacobienne complètement décomposables. C. R. Acad. Sci. Paris **317**, 509–513 (1993)
13. H. Farkas, I. Kra. *Riemann Surfaces*. Grad. Texts in Math., vol. 71 (Springer, 1991)
14. W. Fulton, J. Harris, *Representation Theory, A First Course*. Grad. Texts in Math., vol. 129 (Springer, 1991)
15. P. Griffiths, J. Harris, *Principles of Algebraic Geometry* (Wiley, 1978)
16. A. Grothendieck, Éléments de géométry algébrique II. Inst. Hautes études Sci., 8 (Publ. Math., 1961)
17. E. Kani, Hurwitz spaces of genus 2 covers of an elliptic curve. Collect. Math. **54**, 1–51 (2003)
18. A. Krazer, *Lehrbuch der Thetafunktionen* (Teubner, Leipzig, 1903)
19. A. Krazer, W. Wirtinger, Abelsche Funktionen und allgemeine Thetafuntionen. Enzyklopedia der Mathem. Wiss. II B 7
20. H. Lange, S. Recillas, Prym varieties of pairs of coverings. Adv. Geom. **4**, 373–387 (2004)
21. H. Lange, S. Recillas, Polarizations of Prym varieties of pairs of coverings. Arch. Math. **86**, 111–120 (2006)
22. H. Lange, A. Ortega, Prym varieties of Triple coverings. Geom. Dedicata **150**, 391–403 (2011)

23. H. Lange, A. Ortega, The trigonal construction in the ramified case. Preprint (2019). arXiv:1902.00251

24. H. Lange, C. Pauly, Polarizations of Prym varieties for Weyl groups via abelianization. J. Eur. Math. Soc. **11**, 315–349 (2009)

25. K. Magaard et al., The locus of curves with prescribed automorphism group, Comm. arithm. fund. groups (Kyoto). Surikaisenkikenkyusho Kokyuroko **1267**, 231–244 (2002)

26. D. Mumford, *Abelian Varieties* (Oxford Univ. Press, 1970)

27. D. Mumford, *Prym Varieties I.* Contr. to Analysis (Academic Press, 1974), pp. 325–350

28. S. Pantazis, Prym varieties and the Geodesic flow on $SO(n)$. Math. Ann. **273**, 297–315 (1986)

29. J. Paulhus, A. M. Rojas, Completely decomposable Jacobian varieties in new genera. Experimental Math. **26**(4), 430–445 (2017)

30. S. Recillas, Jacobians of curves with a g_4^1 are Prym varieties of trigonal curves. Bol. Soc. Math. Mexicana **19**, 9–13 (1974)

31. S. Recillas, R. E. Rodríguez, *Jacobians and representations of S_3*. Aportaciones Mathem. 13, Soc. Mat. Mexicana, 117–140 (1998)

32. S. Recillas, R. E. Rodríguez, Prym varieties and fourfold covers. Unpublished preprint

33. A. M. Rojas, Group actions on Jacobian varieties. Rev. Mat. Iberoamericana **23**, 397–420 (2007)

34. J.-P. Serre, *Représentations linéaires des group finis* (Hermann, Paris, 1967)

35. A. Weil, *Variétés abéliennes* (Hermann, Paris, 1948)

36. G. Welters, *Divisor varieties, Prym varieties and a conjecture of Tyurin*. Preprint No. 139 (University Utrecht, 1980)

Index

A

abelian variety, 8
  completely decomposable, 240
  dimension of, 8
  dual, 10
  G-decomposable, 25
  G-simple, 25
  polarized, 9
Abel-Jacobi isomorphism, 46
action of G on an abelian variety by a
      representation, 32
ample, 9
analytic norm, 13
analytic representation, 8
analytic trace, 13
Appell-Humbert Theorem, 9

B

bigonal
  construct, 132
  construction, 131
bigonal cover, 129
  branch point of type (j), 130
  fibre of type (j), 130
branch divisor, 48
branch point, 40

C

character field, 26
characteristic polynomial, 12
complementary abelian subvariety, 16
  pair of, 16

D

decomposition
  group algebra, 33
  Hecke algebra, 38
  isotypical, 32

E

endomorphism
  characteristic polynomial of, 12
  primitive, 16
exponent
  of an abelian subvariety, 15
  of a pair, 17
  of a polarization, 12

F

factor of automorphy, 8
field
  character field, 26
  of definition, 26

correspondence, 129
cover of curves, 40
  bigonal, 129
  cycle structure of, 40
  degree of, 40
  norm map of, 46
criterion
  of Ekedahl-Serre, 240
curve, 40

© The Author(s), under exclusive license to Springer Nature Switzerland AG 2022
H. Lange, R. E. Rodríguez, *Decomposition of Jacobians by Prym Varieties*, Lecture
Notes in Mathematics 2310, https://doi.org/10.1007/978-3-031-10145-8

Printed in the United States
by Baker & Taylor Publisher Services